Crop Protection: Advances in Agricultural Science

Crop Protection: Advances in Agricultural Science

Edited by Harley Wells

SYRAWOOD
PUBLISHING HOUSE

New York

Published by Syrawood Publishing House,
750 Third Avenue, 9th Floor,
New York, NY 10017, USA
www.syrawoodpublishinghouse.com

Crop Protection: Advances in Agricultural Science
Edited by Harley Wells

International Standard Book Number: 978-1-64740-061-3 (Hardback)

Cataloging-in-Publication Data

Crop protection : advances in agricultural science / edited by Harley Wells.
    p. cm.
Includes bibliographical references and index.
ISBN 978-1-64740-061-3
1. Plants, Protection of. 2. Crops. 3. Agricultural pests--Control. 4. Agriculture. I. Wells, Harley.
SB950 .C76 2022
632.9--dc23

TABLE OF CONTENTS

PREFACE

Over the recent decade, advancements and applications have progressed exponentially. This has led to the increased interest in this field and projects are being conducted to enhance knowledge. The main objective of this book is to present some of the critical challenges and provide insights into possible solutions. This book will answer the varied questions that arise in the field and also provide an increased scope for furthering studies.

Crop protection is the practice of managing pests, weeds and diseases that affect agricultural crops. It is generally achieved by the implementation of pesticide-based approaches, barrier-based approaches, biological pest control methods, animal-psychology based approaches and methods of biotechnology. Ploughing and cultivating the soil prior to sowing potentially reduces the risk of pests. Pesticide-based approaches include the use of herbicides, insecticides and fungicides. Biological pest control is a method of pest management that encourages herbivory, predation, parasitism, or other natural mechanisms. Diseases in crops are generally prevented by using pathogen-free seeds, controlling field moisture, or implementing approaches like crop rotation. This book provides comprehensive insights into the latest advances in agricultural science. Most of the topics introduced herein cover new techniques of crop protection. For someone with an interest and eye for detail, this book covers the most significant topics in this domain.

I hope that this book, with its visionary approach, will be a valuable addition and will promote interest among readers. Each of the authors has provided their extraordinary competence in their specific fields by providing different perspectives as they come from diverse nations and regions. I thank them for their contributions.

Editor

Effects of botanical insecticides on the instantaneous population growth rate of *Aphis gossypii* Glover (Hemiptera: Aphididae) in cotton

Lígia Helena de Andrade[1], José Vargas de Oliveira[2*], Mariana Oliveira Breda[2], Edmilson Jacinto Marques[2] and Iracilda Maria de Moura Lima[3]

[1]*Universidade Federal Rural de Pernambuco, Recife, Pernambuco, Brazil.* [2]*Área de Fitossanidade, Departamento de Agronomia, Universidade Federal Rural de Pernambuco, R. Dom Manoel de Medeiros, s/n, 52171-900, Recife, Pernambuco, Brazil.* [3]*Laboratório de Entomologia, Departamento de Zoologia, Centro de Ciências Biológicas, Universidade Federal de Alagoas, Maceió, Alagoas, Brazil. *Author for correspondence. E-mail: vargasoliveira@uol.com.br*

ABSTRACT. Botanical insecticides have been studied aiming the alternative pest control. The present study investigated the effects of these insecticides on the instantaneous population growth rate (ri) of *Aphis gossypii*. Botanical insecticides were tested in the following concentrations: Compostonat®, Rotenat-CE® and Neempro (0, 0.50, 0.75, 1.00, 1.25, 1.50 and 1.75%); Natuneem® and Neemseto® (0, 0.25, 0.50, 0.75 and 1.00%) and essential oils of *Foeniculum vulgare* Mill., *Cymbopogom winterianus* (L.), *Chenopodium ambrosioides* L. and *Piper aduncum* L (0, 0.0125, 0.025, 0.0375 and 0.05%). Cotton leaf discs, CNPA 8H cultivar with 5 cm in diameter were immersed for 30 seconds in products broth and dried for 30 minutes. Eight replicates per concentration were used and each disc was infested with five apterous adult females of *A. gossypii* uniform in size and confined for 10 days. Compostonat®, Rotenat® and Neempro provided negative r_i decreasing *A. gossypii* population. Natuneem® and Neemseto® and the essential oil of *F. vulgare* showed positive r_i increasing the population. The coefficients of determination (R^2) of regression lines ranged from 0.46 to 0.85. The essential oils of *C. winterianus*, *C. ambrosioides* and *P. aduncum* were not statistically significant precluding the establishment of regression lines.

Keywords: cotton aphid, bioactivity, *Gossypium hirsutum*, natural insecticides.

Introduction

Aphis gossypii Glover (Hemiptera: Aphididae) is a pest of great economic importance to cotton crop. Adults and nymphs suck sap from the phloem, inoculate toxins, excrete sugary substances (honeydew), favoring the sooty mold development, and are vectors of virus in cotton, such as the vermilion and vein mosaic form Ribeirão Bonito (MICHELOTTO; BUSOLI, 2003). The rapid multiplication of *A. gossypii* requires the producer to maintain short intervals of insecticide applications, which can lead to selection of resistant insect populations (BARROS et al., 2006).

Due to the concern of researchers, farmers and society as a whole, about the side effects of pesticides overuse, researched on new alternative methods of pest control has a major boost in recent decades, citing as example studies with botanical insecticides (AHMAD et al., 2003; CLOYD et al., 2009; ESTRELA et al., 2006; ISMAN, 2006; MARTINEZ; VAN EMDEN, 2001). These products, derived from secondary metabolism of plants, are composed of complex mixtures of chemical substances such as monoterpenes, sesquiterpenes and flavonoids playing important roles in the processes of tritrophic interactions (plant-insect-natural enemy) and in the control of insects, mites, fungi and nematodes (FAZOLIN et al., 2005; SCHMUTTERER, 1990; TAVARES; VENDRAMIM, 2005). They act on insects by ingestion, contact and fumigation, and can be used as powders, extracts, essential oils and oil emulsion (ABRAMSON et al., 2006; RAJENDRAN; SRIRANJINI, 2008). They are generally biodegradable, low toxicity to vertebrates and may have selectivity for natural enemies (COSME et al., 2007; SILVA; MARTINEZ, 2004). Its effects on insects include mortality, feeding deterrence and oviposition, reductions in fecundity, fertility and growth process (BOEKE et al., 2004; MARTINEZ, 2002; ROEL et al., 2000).

The instantaneous population growth rate (r_i) has been used in the assessment of lethal and sublethal effects of insecticides on pests and natural enemies by having more consistent results, compared with other techniques used in toxicology. This rate is a direct measure of population growth, and also integrates survival and fecundity, as increase intrinsic rate of (r_m), and both provide similar results (WALTHALL; STARK, 1997).

Several authors evaluated the performance of insecticides and acaricides on pests and natural enemies, using the instantaneous population growth rate (ri), among which are the aphid *Acyrthosiphon pisum* (KRAMARZ et al., 2007), *Myzus persicae* and the ladybug *Eriopis connexa* (VENZON et al., 2007), white mite *Polyphagotarsonemus latus* (VENZON et al., 2006), spider mite *Tetranychus urticae* (KIM et al., 2006), and phytoseiid predator *Phytoseiulus persimilis* (TSOLAKIS; RAGUSA, 2008), parasitoids *Ttrychogramma pretiosum* and *Telenomus remus* (TAVARES et al., 2009), and the coffee red mite (*Olygonicus ilicis*) (MOURÃO et al., 2004).

To test a promising alternative tactics that can be used in the integrated management of cotton pests, especially organic and family crops, this study propose to evaluate the r_i of *A. gossypii* under different concentrations of botanical insecticides.

Material and methods

This work was performed at the Agricultural Entomology Laboratory, Department of Agronomy (DEPA), Rural Federal University of Pernambuco (UFRPE), using acclimatized chambers with temperature and relative humidity being monitored, and 12h photoperiod.

***Aphis gossypii* rearing**. Insects were reared according to the technique adapted from the Biology and Rearing Insects Laboratory (LBCI), Department of Fitossanidade, Faculty of Agricultural and Veterinary Sciences of Jaboticabal (FCAVJ), University Estadual Paulista (UNESP), being kept in acclimatized room at 27 ± 1°C, 70 ± 5% RH and 12h photophase.

Cotton seeds (*Gossypium hirsutum* L. race *latifolium* Hutch), cultivar CNPA 8H, were sown in styrofoam cell trays (272 x 280 mm, 64 cells) containing Base Plant® substrate, consisting of pine barks, vermiculite, peat, lime acid correctives and additives, with moisture content between 50 and 55%. Subsequently, styrofoam cell trays were placed inside plastic trays with water maintaining the appropriate level for plant roots absorption, and also contributing to the moisture maintenance in the cultivation environment.

The rearing was initiated with aphids collected on cotton plants seeded in areas of the DEPA. Plants were kept inside cages of germination and infestation, with dimensions of 1.0 x 1.20 x 0.60 m covered with 'voile' fabric. Inside the cages were installed fluorescent 'daylight' and 'Grolux' lamps to stimulate the photosynthetic process. Trays were placed on PVC pipe supports at approximately 60 cm from lamps. Pots with water and detergent were kept at the cages bottom to prevent ants' infestation. Plants remained in germination cages for approximately 20 days and then were transferred to infestation cages by placing leaves with aphids on them. Colonies were periodically observed to prevent the presence of parasitoids, predators and other undesirable insects. The rearing process was established to ensure the adequate supply of plants and aphids for performing experiments.

Botanical insecticides used. The botanical insecticides Compostonat® (essencial oils of Neem, Karanja and Castor beans; Natural Rural Ltda.), Rotenat-CE® (*Derris* sp.; Natural Rural Ltda.) and Neempro (*Azadirachta indica* A. Juss.; Quinabra e Trifolio-M GmbH., Química Natural Brasileira Ltda.) were tested at 0, 0.50, 0.75, 1.00, 1.25, 1.50 and 1.75% concentrations; Natuneem® (*A. indica*; Natural Rural Ltda.) and Neemseto® (*A. indica*; Cruangi Neem do Brasil Ltda.) at 0, 0.25, 0.50, 0.75 and 1.00% concentrations and the essential oils of *Foeniculum* *vulgare* Mill., *Cymbopogom winterianus* (L.), *Chenopodium ambrosioides* L. and *Piper aduncum* L. at 0, 0.0125, 0.0250, 0.0375 and 0.0500% concentrations.

Insecticides were tested in randomized experimental design with eight replications. It was used leaf discs of 5 cm in diameter, obtained from cotton plants, cultivar CNPA-8H with approximately 20 days old, grown in green-house. The disks were immersed in insecticide broods for 30 seconds and dried for 30 minutes. Afterwards they were placed into Petri dishes containing 1% agar-water medium, being each disc infested with five apterous adult females of *A. gossypii* with uniform size. Plates were sealed with PVC film and placed in environmental chambers for 10 days at $25 \pm 1°C$, $67 \pm 5\%$ RH and 12h photophase.

To evaluate insecticides effects on the population growth of *A. gossypii*, the r_i was estimated according to the equation:

$r_i = \ln(N_f / N_0) / \Delta t$, where N_f is the final number of aphids (nymphs and adults); N_0 is in the initial number of aphids transferred and Δt is the change in

time, in the case of ten days (WALTHALL; STARK, 1997). Positive value of r_i indicates population growth increase; $r_i = 0$ means that the population is stable; and negative value of r_i indicates population decline to extinction, when $N_f = 0$ (STARK; BANKS, 2003).

Regression analysis were performed to correlate the concentrations of insecticides tested with the r_i values by the statistical program SAS (SAS, 2001). The regression lines were plotted using the graphics program Sigma Plot (SYSTAT SOFTWARE INC., 2006).

Results and discussion

The regression equations relating to the r_i were significant for the insecticides Compostonat® (F = 313.87, p < 0.001, R^2 = 0.85), Rotenat® (F = 157.43, p < 0.0001, R^2 = 0, 74), Neempro (F = 235.97, p < 0.001, R^2 = 0.81), Natuneem® (F = 167,74,08, p < 0.0001, R^2 = 0.82), Neemseto® (F = 31.79, p < 0.0001, R^2 = 0.46) and the *F. vulgare* essential oil (F = 41.50, p < 0.0001, R^2 = 0.52) (Figures 1, 2, 3).

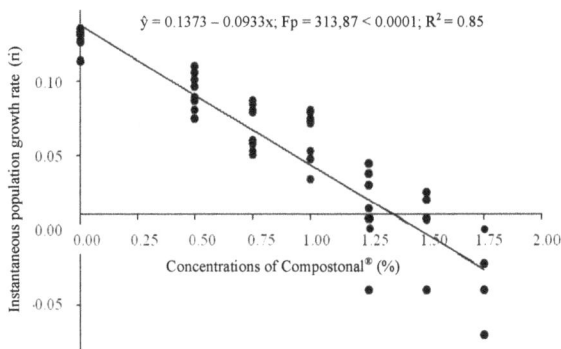

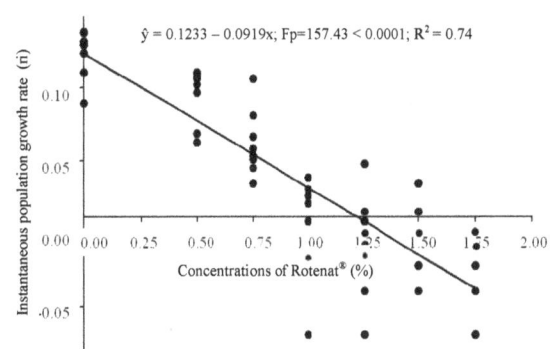

Figure 1. Instantaneous growth rate (ri) observed (•) and estimated (-) to *A. gossypii* in cotton leaf discs treated with Compostonat® and Rotenat®.

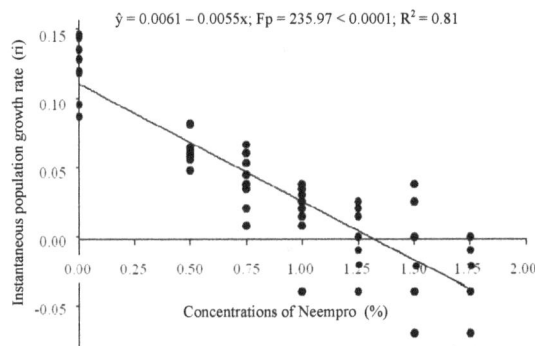

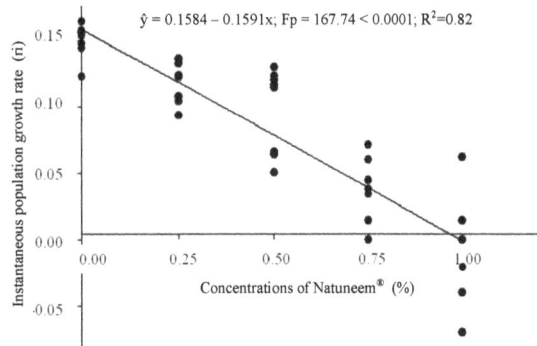

Figure 2. Instantaneous growth rate (ri) observed (•) and estimated (-) to *A. gossypi* Glover in cotton leaf discs treated with Neempro and Natuneem®.

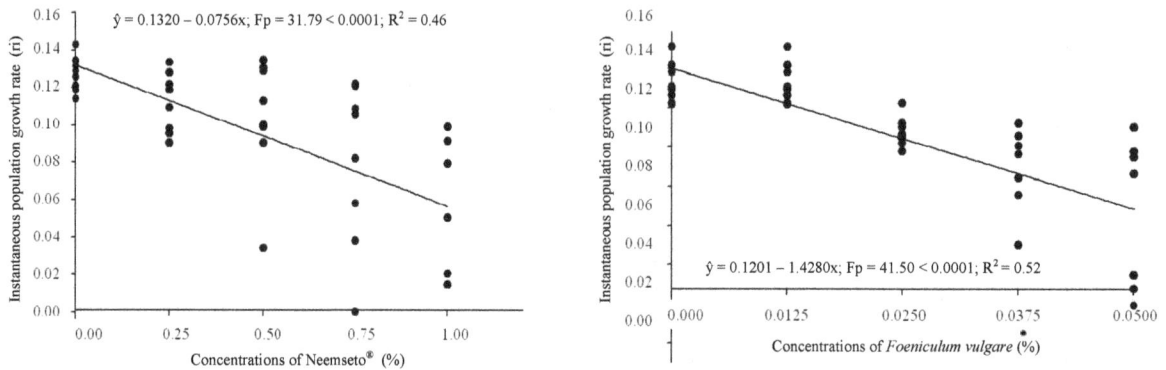

Figure 3. Instantaneous growth rate (ri) observed (●) and estimated (-) to *A. gossypii* Glover in cotton leaf discs treated with Neemseto® and essential oil of *F. vulgare*.

The first three insecticides showed negative r_i, indicating that the *A. gossypii* population has declined towards extinction. Other insecticides and the essential oil obtained positive r_i, confirming that population increased (STARK; BANKS, 2003). Oils of *C. winterianus*, *C. ambrosioides* and *P. aduncum* presented no statistical significance, preventing the establishment of regression lines.

Studies assessing effects of botanical insecticides on *A. gossypii* through the use of r_i are scarce or nonexistent based on the literature searched; but there are similar studies with other pest species and natural enemies. NeemAzal T/S at 0.5 and 1.0% concentrations had reduced population growth of *M. persicae* in laboratory on pepper leaves, with positive r_i. At concentrations of 0.025 and 0.05%, it also had caused lethal and sublethal effects on *Eriopis connexa* (VENZON et al., 2007). Regarding the white mite *Polyphagotarsonemus latus* in Chilli pepper, the r_i values had been negative for lime sulfur brood and 'Calda Viçosa', there had been population decline; a similar result to that found in this study with *A. gossypii* at 1.50 and 1.75% concentrations of Compostonat®, Rotenat® and Neempro. However, positive r_i values had been obtained for mites on plants treated with 'Supermagro' and in the control with water (VENZON et al., 2006). In females of red mites *Oligonychus ilicis*, in coffee trees, the r_i had decreased linearly with increasing concentration of extracts of oil cake, seeds and leaves of neem (MOURÃO et al., 2004).

Considering the good performance of neem-based insecticides and the action mode of these products on pests, it favors its association with biological control. The fact of azadirachtin, the main bioactive compound, to be less effective by contact than by ingestion, favors predators due to the absorption of smaller amounts of active ingredient. Thus, even if preys have been contaminated, the quantity of azadirachtin present would have been quite low, due to its rapid excretion (MARTINEZ, 2002).

According to Venzon et al. (2007), the neem use in the field must be accompanied by regular sampling, being necessary when possible, additional applications in order to reduce the aphid population in case of the remaining population from the first application increase. It is also important to mention the need for research on green-house and field for extension and adjustment of application, validating the use of this alternative tactics as well as assess the effects of botanical insecticides on natural enemies of *A. gossypii*.

Conclusion

Results obtained in this study demonstrate the importance of botanical insecticides in reducing the population of *A. gossypii*, mainly on organic crops, where the synthetic insecticides are not allowed, and family farming due to lack of resources; since in addition to be efficient, they have low toxicity to vertebrates, degrade rapidly, not severely affecting the environment and encourage the population of predators and parasitoids. On the other hand, the use of r_i was very appropriate for its efficiency and speed in obtaining results, eliminating the need for a life table of fertility.

Acknowledgements

To the CNPq and FACEPE for granting the authors' scholarships; to Mauricéa Fidelis Santana for her support in maintaining the aphids' rearing and to Solange Maria de França for her suggestions on statistical analysis.

References

ABRAMSON, C. I.; WANDERLEY, P. A.; WANDERLEY, M. J. A.; MINÁ, A. J. S.; SOUZA, O. B. Effect of essential oil from citronella and alfazema on fennel aphids *Hyadaphis foeniculi* Passerini (Hemiptera: Aphididae) and its predator *Cycloneda sanguinea* L. (Coleoptera: Coccinelidae). **American Journal of Environmental Science**, v. 3, n. 1, p. 9-10, 2006.

AHMAD, M.; OBIEWATSCH, H. R.; BASEDOW, T. Effects of neem-treated aphids as food/hosts on their predators and parasitoids. **Journal of Applied Entomology**, v. 127, n. 8, p. 458-464, 2003.

BARROS, R.; DEGRANDE, P. E.; RIBEIRO, J. E.; RODRIGUES, A. L. L.; NOGUEIRA, R. F.; FERNANDES, M. G. Flutuação populacional de insetos predadores associados a pragas do algodoeiro. **Arquivos do Instituto Biológico**, v. 73, n. 1, p. 57-64, 2006.

BOEKE, S. J.; BOERSMA, M. G.; ALINK, G. M.; VAN LOON, J. J. A.; VAN HUIS, A.; DICKE, M.; RIETJENS, I. M. C. M. Safety evaluation of neem (*Azadirachta indica*) derived pesticides. **Journal of Ethnopharmacology**, v. 94, n. 1, p. 25-41, 2004.

CLOYD, R. A.; GALLE, C. L.; KEITII, S. R.; KALSCIIEUR, N. A.; KEMP, K. E. Effect of commercially available plant-derived essential oil products on arthropod pests. **Horticultural Entomology**, v. 102, n. 4, p. 1567-1579, 2009.

COSME, L. V.; CARVALHO, G. A.; MOURA, A. P. Efeitos de inseticidas botânicos e sintéticos sobre ovos e larvas de *Cycloneda sanguinea* (Linnaeus) (Coleoptera: Coccinellidae) em condições de laboratório. **Arquivos do Instituto Biológico**, v. 74, n. 3, p. 251-258, 2007.

ESTRELA, J. L. V.; FAZOLIN, M.; CATANI, V.; ALÉCIO, M. R. Toxicidade de óleos essenciais de *Piper aduncum* e *Piper hispidinervum* em *Sitophilus zeamais*. **Pesquisa Agropecuária Brasileira**, v. 41, n. 2, p. 217-222, 2006.

FAZOLIN, M.; ESTRELA, J. L. V.; CATANI, V.; LIMA, M. S.; ALÉCIO, M. R. Toxicidade do óleo de *Piper aduncum* L. a adultos de *Cerotoma tingomarianus* Bechyné (Coleoptera: Chrysomelidae). **Neotropical Entomology**, v. 34, n. 3, p. 485-489, 2005.

ISMAN, M. B. Botanical insecticides, deterrents, and repellents in modern agriculture and increasing regulated world. **Annual Review of Entomology**, v. 51, unit number, p. 45-66, 2006.

KIM, M.; SIM, C., SHIM, D.; CHO, E. S. K. Residual and sublethal effects of fenpyroximate and pyridaben on the instantaneous rate of increase of *Tetranychus urticae*. **Crop Protection**, v. 25, n. 6, p. 542-548, 2006.

KRAMARZ, P. E.; BANKS, J. E.; STARK, J. D. Density-dependent response of the pea aphid Hemiptera: Aphididae) to imidacloprid. **Journal of Entomological Science**, v. 42, n. 2, p. 200-206, 2007.

MARTINEZ, S. S. **O nim – *Azadirachta indica***: natureza, usos múltiplos, produção. Londrina: Instituto Agronômico do Paraná, 2002.

MARTINEZ, S. S.; VAN ENDEM, H. F. Growth disruption, abnormalities and mortality of *Spodoptera littoralis* (Boisduval) (Lepidoptera: Noctuidae) caused by Azadirachtin. **Neotropical Entomology**, v. 30, n. 1, p. 113-124, 2001.

MICHELOTTO, M. D.; BUSOLI, A. C. Eficiência de ninfas e adultos de *Aphis gossypii* Glov. na transmissão do vírus do mosaico das nervuras do algodoeiro. **Bragantia**, v. 62, n. 2, p. 255-259, 2003.

MOURÃO, S. A.; ZANUNCIO, J. C.; PALLINI FILHO, A.; GUEDES, R. N. C.; CAMARGOS, A. B. Toxicidade de extratos de nim (*Azadirachta indica*) ao ácaro-vermelho-do-cafeeiro *Oligonychus ilicis*. **Pesquisa Agropecuária Brasileira**, v. 39, n. 8, p. 727-830, 2004.

RAJENDRAN, S.; SRIRANJINI, V. Plant products as fumigants for stored-product insect control. **Journal of Stored Products Research**, v. 44, n. 2, p. 126-135, 2008.

ROEL, A. R.; VENDRAMIM, J. D.; FRIGHETTO, R. T. S.; FRIGHETTO, N. Efeito do extrato acetato de etila de *Trichilia pallida* Swartz (Meliaceae) no desenvolvimento e sobrevivência da lagarta-do-cartucho. **Bragantia**, v. 59, n. 1, p. 53-58, 2000.

SAS-Statistical Analisys System. **SAS/STAT User's guide, version 8.02, TS level 2MO**. Cary: Statistical Analysis System Institute, 2001.

SCHMUTTERER, H. Properties and potential of natural pesticides from the neem tree, *Azadirachta indica*. **Annual Review of Entomology**, v. 35, unit number, p. 271-297, 1990.

SILVA, F. A. C.; MARTINEZ, S. S. Effect of neem seed oil aqueous solutions on survival and development of the predator *Cycloneda sanguinea* (L.) (Coleoptera: Coccinellidae). **Neotropical Entomology**, v. 33, n. 6, p. 751-757, 2004.

STARK, J. D.; BANKS, J. E. Population-level effects of pesticides and other toxicants on arthropods. **Annual Review of Entomology**, v. 48, unit number, p. 505-519, 2003.

SYSTAT SOFTWARE, INC. **SigmaPlot for windows version 10.0**. Copyright©, Port Richmond, CA, 2006.

TAVARES, M. A. G. C.; VENDRAMIM, J. D. Bioatividade da Erva-de-Santa-Maria, *Chenopodium ambrosioides* L., sobre *Sitophilus zeamais* Mots. (Coleoptera: Curculionidae). **Neotropical Entomology**, v. 34, n. 2, p. 319-323, 2005.

TAVARES, W. S.; CRUZ, I.; PETACCI, F.; ASSIS JUNIOR, S. L.; FREITAS, S. S.; ZANUNCIO, J. C.; SERRÃO, J. E. Potencial use of Asteraceae extracts to control *Spodoptera frugiperda* (Lepidoptera: Noctuidae) and selectivity to their parasitoids *Thichogramma pretiosum* (Hymenoptera: Thrichogrammatidae) and *Telenomus remus* (Hymenoptera: Scelionidae). **Industrial Crops and Products**, v. 30, n. 3, p. 384-388, 2009.

TSOLAKIS, H.; RAGUSA, S. Effects of a mixture of vegetables and essential oils and fatty acid potassium salts on *Tetranychus urticae* and *Phytoseiulus persimilis*. **Ecotoxicology and Environmental Safety**, v. 70, n. 2, p. 276-282, 2008.

VENZON, M.; ROSADO, M. C.; PALLINI, A; FIALHO, A.; PEREIRA, C. J. Toxicidade letal e subletal do nim sobre o pulgão-verde e seu predador *Eriopis connexa*. **Pesquisa Agropecuária Brasileira**, v. 42, n. 5, p. 627-631, 2007.

VENZON, M.; ROSADO, M. C.; PINTO, C. M. F.;

DUARTE, V. S.; EUZÉBIO, D. E.; PALLINI, A. Potencial de defensivos alternativos para o controle do ácaro-branco em pimenta "Malagueta". **Horticultura Brasileira**, v. 24, n. 2, p. 224-227, 2006.

WALTHALL, W. K.; STARK, J. D. Comparison of two population-level ecotoxicological endpoints: the intrinsic (r_m) and instantaneous (r_i) rates of increase. **Environmental Toxycology and Chemistry**, v. 16, n. 5, p. 1068-1073, 1997.

2

Description of the application method in technical and scientific work on insecticides

Marcelo Gonçalves Balan[1]*, Otavio Jorge Grigoli Abi Saab[2], Arney Eduardo do Amaral Ecker[1] and Gustavo de Oliveira Migliorini[2]

[1]Departamento de Agronomia, Universidade Estadual de Maringá, Avenida Colombo, 5790, 87020-900, Maringá, Paraná, Brazil. [2]Departamento de Agronomia, Universidade Estadual de Londrina, Londrina, Paraná, Brazil. *Author for correspondence. E-mail: mgbalan2@uem.br

ABSTRACT. Chemical control is a viable and practically indispensable tool in the control and management of cultivated plant pests, but insufficient detail in documenting the methods used for applying phytosanitary products has been reported in the majority of scientific publications dealing with insecticide application. A survey of 200 scientific studies was conducted to examine how much basic information was provided on the application method. The amount of descriptive detail concerning the insecticide application method was found to be below the minimum requirements. In particular, there was insufficient detail concerning the spray droplet spectrum (no information in 173 studies evaluated – 86.5%), operating pressure (38 studies – 19%), solution concentration (52 studies – 26%), distance and position of spray nozzles in relation to the target (114 studies – 57%), temperature (128 studies - 64%), relative humidity (134 studies - 67%) and wind speed (145 studies – 72.5%). All the studies evaluated contained information on the application rate used (L ha^{-1}). To change this situation and reestablish the importance of the application method, we propose a simplified method description for the application of phytosanitary chemicals. Use of the proposed minimum methodological description is practicable for insecticide treatments and will also enable them to be accurately repeated.

Keywords: spray nozzles, droplet spectrum, operating pressure.

Introduction

Chemical control is a viable alternative for the management and control of cultivated plant pests. Chemical control has made a valuable contribution to agriculture and helped boost the potential yield, provided that it is used rationally, minimizing environmental contamination, human health problems, and the appearance of resistant pests.

Phytosanitary defense strategies over the last 30 years have seen significant changes and technological innovations in response to an increase in the number of important pests. In the period from 1992 to 2005, insecticides used for pest management and control represented an average of 25.73% of the total volume of phytosanitary products commercially available in Brazil, in commercial terms amounting to US$ 7.80 billion (Sindicato Nacional da Indústria de Produtos para a Defesa Agrícola [Sindag], 2008).

According to Zambolim, Conceição and Santiago (2008), since the discovery of organochlorine and organophosphate insecticides in the 1940s, the dosage of the active ingredient and the persistence of the insecticides recommended for use in Brazil during the 1960s, 1970s, 1980s and 1990s decreased significantly by approximately 88.69%. Also of note is the development of insecticide molecules. These molecules used to be thought of as highly toxic but are now available at lower toxicity and with different levels of selectivity.

Technology is used in the application of phytosanitary products to deposit the appropriate quantity of the active ingredient on the target with maximum efficiency, in the most economical way and with the lowest possible impact on the environment (Matthews, 2002). The skilled use of this technology is essential to increase insecticide efficiency and minimize contamination of the application operators and the environment, as well as cutting application costs.

Today, there is a tendency to reduce the volume of spray solution to cut costs and increase spraying efficiency. Matthews (2004; 2008) highlights a reduction in spray application rates from 500 to less than 200 L ha^{-1}, underlining the urgent need for improvements in the field application technology.

For Matthews (2008), although phytosanitary products are applied in various situations, each situation must be treated individually to achieve maximum precision and optimum dosage, as well as minimum operator exposure to the active ingredient. He notes that developing new spray nozzles provides users with greater flexibility, improving the distribution of phytosanitary products while reducing spray volumes and influencing the dose transfer.

The author affirms that the dosages of phytosanitary products recommended by the molecule patent-holders are high, based on the results of rigorous and prolonged field trials to ensure that the product is successfully registered with the regulatory agencies. In many cases, the significant drop in the application rates currently used is the result of the end user's decision because, in most cases, the volume to be applied is determined by the user.

The droplet spectrum formed at the hydraulic nozzle is determined by the type of nozzle, orifice size (nominal flow rate), spray discharge angle, operating pressure and formulation of the phytosanitary product. These factors therefore affect the target coverage by the spray because once the volume to be applied and the crop area to be covered have been determined, the coverage can be modified by altering the droplet spectrum, subject to the limits imposed by drift and the run-off point. As a result, it is very important to select appropriate spray nozzles because doing so is crucial in determining the quantity applied per unit area, uniformity of application, coverage obtained and potential drift risk (Lan, Hoffmann, Fritz, Martin & Lopes Jr., 2008; Matthews, 2004, 2008; Zande et al., 2008).

The international classification system for droplet spectra and spray nozzles is based on two components: droplet size distribution and drift risk (Miller, Ellis & Gilbert, 2002). This classification, including the spray nozzle color code, is given in the Asabe/Asae standard S-572 (American Society of Agricultural and Biological Engineers [Asabe], 2004).

The great challenge of research in this field is to obtain ideal coverage of the target, distributing the droplets produced in a uniform way. If the droplets are too large, then there are problems with insufficient coverage of the target, lack of uniformity in the distribution and excess mass, which interferes with adhesion to the target and results in run-off to the soil. However, although using very small droplets would solve all these problems, they are likely to evaporate in low humidity conditions or be carried off by the downwind, aggravating the phenomenon of drift with increased risk of environmental contamination (Fritz, Hoffmann, Martin & Thomson, 2007; Jamar, Mostade, Huyghebaert, Pigeon & Lateur, 2010; Wolf & Daggupati, 2009; Zhu, Dorner, Rowland, Derksen & Ozkan, 2004). There is a strict correlation between the evaporation of the droplet spectrum produced and the weather conditions, such as temperature and relative humidity. Ramos and Pio (2008) reported that an air temperature above 30°C and a relative humidity below 55% are factors that favor this phenomenon and should therefore be monitored. Fritz (2006) and Yu et al. (2009) demonstrated the importance of meteorological factors in the spray and efficacy performance.

Matthews (2002) affirmed that each nozzle has its own volumetric distribution characteristics, specific to the height of the nozzle in relation to the target and the spacing between nozzles on the boom. If the volume applied is not adequate and uniform, then there is a risk that further applications will be needed to compensate for irregular application or untreated swathes (Peressin & Perecin, 2003). In Brazil, the spacing used between nozzle sets is usually 50 cm. Based on knowledge of the spray nozzle to be used and the jet emitted, it is possible to find the best relationship between spacing and minimum boom height over the target to set the parameters so that, depending on the swathe and operating pressure, the spray solution is deposited as uniformly as possible, with the lowest coefficient of variation (Cunha & Ruas, 2006; Peressin & Perecin, 2003).

Despite all these considerations and efforts towards safer and more efficient spraying practices for phytosanitary products, a great deal of importance is still attached to the insecticide but little to the application method (Cunha, Teixeira, Coury & Ferreira, 2003; Hislop, 1991).

The lack of knowledge of the basic concepts involved in applying phytosanitary products is evident in the scientific literature, which frequently indicates that the volume of solution applied is considered an adequate parameter to characterize and allow for repeatability of an application.

Even with the important mission and objective of making a name in the scientific world through publication in high-impact scientific periodicals (Slafer, 2008), insufficient details concerning phytosanitary product application methods have been reported in most of these publications, according to Matthews (2004). In his work ('How was the pesticide applied?'), he notes that the majority of scientific studies do not sufficiently detail the application methods used. Apart from the volume of water used for dilution, which is indicated in all publications, the author highlights a lack of information concerning the types of nozzle used, the application angle, the droplet spectrum category, the spray concentration, the operating pressure and how pressure is maintained, the nozzle position in relation to the crop and information on the weather conditions at the time of application. This lack of information makes it more difficult to judge whether the result of inefficient phytosanitary treatment was caused by the insecticide applied or by an inadequate application method.

The objective of this study was to verify the presence or absence of a basic methodological description, based on the suggestions made by Matthews (2004), of how insecticides were applied in the scientific papers published in Brazil and other countries. We also propose minimum requirements for an application method description to verify the adequacy of the conditions under which the experiments were performed and to afford technical assistance in repeating the insecticide application methods.

Material and methods

Based on the work of Matthews (2004) and bearing in mind his affirmation that scientific documentation contains insufficient basic methodological information on the phytosanitary product application method, we performed a survey of the bibliographical database available for consultation at the periodicals portal of Capes – the Brazilian Higher Education Coordination Agency (Qualis, 2006), journals and bibliographical databases freely accessible over the Internet, and periodicals, journals, and scientific magazines available at the libraries of the State University of Londrina (UEL), the State University of Maringá (UEM), the 'Luiz de Queiroz' Agricultural College (ESALQ/USP) and the Campo Mourão Integrado Faculty. We selected 200 studies involving the application of insecticides, 100 of which were published in Brazil and 100 published abroad. The studies examined were all published after 1990.

In choosing the studies examined, preference was given to those classified during the 2004/2006 triennial period (level A publications) by Qualis, a body that classifies publication media for intellectual output (bibliographical) from post-graduation programs and is used by Capes in its post-graduate assessment process.

We verified the presence or absence of the following information in the methodological description of the scientific work evaluated. Spray nozzle: description (type/model) - Spray discharge angle and droplet spectrum - Operating pressure used and how it was maintained - Application rate: volume of solution applied per unit area - Concentration of solution applied: dosage of the active ingredient used at the application rate - Air temperature at the time of application - Relative humidity at the time of application - Wind speed at the time of application.

The results for the national and international bibliographic databases examined were annotated by a simple 'yes' (present) or 'no' (absent) and presented as charts indicating percentages, arranged to give an overview that could be easily described and analyzed. We chose not to identify the studies evaluated because our objective was simply to indicate the publication medium and the quantitative data (Table 1).

Table 1. Bibliographic database examined and quantitative distribution.

Source	Number of issues evaluated	
	National	International
Agronomy Journal	-	12
Arquivos do Instituto Biológico	9	-
Ciência Rural	18	-
Crop Protection	-	10
Engenharia Agrícola	10	-
International Journal of Agriculture Science	-	12
Journal of Applied Entomology	-	9
Journal of Economic Entomology	-	15
Journal of Stored Products Research	-	4
Neotropical Entomology	-	7
Pesquisa Agropecuária Brasileira	14	-
Pesticide Science	-	6
Phytoparasitica	-	7
Planta Daninha	4	-
Revista Brasileira de Agrociência	8	-
Revista Brasileira de Entomologia	18	-
Revista Brasileira de Fruticultura	8	-
Revista Brasileira de Oleaginosas e Fibrosas	2	-
Scientia Agricola	9	-
Transactions of the American Entomological Society	-	15
Weed Technology	-	3
TOTAL	100	100

Results and discussion

The results for descriptions of the nozzles used (model), the spray angle and the droplet size are shown in Figure 1.

The survey indicates that, regardless of the main focus of the publication, this information, together with the spray angle, must be considered essential in scientific work. Although it may not seem significant because manufacturers' manuals are available for consultation and there is specific work on the development and operation of spray nozzles, the information on the spray droplet spectrum and the correct nozzle set-up in relation to the target is essential because each type of spray nozzle has its own characteristics, as noted by Matthews (2002). The most worrying aspect is the lack of information on the droplet size formed by the spray nozzles. Despite technological advances in the development of spray nozzles and international efforts to standardize them, seeking methods of application that are more technical, accurate, and safe in terms of the environment and human health and economical for phytosanitary products (reduction of potential drift), progress continues to be merely a trend, as affirmed by Matthews (2004). This lack of real progress is highlighted by the fact that 173 (86.5%) of the 200 studies evaluated did not contain this information.

Information on the operating pressure, the basic principle of hydraulic spraying, and the fundamentals for forming and maintaining droplet distribution (Figure 2) were not methodologically described in 19% (38) of the studies evaluated. This situation must be changed, and this information must be made mandatory without exception in studies dealing with the spray application of phytosanitary products. Because these factors vary considerably from one set of equipment to another, improper settings can render the application of insecticides non-viable.

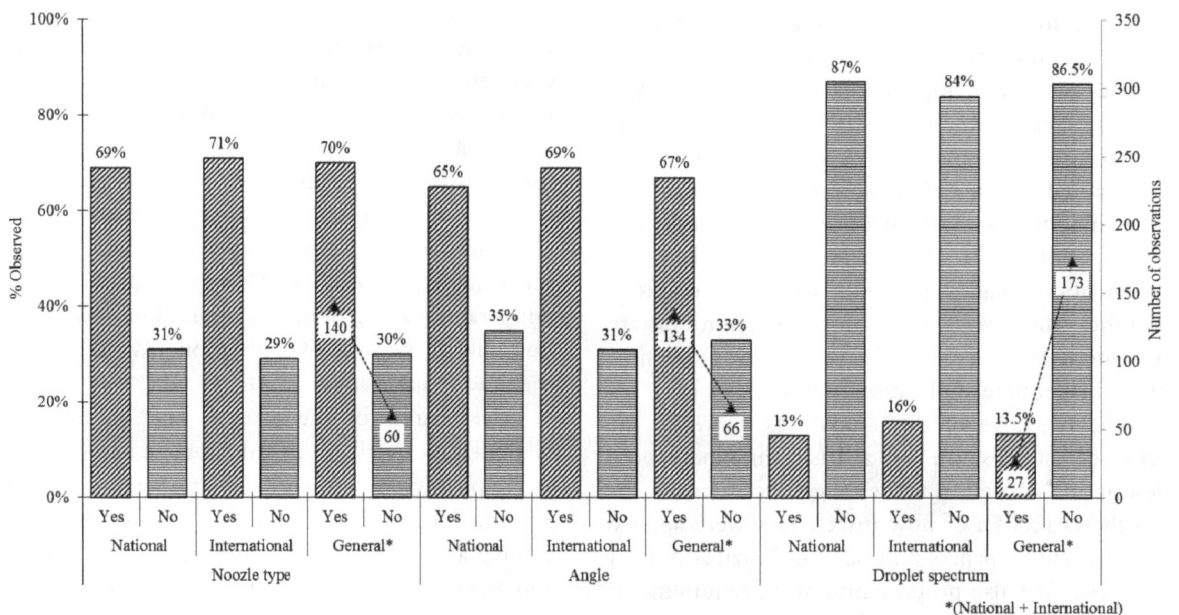

Figure 1. Percentage distribution and total amount of information on the type of spray nozzle used, spray angle and droplet spectrum for 100 national and international scientific studies concerning the application of insecticides.

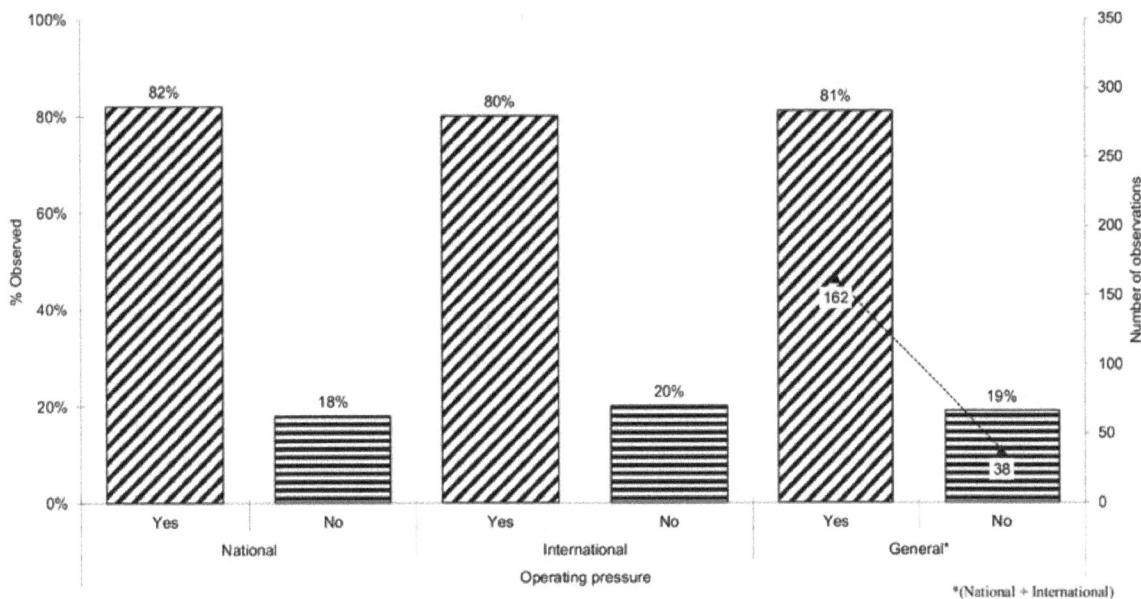

Figure 2. Percentage distribution and total amount of information on operating pressure in 100 national and international scientific studies involving the application of insecticides.

In general, the use of excessive pressure or unsuitable spray nozzles is the greatest cause of drift (Matthews, 2004). In his observations, Matthews affirmed that the volume of solution applied per unit area (application rate) is indicated in the majority of studies, as we ourselves confirmed in the studies we evaluated (Figure 3). However, the insecticide dosage (solution concentration) was not indicated in 26% (52) of the studies evaluated because, in practice, there is a tendency for low rates to be applied (200 L ha^{-1}) and taking into account recommended doses of insecticide per unit area (ha), it is important and indeed essential to indicate the concentration of the active ingredient (dosage) per unit volume applied. Although recommended doses are high, if the application method is unsuitable, then the effect can be attenuated or reinforced to the detriment of dose transfer, and losses are incurred in both cases (Matthews, 2008).

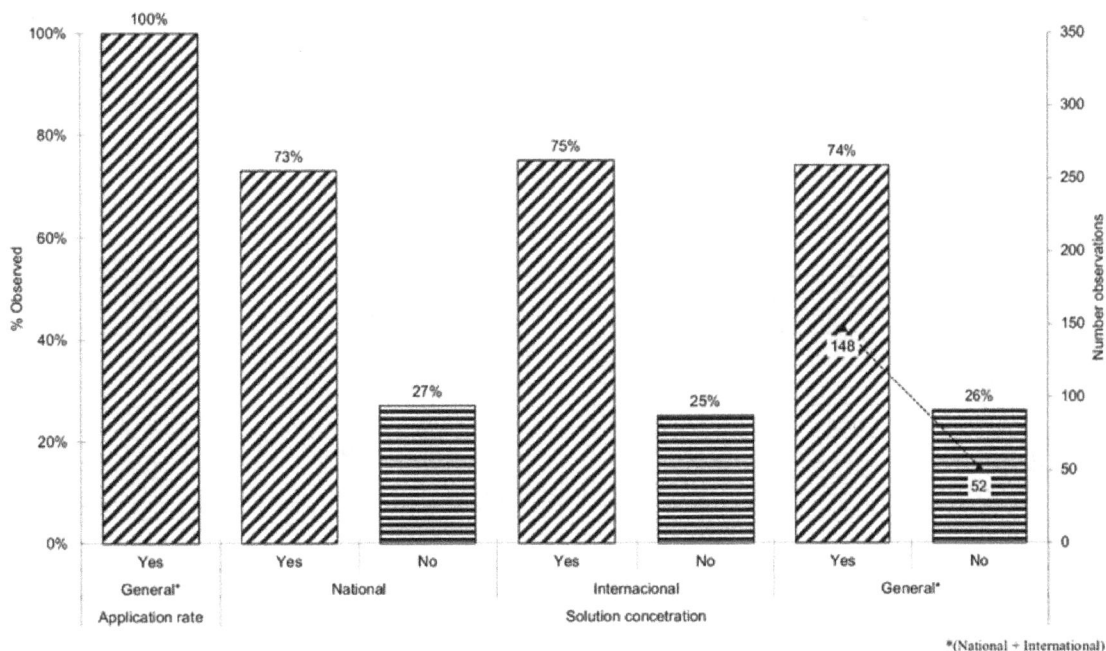

Figure 3. Percentage distribution and total amount of information on application rate and spray solution concentration for 100 national and international scientific studies involving the application of insecticides.

The information on the distance and position of the nozzle relative to the target (Figure 4) was little appreciated in the publications examined, except in specific studies. The fact that this information was omitted in 57% (114) of the studies examined is worrying. The spraying method, with various possibilities in terms of nozzle spacing and the relationship between the spacing and the height (distance) to the target (depending on the nozzle model, technology, operating pressure, type of solution, jet angle, topography, leaf density and crop architecture) can significantly affect the results of spraying, leading to excessive coefficients of variation and the consequent lack of uniformity in the deposition of the insecticide applied. This lack of uniformity can mean that it is necessary to respray some areas (new treatment), which can be directly attributed to the application method, as affirmed by Peressin and Perecin (2003).

The prospects for making progress in the field of science generate significant enthusiasm and concern regarding the objectives and results to be obtained. In some cases, this focus on results can lead to carelessness and negligence in the basic description of the method used in publishing notes, reports and technical and scientific papers in periodicals and journals with significant impact on the scientific community. Slafer (2008) notes the importance for agronomy journals of the number of scientists seeking to publish their most

relevant hypotheses in high-impact journals. However, the importance of publishing this work in the most basic plant science journals should not be underrated.

Following this line of argument, we should have a goal of publishing our work without ignoring the basic methodological aspects. When we evaluated temperature, relative humidity and wind speed (Figure 5), we found the results to be alarming because the results of our survey showed how little importance was attached to them. No fewer than 64 (128), 67 (134) and 72.5% (145) of the studies evaluated contained no information at all concerning temperature, relative humidity and wind speed (in that order). Regarding the use of insecticides, although systemic products can be used with fine, medium and possibly coarse droplets, it is an acknowledged fact that the results do not always match expectations, and notwithstanding the technological options available, inefficient control is the end result (Matthews, 2008). The observations and descriptive criteria for this information are fundamental to safety and efficacy in the application of phytosanitary products, with ample potential for significantly altering application methods in the ongoing search for optimum, homogeneous and safe transfer of the required dose to the direct or indirect target we are attempting to control.

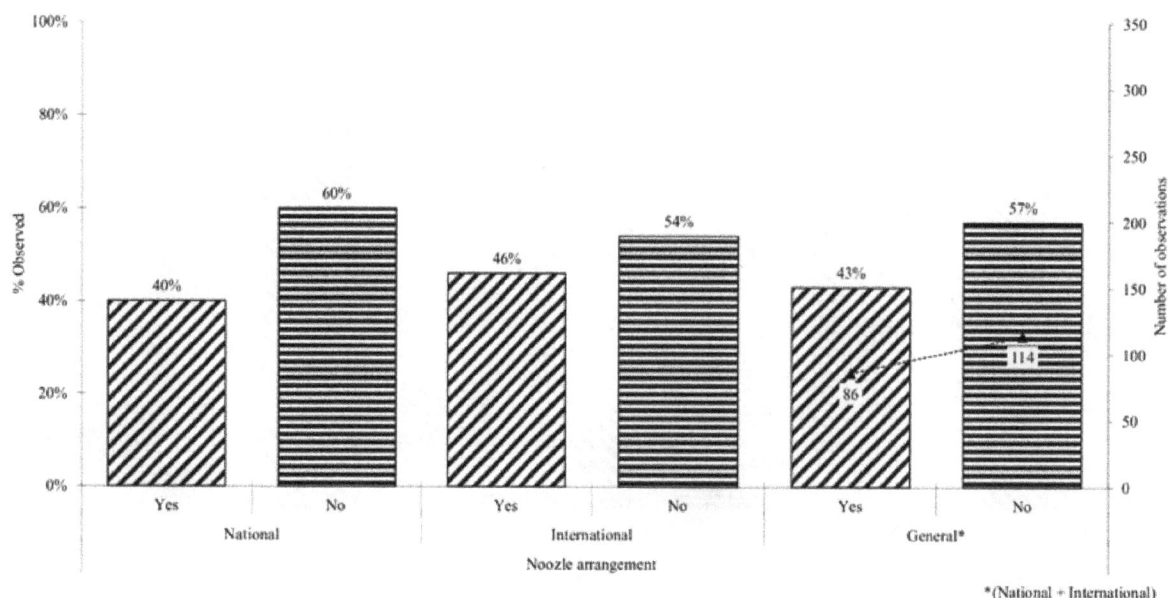

Figure 4. Percentage distribution and total amount of information on the nozzle distance and position in relation to the target in 100 national and international scientific studies on application of insecticides.

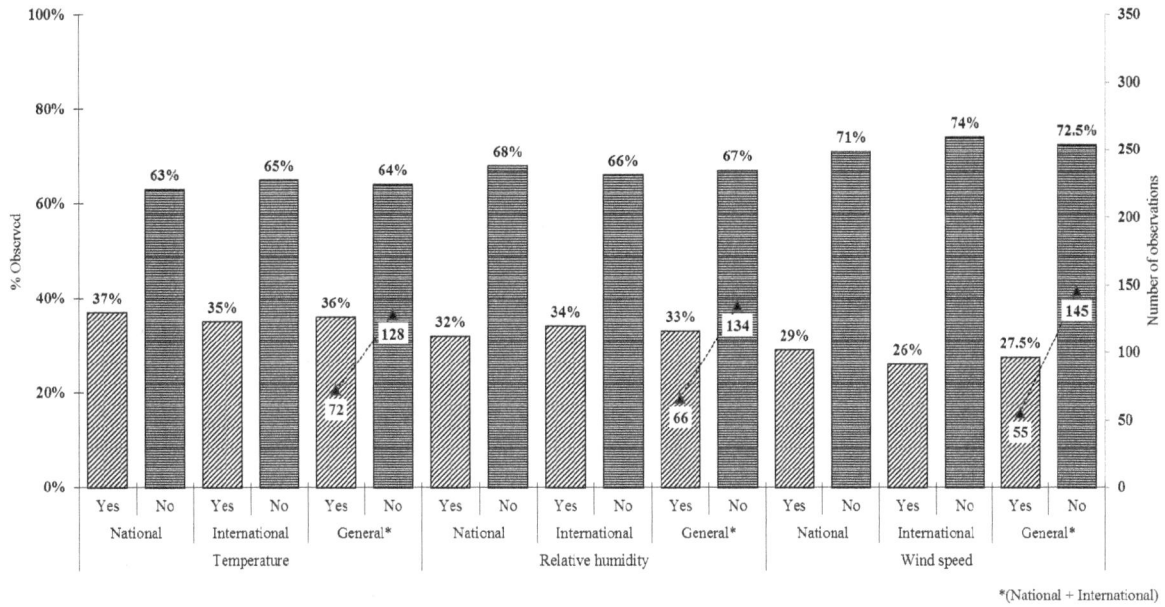

Figure 5. Percentage distribution and total amount of information on weather conditions in 100 national and international scientific studies involving the application of insecticides.

Under adverse weather conditions, with high temperatures, low relative humidity and high wind speed, there is a higher risk of environmental contamination due to drift. The quantity of droplets blown off-course (drift) is directly proportional to the wind speed and the smallness of the spray droplets. On one hand, because water is the dilution medium for most commercially available insecticides, evaporation also plays an important role and should be avoided. On the other hand, producing a spray with larger droplets reduces the risk of drift; however, because of the weight of the droplets, they may not adhere to the leaf surface and end up running off into the soil (Ellis, Webb, & Western, 2004).

In view of our findings, we propose a simplified method description that we hope will help those who are producing scientific work or competing to publish in the scientific periodicals and journals.

Our suggestion is based on the basic principle of the scientific method: repeatability. By providing the information proposed (Table 2), it will be easier to judge and acknowledge whether the application method was efficient or not. Some pieces of information merely give credence to and highlight the importance of the information that is normally provided.

Table 2. Proposed list of items that should be followed and expanded when describing a methodology in technical or scientific articles involving the application of insecticides.

Product description				
Toxicological category	Formulation	Mode of action, selectivity and phytotoxicity	Adjuvant(s)	Recommended dose x Dose used
Toxicology category of the product to be applied	Formulation of the product applied	Systemic or contact – selective or not – phytotoxicity conditions	Description, concentration and characterization (if used)	Indicate the dose recommended by the manufacturer (registered). Indicate the dose used in the application
Application				
Application phase	Recommended application rate (L ha^{-1})	Application rate used (L ha^{-1})	Operating pressure	Dilution agent
Pre-planting; pre-emergence; post-emergence	Solution volume recommended by manufacturer	Solution volume effectively applied	Operating pressure used (kPa)	If possible, describe the quality of the water used (hardness and pH)
Applications performed				
Date and time	Number of applications	Interval between applications	Description of target	Description of protected crop
Date and time of each application	Number of applications	Interval (days) between applications	Age, phenological stage and height of target	Age, phenological stage and height
Spray nozzle				
Model	Spray discharge angle	Flow rate (L min.$^{-1}$)	Droplet spectrum*	Average volumetric diameter (AVD)
Model of nozzle (manufacturer)	If available (degrees)	Nominal flow rate of spray nozzle	Droplet spectrum category at operating pressure used	If available

continue...

... continuation

Equipment				
Model	Type of boom	Nozzle spacing	Nozzle angle	Boom operating height / air assistance
Type of equipment, capacity to produce and maintain operating pressure	Length and type of mounting	Nozzle spacing used for application	Nozzle angle relative to the vertical.	Distance between spray nozzles and target. Type of air assistance used
Weather conditions				
Temperature	Relative humidity	Wind speed	Mist conditions	Rainfall
Temperature at time of application	Relative humidity at time of application	Wind speed at time of and during application	Mist conditions at time of application	If any (before, during and after application)
Safety				
Suggestion: information on the use of individual protective equipment (IPE) when preparing and applying the solution could be given to lay emphasis on the need for operator safety.				

*The international droplet spectrum classification lists droplet categories as very fine, fine, medium, coarse, very coarse and extremely coarse. This classification and the spray nozzle color code are given in Asabe/Asae standard S-572 (Asabe, 2004).

When this minimum information is provided, the researcher is in a better position to address challenges and discuss the results obtained, helping master and improve the techniques used for applying phytosanitary products. In this sense, the production of scientific work will benefit, and it will take less time to reach the professionals and disseminators of technology, who will put the work into practice in the field, thereby benefitting all segments of the production chain.

Conclusion

Currently, the minimum requirements for describing the methods used for applying insecticides are not being met. There are deficiencies in basic information concerning the spray nozzles used and their technical characteristics, spray solution concentration, operating pressure and how pressure is maintained, nozzle position in relation to the target, and information concerning weather conditions at the time of application. The use of the proposed minimum method description is practical for insecticide treatments, facilitating scientific verification through repeatability.

Acknowledgements

To the CNPq (National Council for Scientific and Technological Development).

References

American Society of Agricultural and Biological Engineers. (2004). *Spray nozzle classification by droplet spectra* (p. 437-440). Saint Joseph, MO: Asae.

Cunha, J. P. A. R., & Ruas, R. A. A. (2006). Uniformidade de distribuição volumétrica de pontas de pulverização de jato plano duplo com indução de ar. *Pesquisa Agropecuária Tropical, 36*(1), 61-66.

Cunha, J. P. A. R., Teixeira, M. M., Coury, J. R., & Ferreira, R. L. (2003). Avaliação de estratégias para redução da deriva de agrotóxicos em pulverizações hidráulicas. *Planta Daninha, 21*(2), 325-332.

Ellis, M. C. B., Webb, A., & Western, N. (2004). The effect of different spray liquids on the foliar retention of agricultural sprays by wheat plants in a canopy. *Pest Management Science, 60*(8), 786-794.

Fritz, B. K. (2006). Metereological effects on deposition and drift of aerially applied sprays. *Transactions of the Asabe, 49*(5), 1295-1301.

Fritz, B. K., Hoffmann, W. C., Martin, D. E., & Thomson, S. J. (2007). Aerial application methods for increasing spray deposition on wheat heads. *Applied Engineering in Agriculture, 23*(6), 709-715.

Hislop, E. C. Air assisted crop spraying: an introductory review. In Lavers, A., Herington, P., & Southcombe, E. S. E (p. 3-14). *Air-assisted spraying in crop protection* (Monografia). Swansea, UK: British Crop Protection Council.

Jamar, L., Mostade, O., Huyghebaert, B., Pigeon, O., & Lateur, M. (2010). Comparative performance of recycling tunnel and conventional sprayers using standard and drift-mitigating nozzles in dwarf apple orchards. *Crop Protection, 29*(6), 561-566.

Lan, Y., Hoffmann, W. C., Fritz, B. K., Martin, D. E., & Lopes Jr., J. D. (2008). Spray drift mitigation with spray mix adjuvants. *Applied Engineering in Agriculture, 24*(1), 5-10.

Matthews, G. A. (2002). The application of chemicals for plant disease control. In J. M. Waller, J. M. Lenné, & S. J. Waller. *Plant pathologist's pocketbook* (p. 345-353). London, UK: CAB.

Matthews, G. A. (2004). How was the pesticide applied? *Crop Protection, 23*, 651-653.

Matthews, G. A. (2008). Developments in application technology. *Environmentalist, 28*, 19-24.

Miller, P. C. H., Ellis, M. C. B., & Gilbert, A. J. (2002). Extending the International BCPC spray classification scheme. *Aspects of Applied Biology, 66*, 17-24.

Peressin, V. A., & Perecin, D. (2003). Avaliação do padrão de distribuição de bicos para aplicação de herbicidas: efeitos da altura do alvo nos padrões de distribuição. *Bragantia, 62*(3), 477-497.

Qualis. (2006). *Classificação de periódicos, anais, revistas e jornais.* Recuperado de http://www.qualis.capes.gov.br/webqualis/.

Ramos, H. H., & Pio, L. C. (2008). Tecnologia de aplicação de produtos fitossanitários. In L. Zambolim, M. Z. Conceição, & T. Santiago. *O que engenheiros agrônomos devem saber para orientar o uso de produtos fitossanitários* (p. 133-200). Viçosa, MG: UFV.

Sindicato Nacional da Indústria de Produtos para a Defesa Agrícola. (2008). *ANDEF.* Recuperado de http://www.sindag.com.br/html/estat_dezembro.html

Slafer G. A. (2008). Should crop scientists consider a journal's impact factor in deciding where to publish? *European Journal of Agronomy, 29*, 208-212

Wolf, R. E., & Daggupati, N. P. (2009). Nozzle type effect on soybean canopy penetration. *Applied Engineering in Agriculture, 25*(1), 23-30.

Yu, Y., Zhu, H., Frantz, J. M., Reding, M. E., Chan, K. C., & Ozkan, H. E. (2009). Evaporation and coverage area of pesticide droplets on hairy and waxy leaves. *Biosystems Engineering, 104*(3), 324-334.

Zambolim, L., Conceição, M. Z., & Santiago, T. (2008). *O que os engenheiros agrônomos devem saber para orientar o uso de produtos fitossanitários* (3a ed.). Viçosa, MG: UFV/DFP.

Zande, J. C. D., Huijsmans, J. F. M., Porskamp, E. H. A. J., Michielsen, J. M. G. P., Stallinga, H., Holterman, H. J., & Jong, A. (2008). Spray techniques: how to optimise spray deposition and minimise spray drift. *Environmentalist, 28*, 9-17.

Zhu, H., Dorner, J. W., Rowland, D. L., Derksen, R. C., & Ozkan, H. E. (2004). Spray penetration into peanut canopies with hydraulic nozzle tip. *Biosystems Engineering, 87*(3), 275-283.

3

The absorption and translocation of imazaquin in green manures

Flávia Garcia Florido[1], Patrícia Andrea Monquero[1*], Ana Carolina Ribeiro Dias[2] and Valdemar Luiz Tornisielo[3]

[1]Centro de Ciências Agrárias, Universidade Federal de São Carlos, Rod. Anhanguera, km 174, Cx. Postal 153, Araras, São Paulo, Brazil. [2]Departamento de Produção Vegetal, Escola Superior de Agricultura "Luiz de Queiroz", Universidade de São Paulo, Piracicaba, São Paulo, Brazil. [3]Laboratório de Ecotoxicologia, Centro de Energia Nuclear na Agricultura, Piracicaba, São Paulo, Brazil. *Author for correspondence. E-mail: pamonque@hotmail.com

ABSTRACT. Green manure species that are tolerant to the herbicide imazaquin can be used in crop rotation schemes that aim to reduce herbicide carryover to sensitive plants such as sunflower or corn. Three different doses of imazaquin (0, 0.15 and 0.28 kg ha⁻¹) were applied during the pre-emergence growth stage to *Dolichos lablab, Cajanus cajan, Canavalia ensiformis, Crotalaria juncea, C. breviflora, C. spectabilis, Mucuna deeringiana, M. cinerea, M. aterrima, Lupinus albus, Helianthus annuus, Pennisetum glaucum, Avena strigosa* and *Raphanus sativus*, and the results were evaluated in a greenhouse. *C. ensiformis* and *M. cinerea* were selected from these species for being the most tolerant, and they were then evaluated for absorption and translocation of ¹⁴C-imazaquin in two different growth stages: the cotyledonary stage and the emergence of the first pair of true leaves. *M. cinerea* individuals showed the best potential for translocating imazaquin to the shoot when compared to *C. ensiformes*, which accumulated the herbicide mostly in its roots. These plants had a higher ability to accumulate herbicide during their most advanced stage of development, which demonstrates their potential for use in areas that have residual imazaquin.

Keywords: herbicide, residual, tolerant species.

Introduction

Agrochemicals are classified according to the target organisms they are designed to control (e.g., insects, weeds or fungi). Of all the target organisms, weeds cause by far the greatest economic loss as a consequence of their interference in crop production. It is therefore not surprising that herbicides are the most common class of agrochemicals that are used in Brazil (48% of total expenditures), outstripping insecticides (30%) and fungicides (21%) (PINHEIRO et al., 2011).

Predicting the movement and fate of herbicides in soils is an important step in limiting their environmental impacts (CARABIAS-MARTINEZ

et al., 2000). Herbicides may be sorbed by mineral and organic colloids and, depending on their binding energy, they may become unavailable to the plants (via bond residue fractions) they are intended to kill and can also become unavailable for biodegradation or desorption in soil solution, causing them to be transported or absorbed by plants (HORNSBY et al., 1995).

Herbicide persistence in soil exerts a strong influence on the control of weeds, can cause damage to succession crops and can lead to environmental contamination risks. This persistence varies with the chemical structure of the molecule, the type of soil to which it is applied and the climatic conditions, such as the soil humidity, which in turn affects absorption, leaching and microbial/chemical decomposition (SILVA et al., 1999).

Imazaquin is an imidazolinone herbicide that is widely used for broadleaf and grassy weed control in soybeans and warm-season turf grasses (SEIFERT et al., 2001). The absorption of imazaquin herbicide, also known as 2-[(RS)-4-isopropyl-4-methyl-5-oxo-2-imidazolin-2-yl]quinoline-3-carboxylic acid, occurs through both the roots and the leaves, while translocation occurs through the phloem and xylem, accumulating in the meristems of plants where it produces necrosis (SHANER, 2003). This molecule acts by inhibiting the acetolactate synthase enzyme (ALS), resulting in a blockage of the synthesis of the amino acids valine, leucine and isoleucine. The phytotoxic effect of imidazolinone is caused by deficiency of these amino acids, leading to a decrease in DNA and protein synthesis, which adversely affects cellular division and photosynthate translocation to growing points. These processes cause a reduction in plant growth as well as an elongation of leaves and chlorosis between leaf ribs (TAN et al., 2005).

Imazaquin is an ionizable organic molecule that contains both an ionizable carboxyl group with a pKa of 3.8 and a basic quinoline ring with a dissociation constant of 2.0. At the most common pH range for tropical soils (pH 4.0-6.0), imazaquin predominantly behaves as an organic anion (NOVO et al., 1997), causing low sorption by soil colloids, which are also negatively charged at this pH range. However, organic matter can react with polyvalent cations to form chelates or ionic bridges with acid herbicides, which decreases the pH effect (AICHELE; PENNER, 2005). Several factors, such as speciation, soil solution and sorbent surface pH, charge, ionic strength, and solution composition, must be considered due to the herbicide's amphoteric nature for the user to successfully predict soil sorption (REGITANO et al., 1997, 2001; ROCHA et al., 2003).

The imidazolinone group has a high water solubility and high persistence in the environment, which is predicated on an imazaquin half-life that varies from 16 weeks (AICHELE; PENNER, 2005) to 210 days (VIDAL, 2002). On the one hand, this half-life is good because the herbicide can provide residual weed control during the entire life cycle of the soybean, but on the other hand, it may become a risk to succession crops such as winter maize, cotton, sunflower and brassica crops (ARTUZI; CONTIERO, 2006; DAN et al., 2011; RODRIGUES; ALMEIDA, 2011; SEIFERT et al., 2001; YODER et al., 2000) or lead to environmental

contamination. In the U.S., sixteen active ingredients of herbicides belonging to the sulfonylureas, sulfonamides and imidazolinones were found in samples that were collected from surface and groundwater (BATTAGLIN et al., 2000).

One possible method to assuage the effects of herbicide residues is phytoremediation. This technique aims to decontaminate soil and water by using plants as cleansing agents (NEWMAN et al., 1998). The cleansing action may occur by direct assimilation of the contaminating agents and subsequent accumulation of non-toxic metabolites in the vegetable tissue, such as structural components, or by the stimulation of microbial activity by the plant (SCRAMIN et al., 2001).

The use of phytoremediation is based on the natural or purposefully developed ability that some species exhibit to specific types of compounds or action mechanisms. Phytoremediation is commonly performed with agricultural species and weeds that are tolerant to certain herbicides (PIRES et al., 2003). Plant selection is related to the fact that organic compounds may be translocated by plants to other tissues within the transpiration stream, from which the compounds could be volatilized (GENT et al., 2007). They may also suffer partial or complete degradation or be transformed into less toxic compounds, especially less phytotoxic compounds, when combined and/or connected to plant tissue (via compartmentalization) (ACCIOLY; SIQUEIRA, 2000; SCRAMIN et al., 2001).

A plants' capacity to metabolize pesticides into compounds that are non-toxic (or less toxic) is the principle behind phytodegradation. Another possible strategy is phytostimulation in which plants stimulate microbial activity, promoting the release of root exudates that degrade the compound in the soil, which in some situations, determines the rhizosphere's capacity to engage in the bioremediation of toxic compounds (PIRES et al., 2003).

In Brazil, phytoremediation has been shown to have promise in the decontamination of soils that are contaminated with various herbicides such as tebuthiuron (PIRES et al., 2005), trifloxysulfuron-sodium (SANTOS et al., 2004) and picloram (CARMO et al., 2008). This method has been advantageous when the levels of compound in the soil are high and conditions favor herbicide leaching.

Thus, it is important to select plant species that have phytoremediation potential. It is also important to understand the physiological mechanisms that are responsible for these characteristics in order to create and plan new crop protection strategies, in addition to perfecting already existing treatments. Therefore, this work aims to select species of green manures that are tolerant to imazaquin and to determine the physiological bases for their different responses.

Material and methods

Selection of imazaquin-tolerant plants

This experiment was installed and conducted in a greenhouse. The imazaquin tolerance of 14 species of plants, among which were leguminous, cruciferous, gramineous and composed plants, was evaluated. The evaluated species were *Dolichos lablab, Cajanus cajan, Canavalis ensiformis, Crotalaria juncea, C. breviflora, C. spectabilis, Mucuna deeringiana, M. cinerea, M. aterrima, Lupinus albus, Helianthus annuus, Pennisetum glaucum, Avena strigosa* and *Raphanus sativus*. Samples of a soil classified as Hapludox (EMBRAPA, 1999) without a history of herbicide use were taken from a depth of 0-20 cm for the plantings. Physical and chemical analyses were performed in the Soil Chemistry and Fertility Laboratory at UFSCar [Federal University of São Carlos] (Table 1).

The experiments were set up in randomized blocks with three repetitions. These blocks were arranged in three arrays of 14 x 3 (factorial), which contained 14 species of green manures and three different herbicide doses. The experimental unit was a polyethylene vase that was lined with a plastic bag to avoid the outflow of herbicide, and the vase contained 20 dm³ of soil.

Imazaquin herbicide (at concentrations of 0, 0.15 and 0.28 kg i.a ha^{-1}) was applied during the plant pre-emergence stage with a CO_2-pressurized backpack sprayer that had a sprayer lance with two Teejet AI-110.02 fan-type nozzles and a spray volume of 200 L ha^{-1}.

Seeding was performed the day after herbicide application. The vases were then irrigated frequently using a fixed conventional sprinkler system to maintain the soil's humidity at approximately 80% of field capacity. Visual intoxication that was caused by the herbicide was evaluated at 15, 30 and 45 days after seeding (DAS). Grades from 0 to 100 were given according to the intoxication symptoms demonstrated by the plant shoots, where

0 represents an absence of symptoms and 100 represents the death of the plant. At 46 DAS, the dry matter masses of the shoots and roots were also determined. The shoots and roots were placed in a heater with forced air circulation (60 ± 2°C) for 72 hours to achieve drying.

An analysis of weight averages and variances was performed and compared with the Tukey test at a 5% probability.

Evaluation of the absorption and translocation of imazaquin

This experiment was performed at the Ecotoxicology Laboratory of CENA/USP [*Nuclear Energy in Agriculture Centre/University of São Paulo*]. Based on the previous study, the green manure species *Canavalis ensiformis* (jack bean) and *Mucuna cinerea* (grey mucuna) were selected for their high tolerance to imazaquin herbicide.

Samples of Hapludox were extracted from the 0-10 cm layer (Table 1) from an area without a history of herbicide use and were used as the substrate for the planting of these two species. The soil was dried and passed through a 2-mm sifter to remove part of the organic matter and to increase the soil's homogeneity before planting.

Two vases with 1.5 dm³ of soil were used for each species and phenological stage under study. The soil's density was calculated (1083 g mL^{-1}) with the purpose of estimating the total weight that a vase could support (1270 g). Thus, the necessary weight to fill a vase to a depth of 2-cm of superficial soil (307 g) was calculated. The procedure consisted of contaminating the first soil layer with radiolabeled herbicide and then positioning the seeds between the cold soil layer (that is, the soil without radiolabeled product) and the radiolabeled layer for germination.

The 2-cm vase soil layer was ground prior to the application of ^{14}C-imazaquin solution to ensure homogeneous herbicide application. The herbicide was not homogenized within all the soil inside the vase because the results obtained during the pre-experimentation phase on imazaquin leaching revealed that the herbicide attaches itself strongly to the superficial layer of the soil, and the objective was to evaluate the capacity and behavior of the species that were chosen to tolerate the herbicide under field conditions.

Table 1. Physical and chemical characteristics of the soil that was used in the experiment.

Sample	pH CaCl$_2$	MO g dm^{-3}	P mg dm^{-3}	K	Ca	Mg	H + Al	SB	CTC	V% %	Clay	Silt	Sand
				---------------mmol$_c$ dm^{-3}---------------							---------g kg^{-1}---------		
0-10	6.2	36	14	2.4	29	13	0	44.4	68.4	64.9	560	240	200

Twenty seeds per species were planted (10 seeds per vase). One phenological stage was evaluated per vase (stage 1: cotyledonary leaves and stage 2: first pair of true leaves).

Preparation of radiolabeled imazaquin solution

The herbicide product that was used for experimental purposes was commercial imazaquin (Scepter technical) (2-[4,5-dihydro-4-methyl-4-(1-methylethyl)-5-oxo-1H-imidazol-2-yl]-3-quinolinecarboxylic acid), from the Cyanamid brand (purity = 98.26%). Its respective radioactive isotope (^{14}C-imazaquin, labeled carboxylic group, specific activity = 0.80 MBq mg^{-1}; radiochemical purity = 98%) was used as the radiolabeled product (Figure 1).

For the labeled spray, a solution containing 18,933.73 dpm μL^{-1} of ^{14}C-imazaquin was prepared in which the labeled imazaquin was diluted in a solution containing unlabeled imazaquin (technical product) in a manner that ensured that the final herbicide concentration was equal to the commercial dose of 161 g ha^{-1}, with a total spray volume of 1 L ha^{-1}.

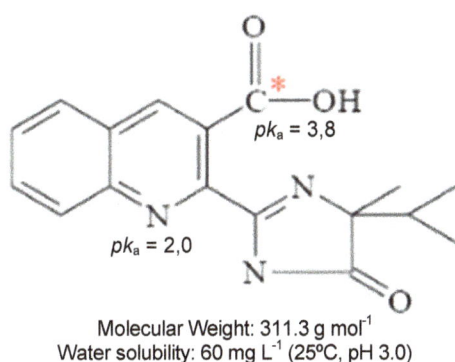

Molecular Weight: 311.3 g mol^{-1}
Water solubility: 60 mg L^{-1} (25°C, pH 3.0)

Figure 1. Imazaquin Molecule. *Radioactively labeled carbon (^{14}C).

Experimental assembly

This experiment was conducted in a randomized block design with ten replications *Canavalis ensiformis* and *Mucuna cinerea* were planted. Sifted dry soil was added to each vase, and its surface was leveled and cleared of possible fragments that could direct the behavior of the herbicide in the soil in a preferential manner.

The green manure seeds underwent a process of mechanical chiseling to break their dormancy. All of the seeds were immersed in 10% sodium hypochlorite for 10 minutes to avoid the proliferation of fungi and bacteria.

Each species' seeds were placed on the 'cold' soil (without radiation), and then the radiolabeled layer was added. The humidity level was reestablished during this first phase with the addition of water to

the vase dishes until the water reached the surface of the soil. The vases were then positioned on a tray to avoid possible radioactive contamination.

To quantify the amount of remaining herbicide, the water and soil that remained after the vase disassembly process were analyzed. The soil was dried at room temperature and the solution of exceeding soil was centrifuged at 4,000 rpm for 15 minutes to remove the suspended soil. The vials were then dried in a heater at 40°C, and the centrifuged solution was added to the dry soil. After being ground, four replicates of 0.2 g of soil each were examined in a biological oxidizer to quantify the remaining herbicide.

The supernatants from the centrifugation step were stored in vials of 500 mL each and filtered with the aid of a vacuum pump and common filter paper to remove the organic matter. The solution that resulted from the disassembly of the vases (SDV) was analyzed by examining 3 aliquots of 0.1 mL each in 10 mL of scintillation solution that were then quantified in a liquid scintillation spectrometer. After these evaluations, the SDV density was 1 g mL^{-1}. The filter paper and the filtered solution were dried in a heater at 40°C, and after drying, the organic matter was incorporated into the soil that was being analyzed.

The absorption and translocation of imazaquin were qualitatively studied by autoradiography and quantitatively evaluated through the combustion of plant tissue. The plants were washed, pressed and dried in a heater with forced air circulation at 45°C for 72 hours. Three specimens from each growth stage were autoradiographed using Crafts and Yamaguchi's protocol (1964) and analyzed in a autoradiography detector.

After drying, the plants were taken out and divided into groups of leaves, roots, stems and cotyledons (whenever present) with the objective of quantifying the radioactivity in each group. The combustion was performed with a biological oxidizer, with six repetitions for each part of the plant. The radioactivity present in all parts of the plants was considered to have been translocated. An average of six repetitions was calculated, and the radioactivity of each plant part was compared to the total radioactivity absorbed by the plant to calculate the translocation.

Results and discussion

Selection of imazaquin-tolerant plants

At 15 days after seeding (DAS), the species that were more sensitive to imazaquin were clearly observable after they received a dose of 0.28 kg ha^{-1}.

The sensitive species were *R. sativus* (with 66.67% of phytotoxicity), *P. glaucum* (73.33%) and *C. juncea* (46.67%), and the more tolerant ones were *C. spectabilis* (0%) and *C. cajan* (13.33%) (Table 2). At 30 DAS, the more tolerant species were *C. cajan* with 0 to 31.67% phytotoxicity at the highest and lower doses, respectively, and *C. ensiformis* with 10% phytotoxicity for both doses (Table 2). Usually the basis for imidazolinone selectivity results from a difference in the nature or rate of herbicide metabolism or through a version of acetolactate synthase (ALS) that is insensitive to inhibition by the herbicide (NEWHOUSE et al., 1992). At 45 DAS, there was a visual observation of green manure responses, with the species *C. ensiformis*, *M. aterrima*, *C. cajan* and *M. cinerea* showing the greatest herbicide tolerance. Seedlings of *R. sativus*, *P. glaucum* and *H. annuus* were the most susceptible, especially at the highest product dose. Imazaquin acts by inhibiting the ALS, which is essential for the synthesis of the amino acids valine, leucine and isoleucine. ALS-inhibiting herbicides are widely used because of their low dose rate, sound environmental properties, low mammalian toxicity, wide crop selectivity and high efficacy (TAN et al., 2005).

C. ensiformis, *C. breviflora* and *M. cinerea* did not present significant shoot biomass differences between the different herbicide doses. However, a higher accumulation of biomass in *C. ensiformis* (10.57 g in the highest dose), *M. cinerea* (6.30 g in the highest dose) and *M. aterrima* (9.64 g in the highest dose) was observed. It was verified that plants with the most significant root production were *C. ensiformis* and *M. cinerea* (Table 3). Although the imidazolinones are weed-selective in some crops such as peanut (*Arachis hypogaea*) and soybean (*Glycine max*), severe injury is normally observed when applied to other crops such as melon (*Cucumis melo*), cucumber (*Cucumis sativus*), sunflower (*Helianthus annuus*) and mustard (*Brassica* sp.) (THOMPSON et al., 2005). Imazaquin carryover was most pronounced after pre-plant incorporation in soils with higher organic matter and clay content (SMITH et al., 2005).

Table 2. Percentage of phytotoxicity from imazaquin herbicide in green manure species at 15, 30 and 45 days after seeding (DAS).

Species	Number of days after seeding (DAS)								
	15			30			45		
	Imazaquin doses kg i.a ha^{-1}								
	0.15	0.28	0	0.15	0.28	0	0.15	0.28	0
D. lablab	6.67 cB	30.00 cdeA	0.00 aC	0.00 dB	36.67cdA	0.00 aB	16.67 cdB	56.67 cdA	0.00 aC
C. cajan	11.67 cA	13.33 efgA	0.00 aB	0.00 dB	31.67 dA	0.00 aB	13.33 cdAB	16.67 fA	0.00 aB
C. ensiformis	0.00 cB	20.00 defA	0.00 aB	10.00 dA	10.00 eA	0.00 aB	0.00 dA	13.33 fA	0.00 aA
C. juncea	6.67 cB	46.67 bA	0.00 aB	28.33 cA	40.00 cdA	0.00 aB	46.67 bB	63.33 cA	0.00 aC
C. breviflora	13.33 cA	16.67 efgA	0.00 aB	28.33cA	40.00 cdA	0.00 aB	0.00 dB	26.67 efA	0.00 aB
C. spectabilis	46.67 bB	73.33 aA	0.00 aC	43.33 bcA	50.00 cA	0.00 aB	33.33 caA	40.00 deA	0.00 aB
M. deeringiana	13.33 cA	16.67 efgA	0.00 aB	6.67 dB	80.0 bA	0.00 aB	13.33 cdB	73.33 bcA	0.00 aC
M. cinerea	65.00 bA	28.33 cdeA	0.00 aC	0.00 dB	38.33 cdA	0.00 aB	0.00 dA	13.33 fA	0.00 aA
M. aterrima	36.67 bA	40.00 bcA	0.00 aB	0.00 dB	40.00 cdA	0.00 aB	0.00 dB	20.00 efA	0.00 aB
L. albus	6.67 cA	0.00 gA	0.00 aA	10.00 dB	93.33 abA	0.00 aB	3.33 dB	90.00 abA	0.00 aB
H. annus	13.33 cA	10.00 fgAB	0.00 aB	55.00 bB	90.00 abA	0.00 aC	83.33 aA	93.67 abA	0.00 aB
P. glaucum	1.67 cB	28.33 cdeA	0.00 aB	56.67 bB	90.00 abA	0.00 aC	93.33 aA	96.67 aA	0.00 aB
A. strigosa	10.0 cAB	16.67 efgA	0.00 aB	38.33 cB	86.67 abA	0.00 aC	40.00 bB	90.00 abA	0.00 aC
R. sativus	66.67 aB	81.67aA	0.00 aC	96.67 aA	100.00 aA	0.00 aB	96.67 aA	100.0 aA	0.00 aB
C.V (%)	35.96			20.89			26.35		

Equal lower-case letters in the column and capital letters in the lines do not differ significantly from each other according to the Tukey test at a 5% probability within each evaluation.

Table 3. Biomass (g) of the shoots and roots of green manure species at 46 days after seeding (DAS).

Species	Imazaquin doses kg i.a ha^{-1}					
	0.15	0.28	0	0.15	0.28	0
	Biomass of the shoot (g)			Biomass of the roots (g)		
D. lablab	3.70 cdB	2.32 bB	6.10 bcA	1.57 abA	0.80 bcB	1.60 bcA
C. cajan	1.07 defA	0.93 bB	2.20 deA	0.79 cdB	0.64 cdB	1.45 bcA
C. ensiformis	10.57 aA	9.24 aA	10.96 aA	1.68 abB	2.10 aAB	2.42 aA
C. juncea	0.92 defB	0.72 bB	3.63 cdA	0.55 cdeB	0.48 deB	1.82 abA
C. breviflora	0.29 fA	0.15 bA	0.37 eA	0.28 deA	0.18 efgB	0.17 hB
C. spectabilis	0.73 defA	0.26 bB	1.20 deA	0.20 eA	0.34 efA	0.36 ghA
M. deeringiana	3.71 bB	1.59 bC	6.27 bcA	0.61 cdeA	0.49 deAB	0.49 fghA
M. cinerea	6.30 bcA	6.87 aA	6.33 bcA	1.74 aA	1.41 abAB	1.08 cdeA
M. aterrima	9.64 aA	2.92 bB	7.89 bA	1.06 aA	1.36 abA	1.03 cdeA
L. albus	8.93 abB	2.10 bC	13.69 aA	1.70 aA	1.20 bB	1.75 abA
H. annus	0.25 fB	0.19 bB	2.39 deA	0.15 eB	0.06 fgB	1.41 bcA
P. glaucum	0.18 fB	0.06 cA	0.41 eA	0.97 cA	0.29 efB	0.84 efgA
A. strigosa	0.27 fB	0.20 bA	1.50 deA	1.55abA	0.67 cdB	1.68 bcA
R. sativus	0.62 efB	0.15 bB	5.54 bcA	0.08 eB	0.01 gB	1.06 cdeA
C.V (%)	30.51			24.40		

Equal lower-case letters in the column and capital letters in the lines do not differ significantly from each other according to the Tukey test at a 5% probability.

Following these data, *M. cinerea* and *C. ensiformis* were selected for further imazaquin absorption and translocation study, as a result of both visual observations and the production of dry mass by shoots and roots in these species.

Absorption and translocation

Canavalis ensiformis

Autoradiography was performed at time 1-T1 (first pair of cotyledonary leaves) and at time 2-T2 (first pair of true leaves) in the species that are shown in Figure 2. One can observe that a higher quantity of radiation is concentrated in the roots. This sequestration resulted from the direct contact between the plants and the radiolabeled herbicide because the imazaquin is absorbed through the roots (SHANER, 2003). Because the entire plant is completely visible in the picture, one can infer that the herbicide that was absorbed by the roots was translocated through all the plant parts because the figure only reveals the locations with observable radiation. A darker outline was also observed in all of the leaves. This indicates a higher quantity of herbicide in these areas, and one possible hypothesis is that the imazaquin is being translocated through the xylem and phloem and is accumulating in the growth points, where there is higher acetolactate synthase enzyme activity (TROXLER et al., 2007).

The values in Table 4 show how the herbicide is translocated inside the plant. At both times, a higher concentration of herbicide was observed in the roots (49.90% and 54.38% in T1 and T2, respectively), which demonstrated that despite the translocation, most of the imazaquin remained in the root system. This same behavior was observed when the weeds *Ipomoea lacunosa* and *I. hederifolia* were used in which the largest percentage of applied ^{14}C imazaquin was found in the roots of the plants (RISLEY; OLIVER, 1992). However, when the plants presented their first pair of true leaves, the percentage of herbicide in the leaves increased to 36.06%. The increased herbicide that was found in the leaf over the course of the plant's life cycle proves that herbicide translocation to the shoot tends to increase in plants during the more developed phenological stages. These results showed an increase in the accumulation of radiation per plant of approximately 2.7 times, or the amount that was absorbed in the cotyledonary leaf stage almost tripled by the true leaf stage. Askew and Wilcut (2002) reported that cotton (*Gossypium hirsutum*), jimsonweed (*Datura stramonium*), peanut (*Arachis hypogaea*), and sicklepod (*Senna obtusifolia*) rapidly absorbed ^{14}C trifloxysulfuron (another ALS inhibitor) between 0 and 4 hours after treatments, but total absorption

after 72 hours varied by species between 30 and 70%. By comparison, tobacco absorbed 40% of the applied ^{14}C-imazaquin at 8 days after treatment and translocated 22% (WALLS et al., 1993). It is important to note that *C. ensiformis* does not have cotyledons when the first pair of true leaves is present.

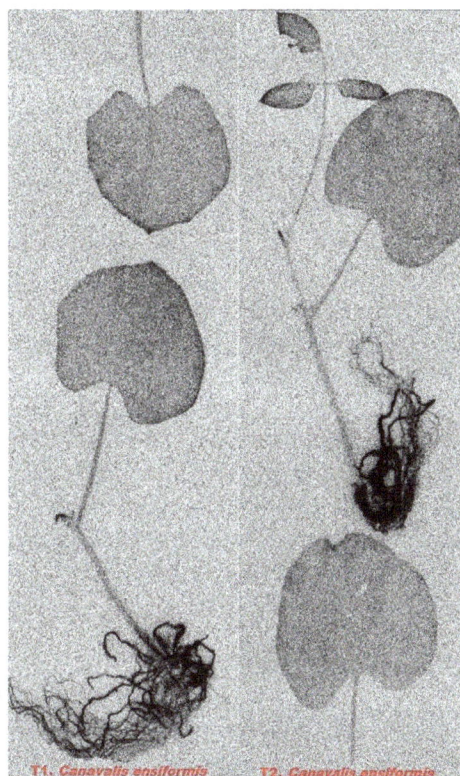

Caption: T1. Jack Bean | T2. Jack bean.

Figure 2. Autoradiography indicating the absorption and translocation of imazaquin herbicide in *C. ensiformis* at time 1-T1 (first pair of cotyledonary leaves) and time 2-T2 (first pair of true leaves).

Out of all the applied radiation, *C. ensiformis* plants absorbed 2.57% and 7.91% during the cotyledonary leaf and first pair of true leaves stages, respectively (Table 5). Thus, one can observe that the plant continued to absorb imazaquin during its growth, increasing absorption as a consequence of the accumulation of dry matter. When observing individual absorption data, an absorption of 0.30% in the cotyledonary stage and 0.98% during the first pair of leaves stage was observed. Marcacci et al. (2006) showed that the high biomass production of these plants is of crucial importance for the progress of phytoremediation. These results provide evidence for the importance of increasing the growth of green manures in areas with herbicide residue, which would increase the potential removal of these products from the soil through the actions of these plants.

Table 4. Translocation of imazaquin herbicide inside *C. ensiformis* during different phenological stages.

Stages	Accumulated radiation (dpm)					Translocation (%)			
	Total	Leaf	Stem	Root	Cotyledons	Leaf	Stem	Root	Cotyledons
Cotyledonary leaves	7647.36	1914.23	927.34	3816.05	989.74	25.03	12.13	49.90	12.94
First pair of true leaves	20383.8	7350.17	1949.89	11083.76	-	36.06	9.57	54.38	-

Table 5. Recovery of the radioactivity of imazaquin in tests performed in different phenological stages of *C. ensiformis*.

Stage	Radioactivity recovered with the solution used in the disassembly of the vases (%)	Radioactivity recovered in the soil (%)	Number of plants in each vase	Radioactivity recovered in the plants (%)	Total of Recovered Radioactivity (%)
Cotyledonary leaves	5.58	88.20	8	2.37	96.16
First pair of true leaves	7.18	69.57	8	7.91	84.67

The total radioactivity that was recovered during the second time period was 7.91%, more than three times that of the cotyledonary leaf stage (2.37%). Pester et al. (2001) reported that the recovery of applied ^{14}C-imazamox (another ALS inhibitor) ranged from 80 to 94% for jointed goatgrass and feral rye. This difference can be explained by his method in which herbicide was applied to the leaves.

Mucuna cinerea

The autoradiography results from the imazaquin herbicide treatment of M. cinerea during the first pair of cotyledonary leaves stage (T1) and the first pair of true leaves stage (T2) are shown in Figure 3. We can observe that the radiation is homogeneously distributed throughout the plant when compared to *C. ensiformes*.

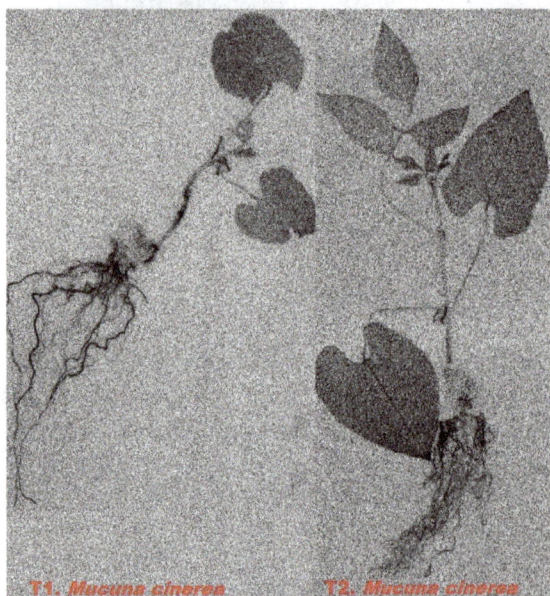

Caption: T1. Grey Mucuna | T2. Grey Mucuna.

Figure 3. Autoradiography indicating the absorption and translocation of imazaquin herbicide in the *M. cinerea* at time 1 (first pair of cotyledonary leaves) and time 2 (first pair of true leaves).

During the cotyledonary leaf stage, it was observed that translocation was higher to the leaves (31.14%), surpassing the values observed for the roots (29.34%). When the plants presented their first pair of true leaves,

an increase in the leaf herbicide concentration (44.54%) and a reduction in the root values (28.45%) and in the cotyledons (18.26%) were observed (Table 6). There was an increase of almost 2000 dpm of total accumulated radiation per plant as the cycle of the cotyledonary stage (5528.93 dpm) advanced toward the first pair of true leaves stage (7289.28 dpm). These results agree with those of Salihu et al. (1998), who showed that the translocation of absorbed radioactivity from roots to shoots increased with time. Once absorbed, herbicide translocation can be affected by many factors, including plant growth stage, photosynthetic rate, phloem mobility, sink strength, and environment (PESTER et al., 2001).

In tobacco plants 40% of the ^{14}C imazaquin that was applied during postemergence was absorbed, 54% remained in the water extract on the leaf surface and 6% stayed in the epicuticular wax layer, and the translocation of herbicide from the treated leaves to the roots was very low (4-5%). By contrast, the application of herbicide to the soil resulted in retention of 40 to 53% of the radiolabeled products in the roots and a translocation of 47 to 60% to the shoots (WALLS et al., 1993).

These results are important because they show that this plant has the potential to accumulate imazaquin in the leaves, which is an advantage because it is easier to remove the shoot of a plant from a location that has residual herbicide than to remove the roots (LEAL et al., 2008).

From the total applied radiation, the *M. cinerea* plants absorbed 1.72% during the first pair of cotyledonary leaves stage and 2.55% in the first pair of true leaves stage (Table 7). In an experiment performed with another ALS-inhibiting herbicide called nicosulfuron, *Elytrigia repens* plants absorbed more ^{14}C-nicosulfuron when the plants presented one leaf than when they had five leaves; however, their translocation rates were similar, regardless of the weed's phenological stage (BRUCE et al., 1996). Differences in imidazolinone absorption and translocation by different plant species have been previously reported (BUKUN et al., 2012; HEKMAT et al., 2008), and have been mainly attributed to different metabolic rates (SHANER; MALLIPUDI, 1991).

Table 6. Translocation of imazaquin herbicide in *M. cinerea* during different phenological stages.

Stage	Accumulated radiation (dpm)					Translocation (%)			
	Total	Leaf	Stem	Root	Cotyledons	Leaf	Stem	Root	Cotyledons
Cotyledonary leaves	5528.93	1721.56	520.03	1622.22	1665.11	31.14	9.41	29.34	30.12
First pair of true leaves	7289.28	3246.42	637.63	2074.08	1331.15	44.54	8.75	28.45	18.26

Table 7. Recovery of radioactivity from imazaquin tests that were performed during different phenological stages of *M. cinerea*.

Stage	Radioactivity recovered with the solution used in the disassembly of the vases (%)	Radioactivity recovered in the soil (%)	Number of plants in each vase	Radioactivity recovered in the plants (%)	Total of Recovered Radioactivity (%)
Cotyledonary leaves	18.12	85.26	8	1.72	100.00
First pair of true leaves	8.48	91.13	8	2.55	100.00

Future studies with more plant growth times could be conducted to build a mathematical model of absorption, translocation and metabolism. It is important to reach the maximum practical potential of *M. cinerea* and *C. ensiformes* for phytoremediation strategies.

Conclusion

The species *M. cinerea* and *C. ensiformis* had the highest herbicide tolerance and were therefore selected for the study of imazaquin absorption and translocation. Considering the absorption and translocation of imazaquin from the cotyledonary stage to the first pair of true leaves stage, *M. cinerea* showed a higher potential for translocating imazaquin to the shoot of the plant when compared to *C. ensiformes*, which accumulated the herbicide mainly in the roots. The plants presented a higher potential for accumulating herbicide during the more advanced development stage, which shows greater potential for their use in areas with imazaquin herbicide residue.

References

ACCIOLY, A. M. A.; SIQUEIRA, J. O. Contaminação química e biorremediação do solo. In: NOVAIS, R. F.; ALVAREZ, V. V. H.; SCHAEFER, C. E. G. R. (Ed.). **Tópicos em ciência do solo**. Viçosa: SBCS, 2000. v. 1, p. 299-352.

AICHELE, J. M.; PENNER, D. Adsorption, desorption, and degradation of imidazolinones in soil. **Weed Technology**, v. 19, n. 1, p. 154-159, 2005.

ARTUZI, J. P.; CONTIERO, R. L. Herbicidas aplicados na soja e produtividade do milho em sucessão. **Pesquisa Agropecuária Brasileira**, v. 41, n. 7, p. 1119-1123, 2006.

ASKEW, S. D.; WILCUT, J. W. Absorption, translocation, and metabolism of foliar-applied CGA 362622 in cotton (*Gossypium hirsutum*), peanut (*Arachis hypogea*), and selected weeds. **Weed Science**, v. 50, n. 5, p. 293-298, 2002.

BATTAGLIN, W. A.; FURLONG, E. T.; BURHARDT, M. R.; PETER, C. J. Ocurrence of sulfonylurea, sulfonamide, imidazolinone, and other herbicides in rivers, reservoirs and ground water in the Midwestern United States, 1998. **Science of the Total Environment**, v. 248, n. 2-3, p. 123-133, 2000.

BUKUN, B.; NISSEN, S. J.; SHANER, D. L.; VASSOS, J. D. Imazamox absorption, translocation, and metabolism in red lentil and dry bean. **Weed Science**, v. 60, n. 3, p. 350-354, 2012.

BRUCE, L. A.; CAREY, B.; PENNER, D.; KELLS, J. J. Effect of growth stage and environment on foliar absorption, translocation, metabolism, and activity of nicosulfuron in quackgrass (*Elytrigia repens*). **Weed Science**, v. 44, n. 3, p. 447-454, 1996.

CARABIAS-MARTÍNEZ, R.; RODRÍGUEZ-GONZALO, E.; FERNÁNDEZ-LAESPADA, M. E.; SÁNCHEZ-SAN ROMÁN, F. J. Evaluation of surface- and ground-water pollution due to herbicides in agricultural areas of Zamora and Salamanca (Spain). **Journal of Chromatography**, v. 869, n. 2, p. 471-480, 2000.

CARMO, M. L.; PROCOPIO, S. O.; PIRES, F. R.; CARGNELUTTI FILHO, A.; BRAZ, G. B. P.; SILVA, W. F. P.; BARROSO, A. L. L.; SILVA, G. P.; CARMO, E. L.; BRAZ, A. J. B. P.; ASSIS, R. L. Influência do período de cultivo de *Panicum maximum* (cultivar Tanzânia) na fitorremediação de solo contaminado com picloram. **Planta Daninha**, v. 26, n. 2, p. 315-322, 2008.

CRAFTS, A. S.; YAMAGUCHI, S. The autoradiography of plant materials. **California Agricultural Experiment Station Manual**, v. 35, n. 1, p. 143, 1964.

DAN, H. A.; DAN, L. G. M.; BARROSO, A. L. L.; PROCÓPIO, S. O.; OLIVEIRA JR., R. S.; ASSIS, R. L.; SILVA, A. G.; FELDKIRCHER, C. Atividade residual de herbicidas pré-emergentes aplicados na cultura da soja sobre o milheto cultivado em sucessão. **Planta Daninha**, v. 29, n. 2, p. 437-445, 2011.

EMBRAPA-Empresa Brasileira de Pesquisa Agropecuária. Centro Nacional de Pesquisa de Solos. **Sistema brasileiro de classificação de solos**. Rio de Janeiro: Embrapa, 1999.

GENT, M. P. N.; WHITE, J. C.; PARRISH, Z. D.; ISLEYEN, M.; EITZER, B. D.; MATTINA, M. I. Uptake and translocation of *p,p*-dichlorodiphenyldichloroethylene supplied in hydroponics solution to *cucurbita*.

Environmental Toxicology and Chemistry, v. 26, n. 12, p. 2467-2475, 2007.

HEKMAT, S.; SOLTANI, N.; SHORSPHERE, C.; SIKKENA, P. H. Effect of imazamox plus bentazon on dry bean. **Crop Protection**, v. 27, n. 12, p. 1491-1494, 2008.

HORNSBY, A. G.; WAUCHOUPE, R. D.; HERNER, A. E. **Pesticide properties in the environment**. New York: Springer-Verlag, 1995.

LEAL, I. G.; ACCIOLY, A. M. A.; NASCIMENTO, C. W. A.; FREIRE, M. B. G. S.; MONTENEGRO, A. A. A.; FERREIRA, F. L. Fitorremediação de solo salino sódico por *atriplex nummularia* e gesso de jazida. **Revista Brasileira de Ciências do Solo**, v. 32, n. 3, p. 1065-1072, 2008.

MARCACCI, S.; RAVETON, M.; RAVANEL, P.; SCHWITZGUÉBEL, J. P. Conjugation of atrazine in vetiver (*Chrysopogon zizanioides* Nash) grown in hydroponics. **Environmental and Experimental Botany**, v. 56, n. 205, p. 205-215, 2006.

NEWHOUSE, K. E.; SMITH, W. A.; STARRETT, M. A.; SCHAEFER, T. J.; SINGH, B. K. Tolarence to imidazolinone herbicides in wheat. **Plant Physiologist**, v. 100, n. 2, p. 882-886, 1992.

NEWMAN, L. A.; DOTY, S. L.; GERY, K. L.; HEILMAN, P. E.; MUIZNIEKS, I.; SHANG, T. Q.; SIEMIENIEC, S. T.; STRAND, S. E.; WANG, X.; WILSON, A. M.; GORDON, M. P. Phytoremediation of organic contaminants: a review of phytoremediation research at the University of Washington. **Journal of Soil Contamination**, v. 7, n. 4, p. 531-542, 1998.

NOVO, M. C.; CRUZ, L. S. P.; PEREIRA, J. C. V. N.; TREMOCOLDI, W. A.; IGUE, T. Persistência de imazaquim em Latossolo Roxo cultivado com soja. **Planta Daninha**, v. 15, n. 1, p. 30-38, 1997.

PESTER, T. A.; NISSEN, S. J.; WESTRA, P. Absorption, translocation and metabolism of imazamox in jointed goatgrass and feral rye. **Weed Science**, v. 49, n. 5, p. 607-612, 2001.

PINHEIRO, A.; MORAES, J. C. S.; SILVA, M. R. Pesticidas no perfil de solos em áreas de plantação de cebolas em Ituporanga, SC. **Revista Brasileira de Engenharia Agrícola e Ambiental**, v. 15, n. 5, p. 533-538, 2011.

PIRES, F. R.; SOUZA, C. M.; SILVA, A. A.; PROCÓPIO, S. O.; FERREIRA, L. R. Fitorremediação de solos contaminados com herbicidas. **Planta Daninha**, v. 21, n. 2, p. 335-341, 2003.

PIRES, F. R.; SOUZA, C. M.; SILVA, A. A.; CECON, P. R.; PROCÓPIO, S. O.; SANTOS, J. B.; FERRREIRA, L. R. Fitorremediação de solos contaminados com tebuthiuron utilizando-se espécies cultivadas para adubação verde. **Planta Daninha**, v. 23, n. 4, p. 711-717, 2005.

REGITANO, J. B.; ALLEONI, L. R. F.; TORNISIELO, V. L. Atributos de solos tropicais e a sorção de imazaquin. **Scientia Agricola**, v. 58, n. 4, p. 801-807, 2001.

REGITANO, J. B.; BISCHOFF, M.; LEE, L. S.; REICHERT, J. M.; TURCO, R. F. Retention of imazaquin in soil. **Environmental Toxicology and Chemistry**, v. 16, n. 3, p. 397-404, 1997.

RISLEY, M. A.; OLIVER, L. R. Absorption, translocation, and metabolism of imazaquin in Pitted (*Ipomoea lacunosa*) and entireleaf (*Ipomoea hederacea* var. integriuscula) Morningglory. **Weed Science**, v. 40, n. 4, p. 503-506, 1992.

ROCHA, W. S. D.; ALLEONI, L. R. F.; REGITANO, J. B. Energia livre da sorção de imazaquin em solos ácricos. **Revista Brasileira de Ciência do Solo**, v. 27, n. 2, p. 239-246, 2003.

RODRIGUES, B. N.; ALMEIDA, F. L. S. **Guia de herbicidas**. Londrina: Edição dos Autores, 2011.

SALIHU, S.; HATZIOS, K. K.; DERR, J. F. Comparative uptake, translocation, and metabolism of root-applied isoxaben in Ajuga (*Ajuga reptans*) and two ornamental. **Pesticide Biochemistry and Physiology**, v. 60, n. 2, p. 119-131, 1998.

SANTOS, J. B.; PROCÓPIO, S. O.; SILVA, A. A.; PIRES, F. R.; RIBEIRO JR., J. I.; SANTOS, E. A.; FERREIRA, L. R. Fitorremediação do herbicida trifloxysulfuron sodium. **Planta Daninha**, v. 22, n. 2, p. 323-330, 2004.

SCRAMIN, S.; SKORUPA, L. A.; MELO, I. S. Utilização de plantas na remediação de solos contaminados por herbicidas – levantamento da flora existente em áreas de cultivo de cana-de-açúcar. In: MELO, J. S. (Ed.). **Biodegradação**. Jaguariúna: Embrapa Meio Ambiente, 2001. p. 369-371.

SEIFERT, S.; SHAW, D. R.; KINGERY, W. L.; SNIPIS, C. E.; WESLEY, R. A. Imazaquin mobility and persistence in a sharkey clay soil as influenced by tillage systems. **Weed Science**, v. 49, n. 2, p. 571-577, 2001.

SHANER, D. L. Imidazolinone herbicides. In: PLUMMER, D.; RAGSDALE, N. (Ed.). **Encyclopedia of agrochemicals**. Hoboke: John Wiley and Sons, 2003. p. 769-784.

SHANER, D. L.; MALLIPUDI, N. M. Mechanisms of selectivity of the imidazolinones. In: SHANER, D. L.; O'CONNER, S. L. (Ed.). **The imidazolinone herbicides**. Boca Raton: CRC Press, 1991. p. 91-102.

SILVA, A. A.; OLIVEIRA, J. R. R. S.; COSTA, E. R.; FERREIRA, L. R. Efeito residual no solo dos herbicidas imazamox e imazethapyr para as culturas de milho e sorgo. **Planta Daninha**, v. 17, n. 3, p. 345-354, 1999.

SMITH, M. C.; SHAW, D. R.; MILLER, D. K. In field bioassay to investigate the persistence of imazaquin and pyriothiobac. **Weed Science**, v. 53, n. 1, p. 121-129, 2005.

TAN, S.; EVANS, R. R.; DAHMER, M. L.; SINGH, B. K.; SHANER, D. L. Imidazolinone tolerant crops: history, current status and future. **Pest Management Science**, v. 61, n. 3, p. 246-257, 2005.

THOMPSON, A. M.; ROSALES-ROBLES, E.; CHANDLER, J. M.; NESTER, P. R.; JINGLE, C. H. Crop tolarence and weed management systems in imidazolinone tolerant corn. **Weed Technology**, v. 19, n. 4, p. 1037-1044, 2005.

TROXLER, S. C.; FISHER, L. R.; SMITH, D. W.; WILCUT, J. W. Absorption, translocation, and metabolism of foliar applied trifloxysulfuron in tobacco. **Weed Technology**, v. 21, n. 2, p. 421-425, 2007.

VIDAL, R. A. **Ação dos herbicidas**: absorção, translocação e metabolização. Porto alegre: Evangraf, 2002.

WALLS, F. R.; CORBIN, F. T.; COLLINS, W. K.; WORSHAM, A. D.; BRADLEY, J. R. Imazaquin absorption, translocation, and metabolism in flue-cured tobacco. **Weed Technology**, v. 7, n. 2, p. 370-375, 1993.

YODER, R. N.; HUSKIN, M. A.; KENNARD, L. M.; ZABIK, J. Aerobic metabolismo of diclosulam on U.S. and South American soils. **Journal of Agricultural and Food Chemistry**, v. 48, n. 1, p. 4335-4340, 2000.

4

Simulate rain about action insecticide flonicamid in the control of the cotton aphid

Roni Paulo Fortunato[*], Paulo Eduardo Degrande and Paulo Rogério Beltramin da Fonseca

*Instituto Federal do Mato Grosso do Sul, Campus Ponta Porá, Rua Intibiré Vieira, s/n, BR-463, km 4,5, 79900-972, Ponta Porá, Mato Grosso do Sul, Brazil. *Author for correspondence. E-mail: roni.fortunato@ifms.edu.br*

ABSTRACT. The cotton production system in Brazil concentrates on the area of the cerrado, characterized by frequent rains that interfere in the effectiveness of the necessary sprays during its cycle. The objective of the work was to evaluate simulate rain of 15 mm in 4 hours after spraying in the control of *Aphis gossypii* with insecticide flonicamid. Plants of *Gossypium hirsutum* were cultivated in pots containing soil as substrate in greenhouse conditions. The pots were arranged in randomized complete design with seven treatments and five replicates, consisting of: test without insecticide spraying, without insecticide spraying with rain, flonicamid spraying with simulate rain of 15 mm after 30 minutes, 1, 2 and 4 hours after spraying. Equivalent insecticide was sprayed 75 g of flonicamid by hectare. The efficiency evaluation was accomplished through the individuals of *A. gossypii* count which started from an artificial infestation 6 days before the application of the treatments. The results were: a 15-mm precipitation during the first four hours after flonicamid spraying interfered negatively in the control of *A. gossypii*.

Keywords: pest management, chemical control, precipitation.

Introduction

The cotton crop (*Gossypium hirsutum* L.) is characterized by suffering from frequent attack of pests during growing season. Among the components of the production costs, the pest control represents approximately 50% of the cost with pesticides in the main areas that produce cotton in Brazil (RICHETTI et al., 2005), which demands discerning analysis for employment of pests chemical control. Among the main pests are the cotton aphids (*Aphis gossypii*) that demand a lot of control interventions. This pest causes indirect damages through the viruses transmission and direct damages, suction sap, contaminating the fibers with sugary excrement in the end of the season.

The climatic conditions of the Brazilian Cerrado, which concentrates the cotton production, are characterized by periods with frequent precipitation during the growing season. This request attention at the moment of the sprays, because the rain can reduce the control efficiency removing part of the product of the leaf, thus reducing your absorption and the protection period (EDWARDS, 1975; PICK et al., 1984; STEFFENS; WIENEKE, 1975).

New insecticides have been developed seeking the aphids control, among them the flonicamid – an insecticide that acts interfering in the alimentary behavior of the insects, quickly inhibiting feeding. According to Morita et al. (2007), about 30 min. are enough to inhibit the feeding of insects submitted to this product. They point out, however, that one possible mechanism is the inhibition of the penetration of the insect estylet in the cells of the plant.

As for handling crop protection management, it esteems that a very small amount of the applied products in the chemical control, less than 0,1%, reach your action site (PIMENTEL, 1995). One of the best tools to increase the efficiency in general of pesticides would be to increase the penetration of the active ingredient inside of the tissues, in the case of systemic products, where foliate absorption has an important factor for their effectiveness. The leaf uptake involves complex processes which depend on the characteristics of the leaf surface, physical-chemical properties of the pesticides, and mainly the environmental conditions (EDWARDS, 1975; WANG; LIU, 2007).

Among the environmental factors, the rain assumes importance as one of the failure factors in the spraying of pesticides, because it is directly related to the mechanical effect because it results in washing leaf surface. Hence, intense rains end up washing the leaf surface, interfering in the uptake of the product (EDWARDS, 1975). The great subject is quantifying the real effect of the precipitation on effectiveness of systemic insecticides application in the control of a given pest.

Steffens and Wieneke (1975) observed that the increase of the relative humidity favors the absorption and metabolism of the insecticides. However, the rain easily removed azinphos-methyl of the bean plant leaves. Furthermore, they point out that removal should be related with the intensity of the rain and the time after the spray. This is similar to the result obtained by Dejonckheere et al. (1982), when they verified that the rain affected the absorption and the final residue of aldicarb in beet leaves. Few studies were accomplished seeking to know the interaction between precipitation and the behavior of insecticides in pest control in the Brazilian Cerrado. Therefore, a comprehensive study is necessary in order to seek the understanding of the insecticide behavior, especially, in what refers to the effects of the rain in the *A. gossypii* control. The objective of the work was to evaluate simulate rain in the *A. gossypii* control with insecticide flonicamid.

Material and methods

The present study was carried out at the greenhouse (28 ± 2°C) and (65 ± 5% RH) of the Agricultural Sciences College (FCA) of the Federal University of Grande Dourados (UFGD), Dourados, State of Mato Grosso do Sul, Brazil, from December 8, 2007 to January 25, 2008.

The experiment consisted of potted cotton plants allocated to seven insecticide spraying treatments, each replicated five times. The treatments included a control with no insecticide application, and four treatments which involved application of the insecticide followed at different intervals by simulated rain. Flonicamid, was applied as the product 'Turbine® 500 WG' at a rate of 150 g ha^{-1} (providing 75 g of flonicamid) with a pressurized CO_2 sprayer containing four Conejeet® nozzles. The volume and operating pressure of the sprayer was 130 L ha^{-1} and 2.5 bar, respectively. The simulated rain was applied to the plants via to provide 15 mm of precipitation, which was simulated 30 min., 1, 2, and 4h after spraying (Figure 1).

The cotton cultivar 'Delta Opal®' was cultivated in pots containing soil as a substrate, and four seeds were placed in the pots at the time of sowing being left 2 uniform plants within a period of 10 days after emergency (DAE).

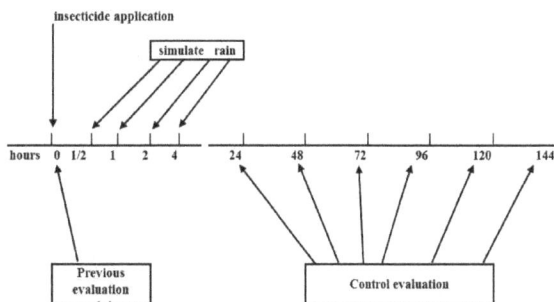

Figure 1. Schematic of the experimental design.

The aphids were collected from the cotton crop in the experimental area at the Agricultural Science College of UFGD. Artificial infestation of the cotton plants was performed when the plants were in stage B2 (MARUR; RUANO, 2001) and possessed an average height of 60 cm. In total, 10 adult individuals of *A. gossypii* were placed by the plant. Six days after infestation, the insecticide was applied, and the plants received 15 mm of simulated rainfall (intensity of 45 mm h^{-1}), according to the time intervals established in the treatments.

The number of individual *A. gossypii* that were observed after 24, 48, 72, 96, 120, and 144h were determined, and the control efficiency was calculated according to the methodology described by Henderson and Tilton (1955). The data were analyzed through an analysis of the variance and were submitted to a Tukey test at a probability level of 5%. Moreover, the standard error for the average control efficiency was determined.

Results and discussion

During the first evaluation period (24h), significant differences among treatments were not observed. However, differences began to appear as the mortality of the insects increased due to the application of the insecticide (Table 1).

Table 1. Statistical analysis of the mean number of aphids per plants 24, 48, 72, 96, 120 and 144 hours after insecticide application.

	Previous evaluation
Variation factors	Prob. (F: α 5%)
Treatments	1.3651 < 2.45 *
	24h
Variation factors	Prob. (F) α = 5%
Treatments	4.1314 > 2.45 **
	48h
Variation factors	Prob. (F) α = 5%
Treatments	9.802 > 2.45 **
	72h
Variation factors	Prob. (F) α = 5%
Treatments	10.1538 > 2.45 **
	96h
Variation factors	Prob. (F) α = 5%
Treatments	8.24 > 2.45 **
	120h
Variation factors	Prob. (F) α = 5%
Treatments	8.361 > 2.45**
	144h
Variation factors	Prob. (F) α = 5%
Treatments	8.361 > 2.45 **

*Not significant (p ≤ 0.05). **Significant (p ≤ 0.05).

When insecticide was not applied to the plants, the presence of rain did not affect the behavior and survival of *A. gossypii*. Thus, the efficiency of the control treatment without insecticide and rain and the efficiency of the control treatment without insecticide in the presence of rain did not differ, indicating that both treatments could be used to calculate the control efficiency (Table 2).

Precipitation that occurred 30 minutes after insecticide application negatively affected the efficacy of the insecticide. Specifically, 48 hours after application, the control efficiency of the 30 min treatment differed from that of the other treatments. After 96 hours, the control efficiency of all of the treatments decreased (Table 2).

The occurrence of rain 1, 2, and 4 hours after flonicamid application negatively affected the control efficiency. However, significant differences were only observed 48 hours after application (Table 2). In a previous study, Pick et al. (1984) observed that 50% of the applied insecticide was removed after 5 mm of precipitation occurred 1 hour after insecticide application. Moreover, the authors concluded that the interval between insecticide application and the

occurrence of rain should be increased as the volume of precipitation increases. Similarly, Mashaya (1993) observed that the biological activity of acephate and monocrotophos was reduced by 85% after 10 mm of precipitation occurred 1 hour after insecticide application.

Table 2. Control efficiency using based on the control treatment without rain (WwR) and control with rain (WR) using the Henderson and Tilton (1955) calculation.

24h after applications			48h after applications		
	Control efficiency (%)			Control efficiency (%)	
Treatments[1]	(WwR)	(WR)	Treatments	(WwR)	(WR)
I + wR	46.6 a*	42.5 a	I + wR	55.3 a	51.8 a
I + R ½ h	34.5 a	30.0 a	I + R ½ h	55.3 a	51.0 a
I + R 1h	40.7 a	36.7 a	I + R 1h	57.8 a	52.1 a
I + R 2h	42.6 a	38.6 a	I + R 2h	57.4 a	52.9 a
I + R 4h	44.1 a	40.2 a	I + R 4h	52.8 a	48.4 a
72h after applications			96h after applications		
	Control efficiency (%)			Control efficiency (%)	
Treatments	(WwR)	(WR)	Treatments	(WwR)	(WR)
I + wR	80.1 a	80.9 a	I + wR	86.8 a	85.1 a
I + R ½ h	55.4 a	57.8 a	I + R ½ h	50.8 a	48.3 a
I + R 1h	56.9 a	59.5 a	I + R 1h	65.5 a	62.0 a
I + R 2h	65.4 a	66.8 a	I + R 2h	69.2 a	63.6 a
I + R 4h	66.3 a	65.2 a	I + R 4h	68.4 a	60.8 a
120h after applications			144h after applications		
	Control efficiency (%)			Control efficiency (%)	
Treatments	(WwR)	(WR)	Treatments	(WwR)	(WR)
I + wR	79.5 a	82.1 a	I + wR	71.6 a	73.1 a
I + R ½ h	51.9 a	54.7 a	I + R ½ h	49.6 a	50.6 a
I + R 1h	67.7 a	68.0 a	I + R 1h	63.2 a	64.4 a
I + R 2h	65.6 a	66.5 a	I + R 2h	63.1 a	64.5 a
I + R 4h	67.0 a	68.8 a	I + R 4h	65.0 a	66.1 a

*Mean in the line followed by the same letter does not differ significantly, Tukey test (p ≤ 0.05%.). 1 I + wR = Insecticide application without rain; I + R ½ h = Insecticide application with rain presence and respective interval time.

Several studies have demonstrated that the rate of absorption of systemic insecticides can favor its effectiveness. Pymetrozine, an insecticide with similar characteristics, was evaluated by Wyss and Bolsinger (1997) in the control of *Myzus persicae* and *Aphis craccivora* in tomatoes, beets, and peas. The results indicated that differences among cultures were related to the distribution of the insecticide in the tissue of the plant.

Under typical cropping conditions, insecticides are applied in the early morning, and daily precipitation occurs in the afternoon; thus, the maximum interval evaluated in the present study was 4 hours. Torres and Silva-Torres (2008) evaluated pimetrozine in the control of *A. gossypii*, and obtained high initial mortality rates, which decreased 6 days after application, demonstrating that the initial biological activity of flonicamid is lower than that of pimetrozine, which is an insecticide with an identical mode of action. Although pimetrozine is a fast-acting insecticide (MORITA et al., 2007), relatively long periods of time on the leaf are required for effective action, which may explain the negative effects of precipitation 4 hours after application.

Conclusion

The occurrence of 15-mm rain up to four hours after flonicamid application interferes negatively in the control of *A. gossypii*.

Acknowledgements

The first author is grateful to Fundect for a research grant.

References

DEJONCKHEERE, W.; MELKEBEKE, G.; STEUBAUT, W.; KIPS, R. H. Uptake and residue of aldicarb in sugarbeet leaves. **Pesticide Science**, v. 13, n. 4, p. 341-350, 1982.

EDWARDS, C. A. Factors that affect the persitence of pesitcides in plant and soil. **Pure and Applied Chemistry**, v. 42, n. 1-2, p. 39-56, 1975.

HENDERSON, C. F.; TILTON, E. W. Tests with acaricides against the brown wheat mite. **Journal of Economic Entomology**, v. 48, n. 1, p. 157-161, 1955.

MARUR, C. J.; RUANO, O. A reference system for determination of developmental stages of upland cotton. **Revista Brasileira de Oleaginosas e Fibrosas**, v. 5, n. 2, p. 313-317, 2001.

MASHAYA, N. Effect of simulated rain on efficacy if insecticide deposits on tobacco. **Crop Protection**, v. 12, n. 1, p. 55-58, 1993.

MORITA, M.; UEDA, T.; YONEDA, T.; KOYANAGI, T.; HAGA, T.; KAISHA, I. S. Flonicamid, a novel insecticide with a rapid inhibitory effect on aphid feeding. **Pest Management Science**, v. 63, n. 10, p. 969-973, 2007.

PICK, F. E.; VAN DYK, L. P.; DE BEER, P. R. The effect of simulated rain on deposits of some cotton pesticides. **Pesticide Science**, v. 15, n. 6, p. 616-623, 1984.

PIMENTEL, D. Amounts of pesticides reaching the target pests: environmental impacts and ethics. **Journal of Agricultural and Environmental Ethics**, v. 8, n. 1, p. 17-29, 1995.

RICHETTI, A.; LAMAS, F. M.; STAUT, L. A. **Estimativa de custo de produção de algodão, safra 2005/06, para Mato Grosso do Sul e Mato Grosso**. Dourados: CPAO, 2005. (Comunicado técnico, 110).

STEFFENS, W.; WIENEKE, J. Influence of humidity and rain on uptake and metabolism of 14C-azinphos-methyl in bean plants. **Archives of Environmental Contamination and Toxicology**, v. 3, n. 3, p. 364-370, 1975.

TORRES, J. B.; SILVA-TORRES, C. S. A. Interação entre inseticidas e umidade do solo no controle do pulgão e da mosca-branca em algodoeiro. **Pesquisa Agropecuária Brasileira**, v. 43, n. 8, p. 949-956, 2008.

WANG, C. J.; LIU, Z. Q. Foliar uptake of pesticides – Present status and future challenge. **Pesticide Biochemistry and Physiology**, v. 87, n. 1, p. 1-8, 2007.

WYSS, P.; BOLSINGER, M. Plant-mediated effects on Pymetrozine efficacy against aphids. **Pest Management Science**, v. 50, n. 3, p. 203-210, 1997.

5

Biological control of phytophagous arthropods in the physic nut tree *Jatropha curcas* L. in Brazil

Flávio Lemes Fernandes[1*], Maria Elisa de Sena Fernandes[1], Elisângela Novais Lopes[2], Madelaine Venzon[3], Juno Ferreira da Silva Diniz[1] and Luís Antônio dos Santos Dias[4]

[1]Instituto de Ciências Agrárias, Universidade Federal de Viçosa, Campus de Rio Paranaíba, M- 230, Km 7, s/n, 38810-000, Rio Paranaíba, Minas Gerais, Brazil. [2]Programa de Pós-graduação em Entomologia, Universidade Estadual Paulista, Jaboticabal, São Paulo, Brazil. [3]Empresa de Pesquisa Agropecuária de Minas Gerais, Viçosa, Minas Gerais, Brazil. [4]Departamento de Fitotecnia, Universidade Federal de Viçosa, Viçosa, Minas Gerais, Brazil. *Author for correspondence. E-mail: flaviofernandes@ufv.br

ABSTRACT. *Jatropha curcas* has a high biofuel oil content, which could replace polluting fuels, and has great potential for large scale monoculture cultivation in the conventional system. We explored the occurrence, spatial distribution and the functional response of the main phytophagous species of this plant and their natural enemies to explore the potential for conservative biological control. We began sampling phytophagous species and predators when *J. curcas* plants were six months old. The most common species of phytophagous insects were nymphs and adults of *Empoasca kraemeri,* followed by *Frankliniella schultzei* and *Myzus persicae.* Among the predators, *Ricoseius loxocheles, Iphiseioides zuluagai,* Araneidae, larvae and adults of *Psyllobora vigintimaculata* and *Anthicus* sp. were the most frequently encountered. The most common parasitoids were the families Encyrtidae and Braconidae. The highest densities of *E. kraemeri* and *F. schultzei* on the edges of the *J. curcas* crop follow spatial patterns similar to those of their natural enemies *I. zuluagai* and *Anthicus* sp. These arthropods can be considered efficient predators of immature stages of *E. kraemeri* and *F. schultzei* on *J. curcas.*

Keywords: predators, parasitoids, phytophagy, spatial distribution, functional response.

Introduction

The physic nut tree (*Jatropha curcas* L.), a native species of Brazil, is considered to be a good crop option for many agricultural regions of the world (LI et al., 2010). This species, which is in the Euphorbiaceae family, has the potential for use in the prevention and control of erosion, can be grown as a hedge due to its rapid growth, has high oil content for biofuels that can replace polluting fuels (ACHTEN et al., 2008; AZAM et al., 2005), is drought tolerant, has low nutrient requirements, has high adaptability to different soil types and emits low levels of gases into the atmosphere. For these reasons, it has great potential to be cultivated in large scale monocultures in the conventional system of cultivation (FRANCIS et al., 2005; LI et al., 2010; PRUEKSAKORN; GHEEWALA, 2008).

A major concern in the conventional system of cultivation is the sustainability of production. Diseases and pests that cause damage to crops compel farmers to use pesticides, thus contaminating the humans, soil, and water and affecting beneficial organisms (DESNEUX et al., 2007). In such cases, biological control may be an alternative to chemical control, especially because there is no insecticide registered to control *J. curcas* pests in Brazil. The practice of conservative biological control (CBC), which involves the manipulation of agricultural habitats to favor the natural enemies of pests (i.e., predators, parasitoids, and pathogens), offers the promise of simultaneously conserving natural enemy biodiversity and reducing pest problems.

To perform CBC in an efficient and planned manner, one needs to know the species of phytophagous pests or those with the potential to become pests of a particular crop, and the main natural enemies of these pests. Additionally, one needs to know about the interactions between phytophagous pests and their natural enemies to develop strategies that conserve the habitat and encourage the maintenance and preservation of natural enemies in the area of the monoculture. Predators, in general, are generalists, feeding on a host of prey, and parasitoids are specialists, completing their life cycle in a single host. Because there is no synthetic insecticide registered to control pests of *J. curcas*, CBC could become a very popular method of control by the producers of this crop.

Understanding the spatial dynamics of insect populations can facilitate the development of CBC strategies, promoting sustainable agroecosystems through functional biodiversity conservation (DIAZ et al., 2010). Knowledge of the spatial distribution pattern of insects allows a focused management of effort in places with the highest densities. Spatial distribution patterns have been determined through the analysis of mathematical models of frequency distribution, where the adjustment of the negative binomial distribution indicates that the insect has an aggregated distribution pattern. This relationship is affected by spatial distribution, but does not represent spatial distribution. A better option for determining the spatial dependence of the density of insects is the use of geostatistics (SCIARRETTA; TREMATERRA, 2008). Geostatistics analyze whether the observed value of a variable for a given location is dependent on the variables of neighboring sites. If there is dependency, the variable exhibits spatial autocorrelation. Thus, geostatistics can be a tool used to study the spatial distribution of pests and natural enemies, which is essential for devising CBC strategies.

Despite the importance of studies of spatial distribution, studies of the functional response of a predator can be used to select key natural enemies. The functional response of a predator is one of the most important factors in the population dynamics of predator-prey systems because it improves the effectiveness of pest biological control (TIMMS et al., 2008). The functional response is based on two basic parameters: the prey handling time (Ht), which involves the gathering, death and ingestion of prey, and the attack rate (a), which represents the efficiency of prey searching. This model evaluates the behavioral aspect of the predator, which can be influenced by its age, its type, the age of the prey, the prey host plant and the climatic conditions.

Little is known about phytophagous species of *J. Curcas*, the primary natural enemies of these phytophagous species, or the interactions between these organisms and their spatial dynamics; this information would be useful for developing agents for natural biological control (SHANKER; DHYANI, 2006). We explored the occurrence, spatial distribution and the functional response of the main phytophagous species and their natural enemies to help develop potential conservative biological control methods.

Material and methods

This work was conducted at the Experimental Station Diogo Vaz de Melo in Viçosa-Minas Gerais State (latitude 20° 45' 54.3" S, longitude 42° 52' 06.07" W and an average altitude of 335 m) at the Federal University of Viçosa (FUV) in an experimental field of physic nut tree *J. curcas* that were one year of age (in the vegetative phase). This area is one hectare, and the trees were planted 2.5 x 2.5 m, with a total of 1,600 plants.

Regarding cultivation, commercial fertilizer 20-5-20 was used during planting and covering, with 0.150 kg per plant divided into three applications. Irrigation was not performed, and weeding of the area was performed when necessary. During the sampling period, no treatment was applied in any area to control pests or diseases. In September, the stems of the plant were trimmed to 0.50 m above the soil and sampling of phytophagous insects and predators stopped; in October, when the plants began sprouting leaves, sampling was resumed.

We selected a subsample of physic nut trees from the North of Minas Gerais State obtained from the Germplasm Bank of the Department of Phytotechny at the Federal University of Viçosa. The subsample studied was 'Filomena', which is currently the most commonly planted variety in Brazil. We started the sampling of phytophagous insects and their predators when the physic nut trees reached six months of age. The evaluation of phytophagous arthropods and their natural enemies was conducted using the technique of leaf beating onto a tray

(0.35 m length x 0.30 m width x 0.05 m of depth) (BACCI et al., 2008). The sampling units consisted of taking a leaf from either the apical, median, or basal third of the plant canopy. All sampled data points were georeferenced using a global positioning system. The samples were taken 10 m apart, and the total area sampled was 800 m².

During sampling, the arthropods found were counted and collected using a fine-bristled brush and stored in 70% alcohol. The mites were taken to the entomology laboratory, Entomologia da Empresa de Pesquisa Agropecuária de Minas Gerais (Epamig), where they were mounted on slides using Hoyer's Medium for later identification by Dr. Manoel GC Gondim Jr. / UFRPE, a specialist of phytoseiid mites. The other arthropods were identified in the taxonomy laboratory at FUV.

The data related to the densities of phytophagous arthropods and their natural enemies were subjected to frequency analysis and presented as the mean ± standard error. To select and study the interactions between variables (densities of phytophagous arthropods and their natural enemies) we used multivariate canonical correlation analysis by implementing the PROC CANCORR command in SAS (2008-2009). Significant variables with the highest correlation coefficient ($p \leq 0.05$) for studies of geostatistics and simple regression analysis ($p \leq 0.05$) were selected to study the causal relations between natural enemies and phytophagous arthropods. We tested the spatial dependence of the Moran Index (I) ($p \leq 0.05$) with the Z t est. This method indicated whether the arthropods were distributed irregularly, uniformly or more aggregated, and values of I near 1 indicated high spatial dependence, while values near zero indicated little or no spatial dependence. If the values of I near 1 were significant ($p \leq 0.05$), linear regression analyses were used to analyze the interaction between the prey and natural enemies' densities using PROC REG.

Arthropods

The phytophagous prey *Frankliniella schultzei* (Trybom) (Thysanoptera: Thripidae) and *Empoasca kraemeri* Ross and Moore (Hemiptera: Cicadellidae) nymphs, and the predators *Iphiseioides zuluagai* (Denmark and Muma) (Acari: Phytoseiidae), *Psyllobora vigintimaculata* (Say) (Coleoptera: Coccinellidae) and *Anthicus* sp. (Coleoptera: Anthicidae) adults were selected for experiments of functional response. Adults of all predators were collected directly from the field using plastic containers and aspirators. These species were collected on random plants growing in the physic nut tree field at the FUV experimental station.

Specimens of arthropods species were stored in 0.004 L vials containing 70% alcohol and sent to taxonomists for identification.

The rearing of *F. schultzei* was performed in the same manner as for *F. occidentalis*. *E. kraemeri* were reared on bean plants (cultivar IAC-Carioca) in 3-l pots filled with the substrate inside wooden cages (1.00 x 0.50 x 0.90 m) covered with white organza. These cages were placed in a greenhouse complex at the Federal University of Viçosa-Rio Paranaíba Campus, at 25 ± 5°C, 50-70% r.h., and L12:D12 photoperiod. Plants were replaced monthly, and the insects were reared for two generations.

Functional response

The experiments were conducted in Petri dishes (0.20 m diameter) lined with a thin layer of solidified agar solution to prevent desiccation of *J. curcas* leaf discs (0.10 m diameter). A single physic nut tree leaf disc was centered upside down on the agar solution. The insects used for the experiments were < 24 hours old and starved for 12 hours before the tests. Nymphs of *F. schultzei* and *E. kraemeri* were offered in densities of 1, 3, 5, 10, 15 and 20 for predators *I. zuluagai*, *P. vigintimaculata* and *Anthicus* sp. The maximum and minimum densities of each prey for all predators were determined from a preliminary study. To check survival of the nymphs of *F. schultzei* and *E. kraemeri* in the absence of the predator, the same number of replicates without predators was setup for each prey density. The predators were added to experimental arenas 1 hour after transferring the prey. The Petri dishes were sealed with Parafilm around the edge to prevent escaping. Seven replicates were conducted for each prey density. The number of consumed prey was counted 5 hours after the release of the predator into the experimental arenas. Consumed prey were not replaced during the experiments. The data from the consumption by the natural enemies were analyzed by one-way univariate ANOVA. For all analyses, the assumptions of normality and homogeneity of variance were verified using PROC UNIVARIATE and PROC GPLOT, and no transformations were necessary.

The functional response type was determined by a logistic regression of the proportion of prey consumed as a function of initial prey number ($p \leq 0.05$). Three basic types of functional response to prey density were identified: linear, convex and sigmoid (HOLLING, 1959). In a linear response (called type I), the number of consumed prey rises linearly up to a plateau. In a convex response (type II), the number of consumed prey rises with prey density but begins to decrease when reaching a maximum point (type III).

Results and discussion

The most common phytophagous species were nymphs and adults of *E. kraemeri,* followed by *F. schultzei* and *Myzus persicae* (Sulzer) (Hemiptera: Aphididae), with densities of 6.29 ± 0.12, 14.71 ± 0.22, 5.57 ± 0.19 and 2.57 ± 0.09/ beating on tray, respectively. Among the predators, *Ricoseius loxocheles* (Denmark and Muma) (Acari: Phytoseiidae), *I. zuluagai,* Araneidae, *P. vigintimaculata* larvae and adults and *Anthicus* sp. were the most common, with average densities of 461.29 ± 3.45, 257.43 ± 1.45, 35.86 ± 0.45, 11.57 ± 0.19, 9.00 ± 0.14 and 4.57 ± 0.14/ beating on tray, respectively. The most common parasitoids were *Oaencyrtus* spp., *Psyllaephagus* spp. and *Hexacladia smithii* (Encyrtidae) (3.00 ± 0.07), and *Aphidius* spp. and *Maecolaspis* sp. (Braconidae) (2.29 ± 0.07) (Table 1).

Table 1. Mean ± standard error and relative frequency of the density of phytophagous arthropods, predators and parasitoids per plant on the cultivated physic nut tree *Jatropha curcas.*

Group	Species	Arthropods per plant Mean ± standard error	Frequency (%)
Phytophagous	*Diabrotica speciosa*	0.71±0.04	0.9
	Cerotoma arcuata	0.57±0.03	0.7
	Empoasca kraemeri (nymph)	6.29±0.12	77*
	Empoasca kraemeri (adult)	14.71±0.22	79*
	Myzus persicae	2.57±0.09	31*
	Frankliniella schultzei	5.57±0.19	68*
Predators	Araneidae	35.86±0.45	43*
	Camponotus spp.	0.14±0.02	0.2
	Chrysopa sp	0.71±0.04	0.9
	Cycloneda sanguinea	0.57±0.04	0.7
	Hippodamia spp.	0.29±0.02	0.4
	Psyllobora vigintimaculata (larvae)	11.57±0.19	41*
	Psyllobora vigintimaculata (adult)	9.00±0.14	10*
	Sciminus sp	1.86±0.06	23*
	Syrphidae	0.86±0.04	10*
	Iphiseioides zuluagai	257.43±1.45	31*
	Ricoseius loxocheles	461.29±3.45	56*
	Anthicus sp.	324.57±0.14	45*
	Reduviidae	0.14±0.02	0.2
	Xylocoris sp	0.43±0.03	0.5
Parasitoids	Braconidae	2.29±0.07	28*
	Encyrtidae	3.00±0.07	37*
	Eulophidae	0.43±0.03	0.05

*Frequency > 10%.

Most phytophagous species found in this study have not yet had their pest status reported for *J. curcas,* and the damages and economic losses generated have not been quantified. However, *E. kraemeri* and *F. schultzei* have been reported to cause damage (OLIVEIRA et al., 2010; SILVA et al., 2008). The natural enemies with higher densities (*R. loxocheles* and *I. zuluagai*) have not been identified previously in the physic nut tree; however, in other crops, their presence is very common because they are generalist predators (SYMONDSON et al., 2002).

There was only one significant canonical axis in the relations between the densities of the phytophagous arthropods with the densities of the natural enemies (Wilk's Lambda = 0.005, F = 111.32, d.f. numerator/density = 5/295; p < 0.001). In this axis, the variation of the densities of the phytophagous nymph and adult stages of *E. kraemeri, M. persicae* and *F. schultzei* explained 85% of the variation in the densities of the natural enemies. The phytophagous insects that most influenced the density of the natural enemies were *E. persicae* nymph (r = 0.88) and *F. schultzei* (r = 0.74) (Table 2). Thus, the variables selected by the canonical correlation analysis studied in the analysis of spatial distribution and tested in the regression models were the phytophagous *E. kraemeri* nymph and *F. schultzei* and the natural enemies *P. vigintimaculata* larvae and adults, *I. zuluagai, R. loxocheles,* Anthicidae and Braconidae (Table 2).

Table 2. Canonical correlation between the densities of the phytophagous arthropods with the densities of the natural enemies in the physic nut tree *Jatropha curcas.*

Density of phytophagous insects	Canonical axis Coefficient	r
Empoasca kraemeri (nymph)	-0.87	0.88
Empoasca kraemeri (adult)	-0.14	0.12
Myzus persicae	1.65	0.45
Frankliniella schultzei	-0.95	0.74
Density of natural enemies		
Araneidae	0.11	0.21
Psyllobora vigintimaculata (larvae)	1.25	0.75
Psyllobora vigintimaculata (adult)	1.74	0.78
Sciminus sp	0.10	0.20
Syrphidae	0.05	0.12
Iphiseioides zuluagai	2.14	0.89
Ricoseius loxocheles	2.74	0.71
Anthicus sp.	2.01	0.87
Braconidae	1.01	0.61
Encyrtidae	0.12	0.21
Characteristics of the axis	$R^2 = 0.85$, F = 9.85, p < 0.01	

The pattern of positive spatial dependence of the density of the phytophagous species *E. kraemeri* and *F. schultzei* and the natural enemies *P. vigintimaculata* larvae and adults, *I. zuluagai, R. loxocheles,* Anthicidae and Braconidae showed that some areas favor the permanence of phytophagous species and their natural enemies. Phytophagous species most likely contain enough food for the natural enemies' diets without competition for resources with other organisms. The natural enemies *P. vigintimaculata* larvae and adults and Anthicidae may have a density dependent relationship with the phytophagous *E. kraemeri* and *F. schultzei.* Density dependent relationships between natural enemies and pests are often cited in the literature (BYRNES et al., 2006; CARDINALE et al., 2003; HOUGARDY; MILLS, 2008; SNYDER et al., 2006; SNYDER et al., 2008), especially when these relationships involve interactions between prey and natural enemies in a spatial scale.

There was a positive spatial dependence of the density of phytophagous *M. persicae* and *F. schultzei* and their natural enemies *P. vigintimaculata* larvae and adults, *I. zuluagai*, *R. loxocheles*, Anthicidae and Braconidae (I = 0.71, Z = 3.37, p≤ 0.05; I = 0.88, Z = 5.52, p≤ 0.05; I = 0.65, Z = 4.23, p ≤ 0.05; I = 0.55, Z = 3.11; p≤ 0.05, I = 0.68; Z = 3.08, p ≤ 0.05, I = 0.61, Z = 3.43, p ≤ 0.05; I = 0.64, Z = 3.34, p ≤ 0.05 and I = 0.84, Z = 4.98, p ≤ 0.05, respectively). This spatial dependence is consistent with the patchy distribution of these arthropods. In addition, there was a trend of early attack and action of natural enemies on the sides of the region evaluated at the *J. curcas* farm, where the adult density is markedly higher (Figure 1).

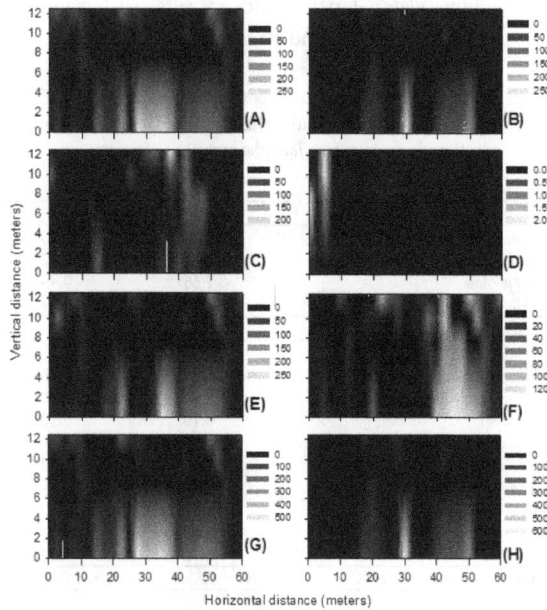

Figure 1. Maps of spatial distributions of natural enemies (A) *Iphiseioides zuluagai,* (B) *Ricoseius loxocheles,* (C) *Psyllobora vigintimaculata* larva, (D) *Psyllobora vigintimaculata* adult, (E) *Anthicus* sp., (F) Braconidae, (G) *Frankliniella schultzei* and (H) *Empoasca kraemeri* nymph. The colors on the maps show the different densities of arthropods per 100 plants.

In this study, the distributions of *E. kraemeri* and *F. schultzei* follow spatial patterns similar to those of the natural enemy *P. vigintimaculata*, both larvae and adults. The causal relationship between the phytophagous *E. kraemeri* and *F. schultzei* with the larvae and adults of *P. vigintimaculata* and adult Anthicidae explains the clustered distribution of these arthropods. These predators have been reported to feed on *E. kraemeri* and *F. schultzei*. Coccinellids have been observed feeding on aphids, thrips, mites and eggs of lepidoptera (ONKAR, 2014; SARMENTO et al., 2007). Because *Anthicus* are coleopterous, according to (MAES; CHANDLER, 1994), they feed on Lepidoptera eggs and first stage

larvae and the pupae of other insects. Given the small sizes of the leafhopper *E. kraemeri* and the thrips *F. schultzei* associated with one place of occurrence of predators (vicinity of niches), they could serve as a suitable food source for these predators. Bastos et al. (2003), studying the diversity of phytophagous arthropods and predators, found high densities of *E. kraemeri* (Ross and Moore) and the predator *Anthicus* sp. in beans (*Phaseolus vulgaris*).

The highest densities of phytophagous *E. kraemeri* and *F. schultzei* on the edges of the *J. curcas* crop follow spatial patterns similar to those of the natural enemies *I. zuluagai* and *Anthicus* sp. (Figure 1A and E). This trend may be because of a causal relationship between phytophagous densities and these natural enemy densities (Figure 2A and E).

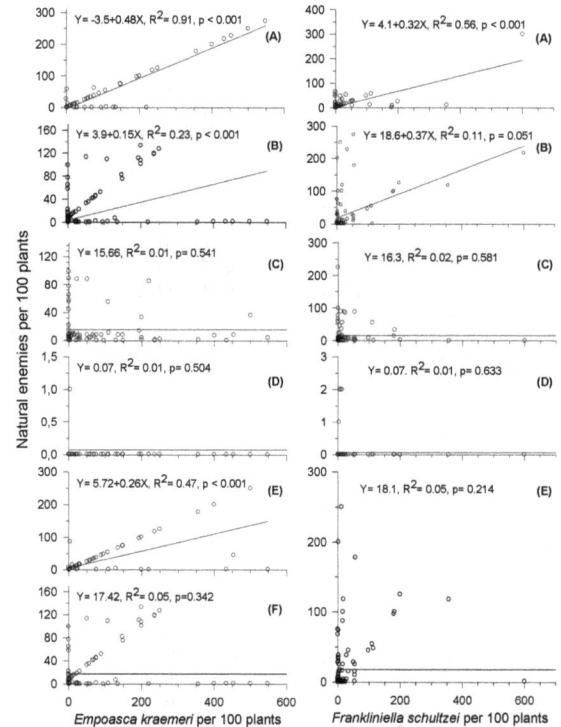

Figure 2. Causal relationship between phytophagous (independent variable) *Empoasca kraemeri* nymph and *Frankliniella schultzei* with (dependent variable) natural enemies (A) *Iphiseioides zuluagai,* (B) *Psyllobora vigintimaculata* adults, (C) *Psyllobora vigintimaculata* larvae, (D) *Ricoseius loxocheles,* (E) *Anthicus* sp. and (F) Braconidae.

The non-causal relationship between the other natural enemies with phytophagous insects may be due to a greater dispersal to other locations in the area. This behavior may have reduced the chances of an encounter between natural enemies and phytophagous insects. Relations between phytophagous insects, predators and parasitoids on a spatial scale are important in establishing the rates of dispersal of species. These rates may be higher or lower depending on the available resources.

Functional response

The mean numbers of *F. schultzei* and *E. kraemeri* nymphs consumed were significantly different among the predators *P. vigintimaculata* (F = 60.48; d.f. = 1,39; p < 0.0001), *I. zuluagai* (F = 32.44; d.f. = 1,39; p < 0.0001) and *Anthicus* sp. (F = 28.57; d.f. = 1,39; p < 0.0001), respectively. The predators *P. vigintimaculata*, *I. zuluagai* and *Anthicus* sp. consumed 1.00 ± 0.41, 5.55 ± 1.06 e 4.98 ± 1.30 *F. schultzei* nymphs, and 0.28 ± 0.15, 5.39 ± 1.05 and 3.14 ± 0.90 nymphs of *E. kraemeri*, respectively.

No relation was observed between the number of *F. schultzei* and *E. kraemeri* consumed by *P. vigintimaculata* (Figure 3A). However, the number of *F. schultzei* and *E. kraemeri* consumed by *I. zuluagai* and *Anthicus* sp. increased as a function of the density, with a positive and highly significant correlation (Figure 3B and C). Thus, the functional response to prey density of *I. zuluagai* and *Anthicus* sp. were type II, where the number of killed prey rises linearly up to a plateau (Figure 3). It was also established that the three predators found and captured prey even at the lowest densities, killing them even after satiation.

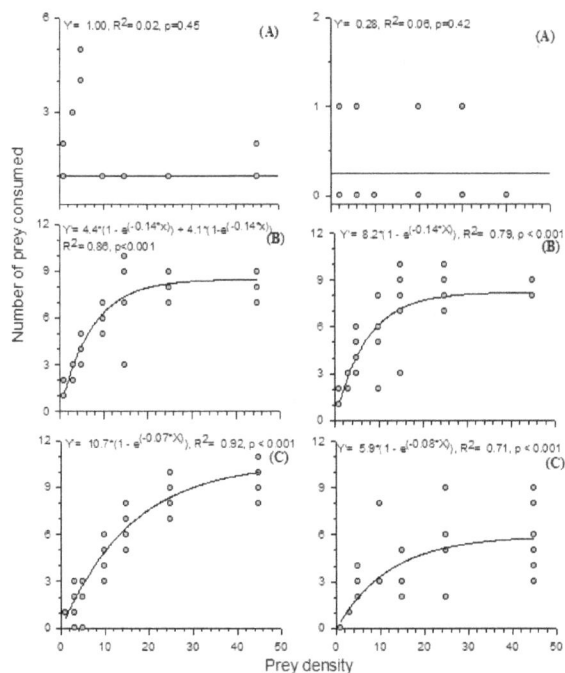

Figure 3. The functional response of *Psyllobora vigintimaculata* adult (A), *Iphiseioides zuluagai* (B) and *Anthicus* sp. (C) to *Frankliniella schultzei* and *Empoasca kraemeri* nymphs at temperature: 25.21±1.05°C, relative humidity: 65.32±2.12%, photoperiod: 12 hours.

The stabilization of type II functional response is derived from a limitation of the predators; they cannot increase their rate of predation due to the limitation of time for searching, handling prey, and their satiation (see Figure 3). It was established that the three predators found and captured prey even at the lowest densities, killing prey even after satiated. The nature of the functional response of *I. zuluagai* and *Anthicus* sp. to *F. schultzei* and *E. kraemeri* is expected because the type II response is common in many predatory mites and Anthicidae: *Phytoseiulus persimilis* Athias-Henriot, *Neoseiulus californicus* (McGregor), *Phytoseiulus macropilis* (Banks) and *Orius insidiosus* (Say) (GOTOH et al., 2004; POLETTI et al., 2007; RUTLEDGE; O'NEIL, 2005; XIAO; FADAMIRO, 2010).

Considering the obtained results, it is possible that *I. zuluagai* and *Anthicus* sp. can reduce the population of *E. kraemeri* and *F. schultzei* nymphs in field conditions of *J. curcas*, even if they reach high densities. The fact that *P. vigintimaculata* needs less prey than *I. zuluagai* and *Anthicus* sp. suggests that its survivorship will be better than predators with a low prey density in the field. The high aggressiveness of *I. zuluagai* was also recorded by Reis et al. (2000).

Knowledge of spatial distribution is important because dispersal behavior, such as immigration and inter- and intra-field crop movement, must be precisely monitored to understand the population dynamics of insect predators and prey. Thus, creating ideal conditions for natural enemies in physic nut tree crops can be very important in reducing dispersion and increasing the encounter rates between predators and prey.

Conclusion

In conclusion, the phytoseiid *I. zuluagai* and the anthicid *Anthicus* sp. can be considered efficient predators of immature stages of *E. kraemeri* and *F. schultzei* at different densities, and they contribute to the reduction of the prey population on *J. curcas*, where they are most commonly present. These data are extremely important for developing CBC measures for this crop of great economic value because there are still no insecticides registered for the control of its pests.

Acknowledgements

The funding and fellowships provided by the following Brazilian agencies were greatly appreciated: Ministério da Agricultura e Abastecimento (Mapa), Coordenação de Aperfeiçoamento de Pessoal de Nível Superior (Capes), Conselho Nacional de Desenvolvimento

Científico e Tecnológico (CNPq), and Fundação de Amparo à Pesquisa do Estado de Minas Gerais (Fapemig).

References

ACHTEN, W. M. J.; VERCHOT, L.; FRANKEN, Y. J.; MATHIJS, E.; SINGH, V. P.; AERTS, R.; MUYS, B. *Jatropha* bio-diesel production and use. **Biomass and Bioenergy**, v. 32, n. 12, p. 1063-1084, 2008.

AZAM, M. M.; WARIS, A.; NAHAR, N. M. Prospects and potential of fatty acidmethyl esters of some non-traditional seed oils for use as biodiesel in India. **Biomass and Bioenergy**, v. 29, n. 4, p. 293-302, 2005.

BACCI, L.; PICANÇO, M. C.; MOURA, M. F.; SEMEÃO, A. A.; FERNANDES, F. L.; MORAIS, E. G. F. Sampling plan for thrips (Thysanoptera: Thripidae) on cucumber. **Neotropical Entomology**, v. 37, n. 5, p. 582-590, 2008.

BASTOS, C. S.; PICANÇO, M. C.; GALVÃO, J. C. C.; CECON, P. R.; PEREIRA, P. R. G. Incidência de insetos fitófagos e de predadores no milho e no feijão cultivados em sistema exclusivo e consorciado. **Ciência Rural**, v. 33, n. 3, p. 391-397, 2003.

BYRNES, J. E.; STACHOWICZ, J. J.; HULTGREN, K. M.; RANDALL, A. H.; OLYARNIK, S. V.; THORNBER, C. S. Predator diversity strengthens trophic cascades in kelp forests by modifying herbivore behaviour. **Ecology Letters**, v. 9, n. 1, p. 61-71, 2006.

CARDINALE M.; CASINI, M.; ARRHENIUS, F.; DIEL, H. N. Spatial distribution and feeding activity of herring (*Clupea harengus*) and sprat (*Sprattus sprattus*) in the Baltic Sea. **Aquat Living Resources**, v. 16, n. 3, p. 283-292, 2003.

DESNEUX, N.; DECOURTYE, A.; DELPUECH, J. M. The sublethal effects of pesticides on beneficial arthropods. **Annual Review of Entomology**, v. 52, n. 6, p. 81-106, 2007.

DIAZ, B. M.; LEGARREA, S.; MARCOS-GARCÍA, M. Á.; FERERES, A. The spatio-temporal relationships among aphids, the entomophthoran fungus, *Pandora neoaphidis*, and aphidophagous hoverflies in outdoor lettuce. **Biological Control**, v. 53, n. 3, p. 304-311, 2010.

FRANCIS, G.; EDINGER, R.; BECKER. K. A concept for simultaneous wasteland reclamation, fuel production, and socio-economic development in degraded areas in India: Need, potential and perspectives of *Jatropha plantations*. **Natural Resources**, v. 29, n. 1, p. 12-24, 2005.

GOTOH, T.; MITSUYOSHI, N.; YAMAGUCHI, K. Prey consumption and functional response of three acarophagous species to eggs of the two-spotted spider mite in the laboratory. **Applied Entomology and Zoology**, v. 39, n. 1, p. 97-105, 2004.

HOLLING, C. S. The components of predation as revealed by a study of small-mammal predation of the European pine sawfly. **Canadian Entomologist**, v. 91, n. 5, p. 293-329, 1959.

HOUGARDY, E.; MILLS, N. J. Comparative life history and parasitism of a new color morph of the walnut aphid in California. **Agricultural Forest Entomology**, v. 10, n. 2, p.137-146, 2008.

LI, Z.; LIN, B. L.; ZHAO, X.; SAGISAKA, M.; SHIBAZAKI, R. System approach for evaluating the potential yield and plantation of *Jatropha curcas* L. on a global scale. **Environmental Science Technology**, v. 44, n. 6, p. 2204-2209, 2010.

MAES, J. M.; CHANDLER, D. S. Catálogo de los Meloidea (Coleoptera) de Nicaragua. **Revista Nicaraguence de Entomologia**, v. 28, n. 1, p. 31-42, 1994.

OLIVEIRA, H. N.; SILVA, C. J.; ABOT, A. R.; ARAÚJO, D. I. Cigarrita verde en cultivos de *Jatropha curcas* en el Estado de Mato Grosso do Sul, Brasil. **Revista Colombiana de Entomologia**, v. 36, n. 1, p. 52-53, 2010.

ONKAR, M. B. Consumption, developmental and reproductive attributes of two con-generic ladybird predators under variable prey supply. **Biological Control**, v. 74, n. 1, p. 36-44, 2014.

POLETTI, M.; MAIA, A. H. N.; OMOTO, C. Toxicity of neonicotinoid insecticides to *Neoseiulus californicus* and *Phytoseiulus macropilis* (Acari: Phytoseiidae) and their impact on functional response to *Tetranychus urticae* (Acari: Tetranychidae). **Biological Control**, v. 40, n. 1, p. 30-36, 2007.

PRUEKSAKORN, K.; GHEEWALA, S. H. Full chain energy analysis of biodiesel from *Jatropha curcas* L. in thailand. **Environmental Science and Technology**, v. 42, n. 9, p. 3388-3393, 2008.

REIS, P. R.; TEODORO, A. V.; PEDRO NETO, M. Predatory activity of phytoseiid mites on the development stages of coffee ringspot mite (Acari: Phytoseiidae, Tenuipalpidae). **Anais da Sociedade Entomológica do Brasil**, v. 29, n. 3, p. 547-553, 2000.

RUTLEDGE, C. E.; O'NEIL, R. J. *Orius insidiosus* (Say) as a predator of the soybean aphid, *Aphis glycines* Matsumura. **Biological Control**, v. 33, n. 1, p. 56-64, 2005.

SARMENTO, R. A.; VENZON, M.; PALLINI, A.; OLIVEIRA, E. E.; ARNE, J. Use of odours by *Cycloneda sanguinea* to assess patch quality. **Entomologia Experimentalis et Applicata**, v. 124, n. 3, p. 313-318, 2007.

SAS-Statistical Analisys System. **System for Microsoft Windows**, release 9.2. Cary: Statistical Analysis System Institute, 2008-2009. (CD-ROM).

SCIARRETTA, A.; TREMATERRA, P. Geostatistical tools for the study of insect spatial distribution: practical implications in the integrated management of orchard and vineyard pests. **Plant Protection Science**, v. 50, n. 2, p. 97-110, 2008.

SHANKER, C.; DHYANI, S. K. Insect pests of *Jatropha curcas* L. and the potential for their Management. **Current Science**, v. 91, n. 2, p. 162-163, 2006.

SILVA, P. H. S.; CASTRO, M. J. P.; ARAÚJO, E. C. A. Tripes (insecta: tripidae) associados ao pinhão-manso no estado do Piauí, Brasil. **Revista Brasileira de Oleaginosas e Fibrosas**, v. 12, n. 3, p. 125-127, 2008.

SNYDER, G. B.; FINKE, D. L.; SNYDER, W. E. Predator biodiversity strengthens aphid suppression across single- and multiple-species prey communities. **Biological Control**, v. 44, n. 1, p. 52-60, 2008.

SNYDER, W. E.; SNYDER, G. B.; FINKE, D. L.; STRAUB, C. S. Predator biodiversity strengthens herbivore suppression. **Ecological Letters**, v. 9, n. 7, p. 789-796, 2006.

SYMONDSON, W. O. C.; SUNDERL, K. D.; GREENSTONE, M. H. Can generalist predators be effective biocontrol agents? **Annual Review of Entomology**, v. 47, n. 1, p. 561-594, 2002.

TIMMS, J. E.; OLIVER, T. H.; STRAW, N. A.; LEATHER, S. R. The effects of host plant on the coccinellid functional response: Is the conifer specialist *Aphidecta obliterate* (L.) (Coleoptera: Coccinellidae) better adapted to spruce than the generalist *Adalia bipunctata* (L.) (Coleoptera: Coccinellidae)? **Biological Control**, v. 47, n. 3, p. 273-281, 2008.

XIAO, Y.; FADAMIRO, H. Y. Functional responses and prey-stage preferences of three species of predacious mites (Acari: Phytoseiidae) on citrus red mite, *Panonychus citri* (Acari: Tetranychidae). **Biological Control**, v. 53, n. 3, p. 345-352, 2010.

6

Is curtobacterium wilt biocontrol temperature dependent?

Samuel Julio Martins[1], Flávio Henrique Vasconcelos Medeiros[1*], Ricardo Magela Souza[1] and Laíze Aparecida Ferreira Vilela[2]

[1]Departamento de Fitopatologia, Universidade Federal de Lavras, Campus Universitário, 3037, 37200-000, Lavras, Minas Gerais, Brazil.
[2]Departamento de Ciências do Solo, Universidade Federal de Lavras, Lavras, Minas Gerais, Brazil. *Author for correspondence.
E-mail: flaviomedeiros@dfp.ufla.br

ABSTRACT. Abiotic stress interferes with plant-microbial interactions, but some microorganisms may buffer this interference. We investigated the interaction between temperature and bacterial wilt (*Curtobacterium flaccumfaciens* pv. *flaccumfaciens* - *Cff*) biocontrol and the ability of *Bacillus subtilis* strain ALB629 to colonize bean seedlings, to inhibit pathogen growth and to use different C and N sources. *B. subtilis* ALB629^{rif}, a mutant selected from the wild population of ALB629, was used to monitor plant colonization at 20°C and 30°C. ALB629^{rif} was detected only in the plant roots ($10^{3.22}$ CFU g^{-1}) at the lower temperature but colonized the roots, stems, and leaves ($10^{5.85}$, $10^{4.48}$, and $10^{4.01}$ CFU g^{-1}), respectively, at 30°C. The area under the disease progress curve was also different at the two tested temperatures ($p < 0.01$). Nevertheless, the disease reduction using ALB629^{rif} - treated seeds was similar: 71% and 75%, respectively, at 20 and 30°C ($p < 0.01$). A higher efficiency of C and N source utilization was observed at the higher temperature, but the antagonist inhibited *Cff* growth equally at either temperature *in vitro*. Based on our results, temperature interferes with pathogen and antagonist plant colonization, but the overall suppression of bacterial wilt appears to be stabilized by ALB629.

Keywords: seed treatment, *Phaseolus vulgaris*, antibiosis, temperature, rhizobacteria, PGPR.

Introduction

Common bean (*Phaseolus vulgaris* L.) is the most important legume in Brazil; the crop is cultivated throughout the year, mainly by small growers, with an estimated 10% utilizing certified seeds (SENA et al., 2008). Although various bean seed-transmitted pathogens exist, the bacterial pathogens are the most difficult to control, particularly if no resistant commercial cultivar is available. Presently, the most serious bean bacterial disease is bacterial wilt, which is caused by *Curtobacterium flaccumfaciens* pv. *flaccumfaciens* (*Cff*) (Hedges) Collins and Jones (VALENTINI et al., 2010). The pathogen was reported for the first time in South Dakota, United States by Heges in 1922 (VENETTE et al., 1995) and rapidly spread worldwide. In Brazil, the disease was first reported in 1997 by Maringoni and Rosa (1997) and, thereafter, became a major problem of common beans in different regions. In spite of its relatively recent discovery in Brazil, the pathogen has already spread throughout most of the bean-producing fields (HERBES et al., 2008), causing high economically important losses.

However, disease control by employing resistant cultivars is not possible because there is no commercially available resistant cultivar to date, though there are some germplasm lines of common bean that have demonstrated different levels of resistance to the yellow and orange variants of *Cff* (HSIEH et al., 2005; THEODORO et al., 2007). Infected seeds represent an important source of inoculum for bean bacterial wilt (HSIEH et al., 2003), yet no chemical seed treatment is registered (MARTINS et al., 2013; HSIEH et al., 2004). Therefore, other methods, such as biological control, need to be employed as potential tools for the management of bacterial wilt in common bean, particularly if used as a treatment of contaminated seeds. Plant beneficial bacteria (PBB), such as *Bacillus subtilis* strains, are among the disease control strategies that could potentially be useful to control plant pathogenic bacteria (MEDEIROS et al., 2009). Once an antagonist colonizes the plant tissue, the biological control of pathogens can result from a combination of mechanisms, including the competition for nutrients and space, antibiosis, and induced systemic resistance, and can sometimes result in beneficial effects to the host plant (HARMAN et al., 2010; MEDEIROS et al., 2012). Considering the importance of seeds in the transmission of pathogens and the need to reduce fungicide loads in the environment, seed treatment may provide a practical and cost-effective strategy to reduce such seed-born pathogens as *Cff* (HUANG et al., 2007; MARTINS et al., 2013). Unfortunately, the efficacy of biological control may be strongly influenced by sudden changes in temperature, and its efficacy may be dependent on the stability of such environmental variables as temperature (ASHRAFUZZAMAN et al., 2009). Indeed, both microorganisms and plants are influenced by temperature changes, and the root exudates that support rhizobacterial growth are more rapidly or slowly released at a higher or lower temperature, respectively, thus impacting the growth of bacteria in the rhizosphere (KATO et al., 2005; NIHORIMBERE et al., 2009).

This work aimed to verify the potential of *B. subtilis* ALB629 seed treatment to prevent bacterial wilt at 20 and 30°C and to study the relationship between these temperatures on the ability of ALB629 to colonize bean seedlings, to inhibit pathogen growth and to use different carbon and nitrogen sources, the major components of bean root exudates.

Material and methods

Selection of the B. *subtilis* ALB629[rif] mutant

To study the antagonist colonization of bean seedlings, a naturally occurring mutant, ALB629[rif], was selected from a *B. subtilis* strain (ALB629) and comparatively tested for growth and biocontrol capacity with its wild-type counterpart. To generate a mutant that is resistant to rifampicin, increasing amendments of this antibiotic were added to the nutrient agar (NA) medium up to 100 ppm at each bacterial culture, similar to a previously described method (MEDEIROS et al., 2009). Both ALB629 and ALB629[rif] were preserved in peptone glycerol at -80°C until use.

Biocontrol of bacterial wilt by B. *subtilis* wild-type and mutant strains at two temperatures

In this first experiment, *B. subtilis* ALB629 and its mutant, *B. subtilis* ALB629[rif], were used as biocontrol agents against bacterial wilt at of 20 and 30°C to ascertain the stability of the biological control activity of the mutant strain.

Artificial seed inoculation

Prior to inoculation, commercially available bean seeds (cv. Pérola) were disinfected by soaking in alcohol (70% ethanol) for 30 s, followed by sodium hypochlorite (0.5% active chloride) for 10 min. and sterile distilled water (SDW).

The *Cff* strain was isolated from dried infected leaves and cultivated on 523 Kado and Heskett medium in Petri dishes at room temperature (28°C) for 48 hours. The bacterial cells were then transferred to 523 liquid medium in test tubes and cultivated for 48 hours at room temperature (28°C) with shaking at 150 rpm. A 100 μL aliquot of the bacterial suspension was spread using a *Drigalski* spatula onto the surface of CNS medium amended with mannitol. The Petri dishes were placed at room temperature (28°C) for 72 hours. The disinfected seeds were artificially inoculated with *Cff* by the physiological conditioning technique of Deuner et al. (2011). The seeds were air-dried in a laminar flow cabinet for 8 hours, and fifty seeds/Petri dish were arranged over the *Cff* growth for 48 hours at room temperature (28°C), scraped from the medium and immediately treated as detailed below. Non-inoculated controls were incubated in the same medium under the same conditions without the pathogen.

Biological seed treatment to control bacterial wilt at two temperatures

B. subtilis ALB629 and its mutant ALB629[rif] were cultivated on NA medium in Petri dishes at room temperature (28°C) for 48 hours. Cells were subsequently transferred to a nutrient stock medium and cultivated for 48 hours at room temperature (28°C) with shaking at 150 rpm. The endospore concentration of the bacterial suspension was adjusted using a Neubauer chamber to 1×10^8 CFU mL^{-1} by dilution plating and then used for the seed treatment. The inoculated seeds were individually treated with the *B. subtilis* wild strain (ALB629) and its mutant (ALB629^{rif}) in a suspension of 10^8 CFU mL^{-1}, copper oxychloride (2 g a.i. L^{-1}), or water by imersion at the ratio of 2 mL per gram of seed for 30 min. Seeds without *Cff* inoculation were immersed in water solution as a control. The suspension was rinsed off, and the seeds were sown in 3 L pots filled with Argissoil, with 6 seeds per pot. A portion of the pots were maintained in growth chambers at 20°C and another portion at 30°C under a 12 hours photoperiod. The experiment was arranged in randomized block, with four replicates, and the plants were watered to maintain the soil moisture at field capacity.

Assessment of the analyzed variables

The curtobacterium wilt severity was evaluted according to Hsieh et al. (2003) at 12, 15, 18, 21 and 24 days after sowing (DAS). A 0–5 scale was used, as follows: 0 = no wilt symptoms; 1 = wilt on one of the primary leaves; 2 = wilt on both primary leaves but not on the first trifoliolate leaf; 3 = wilt on the first trifoliolate leaf; 4 = death of the seedling after wilt development on the primary leaves and 5 = unmerged seedling or death of seedling before wilt development on the primary leaves. The data were used to calculate the area under the disease progress curve (AUDPC) according to Shaner and Finney (1977). The experiment was arranged in a randomized block design. The data were subjected to a variance analysis (ANOVA); for significant effects, the data were compared using Tukey's test (p = 0.05). The experiment was repeated twice.

Colonization of bean seedlings by ALB629^{rif}

In this second experiment, the artificial seed inoculation and the biological seed treatment with ALB629^{rif} were performed, as described previously. To track plant colonization by the biocontrol agent, two plants per pot were randomly sampled at 24 DAS and separated into roots, stems and leaves. The cotyledons were used as the leaf samples, the stem samples consisted of the region between cotyledon insertion and the first true leaf, and the root samples consisted of the entire root system. Each assessed plant tissue was weighed, surface disinfected with alcohol (70%) for 30 s, followed by hypochlorite (0.5% active chloride) for 10 min. and washed twice

with SDW. The tissue was then transferred to 10 mL tubes containing 5 mL of SDW; the tissue was crushed, vortexed for 1 min. and serially diluted to 10^{-5}. The crude extract and the 10^{-1} to 10^{-5} dilutions were then plated by spreading on NA medium with 100 ppm rifampicin and incubated at room temperature (28°C). Colonies were counted after 2 days of incubation, and the data were transformed to log10 CFU g^{-1} of fresh tissue. The experiment was performed in a randomized block design. The data were subjected to an ANOVA; for significant effects, the data were compared using Tukey's test (p = 0.05). The experiment was repeated twice.

Antibiosis *in vitro*

The stability of the *Cff* inhibition *in vitro* was tested at two temperatures: 20 and 30°C. The bacterial pathogen was cultivated in liquid medium 523 for 48 hours, and 100 µL of the suspension was spread on CNS medium. The antagonist was grown in liquid medium MCF, and a 10 µL aliquot of 10^8 CFU mL^{-1} of ALB629 and a 10 µL aliquot of MCF medium was transferred to plates containing the *Cff* suspension. Five plates were incubated at 20°C and another five at 30°C. The zone of inhibition was recorded 24 hours afterwards. The experiment was performed in a factorial scheme using a randomized block design. The data were subjected to an ANOVA, and the significant effects were compared using Tukey's test (p = 0.05). The experiment was repeated twice.

B. *subtilis* ALB629 growth using different C and N sources

In this experiment, *B. subtilis* ALB629 was tested for its ability to use different nutritional sources of carbon and nitrogen under at 20 and 30°C. Thus, ALB629 was grown in four different culture media for 12, 24, 36, and 48 hours to create a growth curve. The different culture media differed from each other by the following substrates: glucose, sodium citrate, aspartate, and glutamate, with the two first two being used as carbon sources and the last two used as nitrogen sources. These four different nutrients were used at 10 g L^{-1} concentration except for glutamate (15 g L^{-1}). Each of these nutrients was added to another solution composed of 0.5 g L^{-1} $MgSO_4$ ($7H_2O$), 1.0 g L^{-1} K_2HPO_4, 0.5 g L^{-1} KCl, 1.0 g L^{-1} yeast extract, 1.2 mg L^{-1} $FeSO_4$ ($7H_2O$), 0.4 mg L^{-1} $MnSO_4$ (H_2O), 1.6 mg L^{-1} $CuSO_4$, and 1 g L^{-1} $(NH_4)_2SO_4$. The pH of each medium was adjusted to 7 with NaOH or 0.1 N HCl before sterilization, as described previously (NIHORIMBERE et al., 2009).

The experimental plot consisted of a test tube with 3 mL of each of the above 4 culture media, with three replicates per plot x time x temperature. To evaluate ALB629 growth, the strain was cultivated in a nutrient broth composed of 13.8 g L^{-1} yeast extract, 2.5 g L^{-}K$_2$HPO$_4$, 1.0 g L^{-1} KH$_2$PO$_4$ (anhydrous), 2.5 g L^{-1} NaCl, 6.5 g L^{-1} sucrose, 0.25 g L^{-1} magnesium sulfate and 0.1 g L^{-1} manganese sulfate for 48 hours at 28°C under stirring at 150 rpm. Every 12 hours (for 48 hours), a 150 μL aliquot of the antagonist suspension was transferred to each tube containing 3 mL of each previously described medium. The test tubes were placed in two different orbital shakers at different temperatures, 20 and 30°C, both at 150 rpm. After 48 hours from the first transfer, for each incubation temperature (20 and 30°C) and each sampled time point (12, 24, 36, and 48 hours after the onset of the experiment), three replicates of one tube each were individually analyzed by measuring the absorbance using a light spectrophotometer DU® 640B at λ = 600 nm. The experiment was performed in a factorial scheme (2 temperatures x 4 sampled time points x 4 nutrient sources) using a randomized block design. The data were subjected to a variance analysis (ANOVA); for significant effects, the temperature and nutrient factors were compared using Tukey's test (p = 0.05) and the sampled time points using a regression analysis. The linear regression equations were subjected to a parallelism (F-test) test to verify the null hypothesis that the slopes of the equations were statistically equal. The goodness of fit of the models was tested at 0.05 significance and evaluated by its coefficient of determination (R^2). The data analyses were performed using SISVAR software (FERREIRA, 2011).

Results and discussion

Disease control was achieved by seed treatment with *B. sutilis* ALB629 at approximately. 70% compared to the water control; both wild-type and mutant ALB629 displayed the same growth rate. Moreover, based on the area under the disease progress curve (AUDPC), the biological control of the bacterial wilt of bean was not different between the wild-type and mutant strains at 30°C (p = 0.8596) or 20°C (p = 0.1572), and both were effective in reducing the common bean bacterial wilt severity progress (AUDPC) at either temperature.

Because the wild-type and mutant strains showed the same disease reduction, all of the subsequent experiments were performed using only the mutant. Although the AUDPC was different

between both tested temperatures (p < 0.01), the disease reduction obtained by the ALB629^{rif} seed treatment was similar at 71 and 75%, respectively, for 20 and 30°C when compared with the non-inoculated untreated seeds (control) (p < 0.01) (Table 1).

Table 1. Effect of seed treatment with the rifampicin-resistant strain of *Bacillus subtilis* ALB629 (ALB629^{rif}), copper oxychloride and water on the area under the disease progress curve (AUDPC)* in the presence of *Curtobacterium flaccumfaciens* pv. *flaccumfaciens* (*Cff*).

Treatment	Temperature	
	20°C	30°C
ALB629^{rif}	64.95 bB**	136.60 bA
Copper oxychloride	175.84 aB	497.40 aA
Cff-inoculated control	227.48 aB	554.35 aA
Non-inoculated, untreated control	0.00 bA	0.0 cA
^{1}C.V (%) = 19.39		

*Calculated based on the disease score according to Hsieh et al. (2003) at 12, 15, 18, 21, and 24 days after sowing; **The means written with a common letter do not differ significantly according to Tukey's test (p ≤ 0.01); the lowercase refers to the column, and the uppercase to the row. ^{1}C.V. - Coefficient of variation. Means of two experiments of four replicates of six seedlings each.

Although *B. subtilis* ALB629^{rif} could efficiently reduce the disease severity at either temperature, it had a direct effect on the overall AUDPC (Table 1). At the higher temperature, a higher AUDPC was observed for all the treatments, and this corroborates previous findings for the influence of temperature on bacterial wilt severity (KRAUSE et al., 2009). Nevertheless, the biocontrol agent ALB629 induced a consistent disease reduction of approximately. 72%, regardless of the temperature. The same temperature-independent trend was not observed for the seeds treated with copper oxychloride (p < 0.001), the putative positive control.

The mutant strain, which showed consistent disease control, was used to monitor plant colonization. The plants were colonized by the bacterium, regardless of the temperature; however, the bacterium colonized the roots (10$^{5.85}$ CFU g^{-1}), stems (10$^{4.48}$ CFU g^{-1}) and leaves (10$^{4.01}$ CFU g^{-1}) at 30°C, whereas it remained confined to the roots (10$^{3.22}$ CFU g^{-1}) at 20°C, with a population approx. 100 times smaller. Because the pathogen was not as damaging at the lower temperature, the antagonist was able to control the disease, even at a less robust plant colonization (Figure 1).

ALB629 was confined to the roots when the plants were grown at 20°C and colonized the entire plant at 30°C, suggesting that the induced resistance might play a pivotal role as a control mechanism in this plant-pathogen interaction (ONGENA et al., 2007). Future experiments will investigate the activity of key defense-related enzymes in disease control; although it might also be present at the higher temperature, direct antibiosis will still have to be considered as a key mechanism to cope with more aggressive plant colonization under this condition. Another hypothesis that may be drawn to explain the observed disease control at the lower temperature is that because both the pathogen and biocontrol agent were introduced via seeds, both shared the same niche at the early stages of plant development. Thus, an initial direct antibiosis may have occurred, reducing the initial inoculum of the pathogen, and, even when ALB629 was confined to the roots, the possible induced resistance triggered by the biocontrol agent or simply by the low pathogen inoculum would be sufficient to cope with the disease. We are presently investigating each of hypotheses to explain the observed disease control.

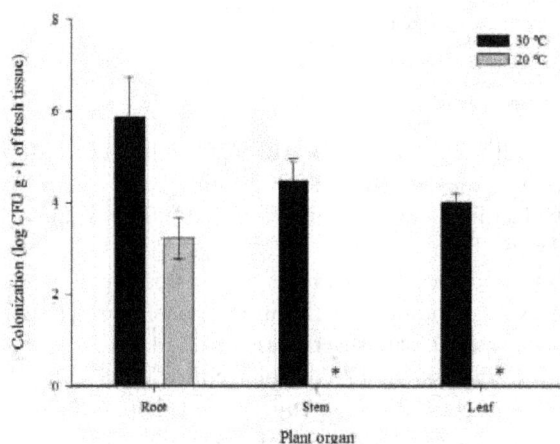

Figure 1. Colonization of bean seedlings by *Bacillus subtilis* ALB629[rif] inoculated by seed immersion at a concentration of 10^8 CFU mL^{-1} for 30 min. *n.d.: not detected by the plate dilution method. The vertical bars represent the standard error of the mean. Means of two experiments of four replicates of six seedlings each.

The diversity of the root exudate secreted at each temperature may also regulate the rate at which ALB629 was able to grow. Bean roots release large amounts of amino acids and sugars (ODUNFA, 1979), which supply the demands of carbon and nitrogen for rhizosphere microorganisms (YOUSSEF; MANKARIOS, 1968).

Based on the spectrophotometric absorbance reading ($\lambda = 600$ nm), a different pattern of ALB629 growth was observed when it was cultivated using different nutrient sources ($p < 0.01$). Moreover, the bacterial growth at 30°C was higher than at 20°C ($p < 0.01$) (Table 2).

The nutrient sources supported different bacterial growth at each temperature. Glutamate supported the highest growth, which was more than two times higher than the growth observed

for aspartate, the nutrient source that supported the second-highest growth. In contrast, sodium citrate supported a comparatively lower growth, but the growth was higher than with glucose. For all of the nutrient sources except aspartate, the growth observed at 30°C was higher than that observed at 20°C.

Table 2. *Bacillus subtilis* ALB629 growth in media with different C and N nutrient sources at temperatures of 20°C and 30°C.

Treatment	Temperature	
	20°C	30°C
Glucose	0.15 d B*	0.45 d A
Sodium citrate	0.40 c B	0.55 c A
Glutamate	1.25 a B	1.36 a A
Aspartate	0.59 b A	0.58 b A
[1]C.V (%) = 2.09		

*The means of absorbance followed by the same uppercase letter in a row and lowercase letter in a column are statistically similar according to Tukey's test ($p \leq 0.01$).
[1]C.V. - Coefficient of variation.

Moreover, the bacterial growth differed over time ($p < 0.01$), with the growth increasing (12, 24, 36, and 48 hours) when the times were compared using Tukey's test (Figure 2A, B, C and D).

Except for sodium citrate and aspartate at 30°C, all of the other substrates differed significantly according to the F-test ($p < 0.05$) including sodium citrate and aspartate at 20°C. All the tested amino acids are components of bean root exudates and, compared to other sugars and amino acids, support higher surfactin, a lipopeptide that acts in the induction of resistance, production by *Bacillus subtilis* (NIHORIMBERE et al., 2009). It remains unknown whether the molecules produced by ALB629 directly or indirectly inhibit the growth of the pathogen. However, with regard to disease control independently of the temperature, the growth using aspartate was similar at either tested temperature, and the bacterial growth displayed a standard characteristic of a probable phase adaptation of twelve hours, a logarithmic phase at 12-24 hours, a stationary phase at 24-36 hours, and a decline phase thereafter. These results suggest that resistance induction plays an important role in the protection of bean against bacterial wilt triggered by ALB629. Future studies will determine the induction of resistance at different temperatures and lipopeptide production under these conditions. Furthermore, induced resistance elicitors may be responsible for the observed biological control because antibiosis at either temperature displayed similar results ($p < 0.05$).

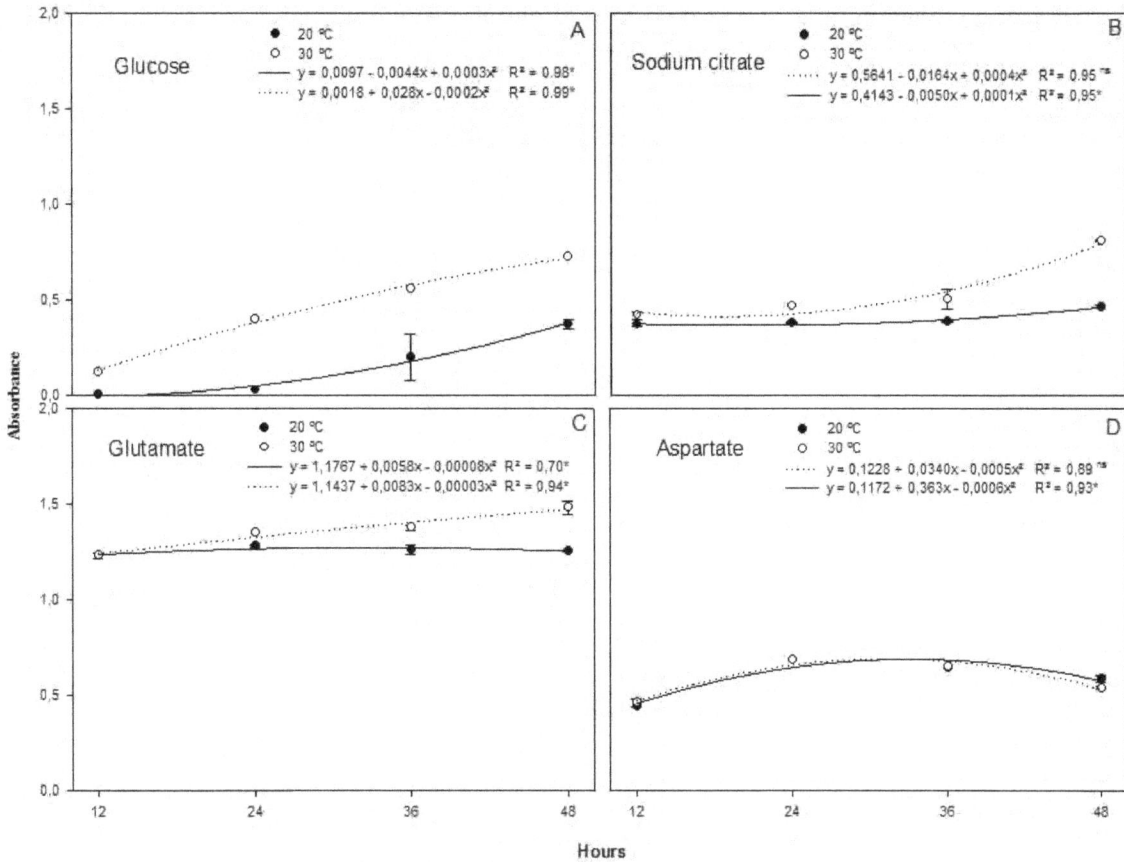

Figure 2. *F*-test comparing the linear regression angular coefficients derived from the *Bacillus subtilis* ALB629 growth in media with different nutrient sources and at temperatures of 20 and 30°C. The graphs in 2A, B, C, and D represent media with glucose, sodium citrate, glutamate, and aspartate, respectively. The vertical bars represent the standard error of the mean. [ns]: angular coefficients for ALB629 growth do not differ significantly. ★Angular coefficients for ALB629 growth differ significantly ($F_{calculated} > F_{tabulated}$).

Therefore, *B. subtilis* ALB629 controls bacterial wilt in bean via treatment of infected seeds, and the stability of the disease control at different temperatures reinforces the potential practical use of this biological control agent, particularly because bean is cultivated throughout the year at different locations where different temperature conditions are found.

Conclusion

The stability of bacterium wilt control, the root colonization, the use of different nutrient sources and antibiosis at 20 and 30°C implies ALB629 as a promising tool to be used in the disease management in the field avoiding important bacterial wilt outbreaks that have been frequently observed in Brazil.

Acknowledgements

We thank Conselho Nacional de Desenvolvimento Cientifico Cultural (CNPq), Fundação de Apoio à Pesquisa do Estado de Minas Gerais (FAPEMIG) and Programa de Apoio a Primeiros Projetos (PAPP/UFLA) for providing the financial support necessary for the development of this work. We thank CNPq for providing assistantship for the second and third authors.

References

ASHRAFUZZAMAN, M.; HOSSEN, F. A.; ISMAIL, M. R.; HOQUE, M. D. A.; ISLAM, M. Z.; SHAHIDULLAH, S. M.; MEON, S. Efficiency of plant growth-promoting rhizobacteria (PGPR) for the enhancement of rice growth. **African Journal of Biotechnology**, v. 8, n. 7, p. 1247-1252, 2009.

DEUNER, C. C.; SOUZA, R. M.; ISHIDA, A. K. N.; ZACARONI, A. B.; VON PINHO, E. V. R.; MACHADO, J. C.; CAMERA, J. N. Inoculação de *Curtobacterium flaccumfaciens* pv. *flaccumfaciens* em sementes de feijão por meio da técnica de condicionamento fisiológico. **Revista Brasileira de Semente**, v. 33, n. 2, p. 9-20, 2011.

FERREIRA, D. F. SISVAR: a computer statistical analysis system. **Ciência e Agrotecnologia**, v. 35, n. 6, p. 1039-1042, 2011.

HARMAN, G.; OBREGON, M. A.; SAMUELS, G. J.; LORITTO, M. Changing models for commercialization and implementation of biocontrol in the developing and the developed world. **Plant Disease**, v. 94, n. 8, p. 928-938, 2010.

HERBES, D. H.; THEODORO, G. F.; MARINGONI, A. C.; DAL PIVA, C. A.; ABREU, L. Detecção de *Curtobacterium fl accumfaciens* pv. *fl accumfaciens* em sementes de feijoeiro produzidas em Santa Catarina. **Tropical Plant Pathology**, v. 33, n. 1, p. 53-156, 2008.

HSIEH, T. F.; HUANG, H. C.; CONNER, R. L. Bacterial wilt of bean: Current status and prospects. **Recent Research Developments in Plant Science**, v. 2, p. 181-206, 2004.

HSIEH, T. F.; HUANG, H. C.; MÜNDEL, H. H.; ERICKSON, R. S. Arapid indoor technique for screening common bean (*Phaseolus vulgaris* L.) for resistance to bacterial wilt [*Curtobacterium flaccumfaciens* pv. *flaccumfaciens* (Hedges) Collins and Jones]. **Revista Mexicana de Fitopatología**, v. 21, n. 3, p. 370-374, 2003.

HSIEH, T. F.; HUANG, H. C.; MÜNDEL, H. H.; CONNER, R. L.; ERICKSON, R. S.; BALASUBRAMANIAN, P. M. Resistance of common bean (*Phaseolus vulgaris*) to bacterial wilt caused by *Curtobacterium flaccumfaciens* pv. *flaccumfaciens*. **Journal of Phytopathology**, v. 153, n. 4, p. 245-249, 2005.

HUANG, H. C.; ERICKSON, R. S.; HSIEH, T. F. Control of bacterial wilt of bean (*Curtobacterium flaccumfaciens* pv. *flaccumfaciens*) by seed treatment with Rhizobium leguminosarum. **Crop Protection**, v. 26, n. 7, p. 1055-1061, 2007.

KATO, K.; WATANABE, K.; ARIMA, Y. *Rhizobium* Proliferation-Supporting Substances in Seed Exudates of Common Bean (*Phaseolus vulgaris* L.). **Soil Science and Plant Nutrition**, v. 51, n. 6, p. 905-910, 2005.

KRAUSE, W.; RODRIGUES, R.; GONÇALVES, L. S. A.; BEZERRA NETO, F. V.; LEAL, N. R. Genetic divergence in snap bean based on agronomic traits and resistance to bacterial wilt. **Crop Breeding and Applied Biotechnology**, v. 9, n. 3, p. 246-252, 2009.

MARINGONI, A. C.; ROSA, E. F. Ocorrência de *Curtobacterium flaccumfaciens* pv. *flaccumfaciens* em feijoeiro no Estado de São Paulo. **Summa Phytopathologica**, v. 23, n. 2, p. 160-162, 1997.

MARTINS, S. J.; MEDEIROS, F. H. V.; SOUZA, R. M.; RESENDE, M. L. V.; RIBEIRO JUNIOR, P. M. Biological control of bacterial wilt of common bean by plant growth-promoting rhizobacteria. **Biological Control**, v. 66, n. 1, p. 65-71, 2013.

MEDEIROS, F. H. V. Management of melon bacterial blotch by plant beneficial bactéria. **Phytoparasitica**, v. 37, n. 5, p. 453-460, 2009.

MEDEIROS, F. H. V.; MARTINS, S. J.; ZUCCHI, T. D.; DE MELO, I. S.; BATISTA, L. R.; MACHADO, J. C. Biological control of mycotoxin–producing molds. **Ciência e Agrotecnologia**, v. 36, n. 5, p. 483-497, 2012.

NIHORIMBERE, V.; FICKERS, P.; THONART, P.; THONART, P.; ONGENA, M. Ecological fitness of *Bacillus subtilis* BGS3 regarding production of the surfactin lipopeptide in the rhizosphere. **Environmental Microbiology Reports**, v. 1, n. 2, p. 124-130, 2009.

ODUNFA, V. S. A. Free amino acids in the seed and root exudates in relation to the nitrogen requirements of rhizosphere soil Fusaria. **Plant and Soil**, v. 52, n. 4, p. 491-499, 1979.

ONGENA, M.; ADAM, A.; JOURDAN, E.; PAQUOT, M.; BRANS, A.; JORIS, B.; ARPIGNY, J. L.; THONART, P. Surfactin and fengycin lipopeptides of *Bacillus subtilis* as elicitors of induced systemic resistance in plants. **Environmental Microbiology**, v. 9, n. 4, p. 1084-1090, 2007.

SENA, M. R.; ABREU, A. F. B.; RAMALHO, M. A. P.; BRUZI, A. T. Envolvimentoe de agricultores no processo seletivo de novas linhagens de feijoeiro. **Ciência e Agrotecnologia**, v. 32, n. 2, p. 407-412, 2008.

SHANER, G.; FINNEY, R. The effect of nitrogen fertilization on the expression of slow mildewing resistance in Knox Wheat. **Phytopathology**, v. 67, n. 8, p. 1051-1056, 1977.

THEODORO, G. F.; HERBES, D. H.; MARINGONI, A. C. Fontes de resistência à murcha de curtobacterium em cultivares locais de feijoeiro, coletadas em Santa Catarina. **Ciência e Agrotecnologia**, v. 31, n. 5, p. 333-1339, 2007.

VALENTINI, G.; GUIDOLIN, A. F.; BALDISSERA, J. N. C.; COIMBRA, J. L. M. *Curtobacterium flaccumfaciens* pv. *flaccumfaciens*: etiologia, detecção e medidas de controle. **Revista Biotemas**, v. 23, n. 1, p. 1-8, 2010.

VENETTE, J. R.; LAMPRA, R. S.; GROSS, P. L. First report of bean bacterial wilt caused by *Curtobacterium flaccumfaciens* subsp. *flaccumfaciens* in North Dakota. **Plant Disease Note**, v. 79, n. 9, p. 966, 1995.

YOUSSEF, Y. A.; MANKARIOS, A. T. Studies on the rhizosphere mycoflora of broad bean and cotton: II., seed and root exudates and their effects on spore germination and growth of the prevalent fungi isolated from the rhizosphere. **Mycopathologia**, v. 38, n. 3, p. 257-269, 1968.

7

Effect of the flavonoid rutin on the biology of *Spodoptera frugiperda* (Lepidoptera: Noctuidae)

Talita Roberta Ferreira Borges Silva[1], André Cirilo de Sousa Almeida[1], Tony de Lima Moura[1], Anderson Rodrigo da Silva[1], Silvia de Sousa Freitas[2] and Flávio Gonçalves Jesus[1*]

[1]*Instituto Federal Goiano, Campus Urutaí, Rodovia Professor Geraldo Silva Nascimento, km 2,5, 75790-000, Urutaí, Goiás, Brazil.*
[2]*Departamento de Química, Universidade Federal de Goiás, Campus de Catalão, Catalão, Goiás, Brazil. *Author for correspondence.*
E-mail: fgjagronomia@zipmail.com.br

ABSTRACT. The fall armyworm *Spodoptera frugiperda* (J.E. Smith) (Lepidoptera: Noctuidae) is a major pest of maize crops in Brazil. The effects of plant metabolites on the biology and behavior of insects is little studied. The aim of the study was to evaluate the activity of rutin on the biology of the *S. frugiperda* by using artificial diets containing rutin. The study evaluated four treatments: regular diet (control group) and diets containing 1.0, 2.0 and 3.0 mg g^{-1} of rutin. The following biological variables parameters of the larvae were evaluated daily: development time (days), larval and pupal weight (g) and viability (%), adult longevity and total life cycle (days). A completely randomized experimental design was used with 25 replication. The rutin flavonoid negatively affected the biology of *S. frugiperda* by prolonging the larval development time, reducing the weight of larvae and pupae and decreasing the viability of the pupae. The addition of different concentrations of rutin prolonged the *S. frugiperda* life cycle. The use of plant with insecticidal activity has the potential with strategy in IPM.

Keywords: fall armyworm, insecticide plant, secondary metabolite, flavonoid.

Introduction

The fall armyworm *Spodoptera frugiperda* (J.E Smith) (Lepidoptera: Noctuidae) is a major pest species of maize crops (Cruz, Figueiredo, Oliveira, & Vasconcelos, 1999; Pereira et al., 2002). *Spodoptera frugiperda* is a polyphagous species that infests cotton (Campos, Boiça-Júnior, Valério Filho, Campos, & Campos, 2012; Jesus, Boiça Junior, Alves, & Zanuncio, 2014), rice, millet, sorghum (Busato et al., 2004), soybean (Boiça Junior, Souza, Neves, Ribeiro, & Stout, 2015) and other crops. In Brazil, one of the factors that contribute to failures in the control of *S. frugiperda* is the large number of hosts caused by the succession of crops with different phenologies (Sá, Fonseca, Boregas, & Waquil, 2009; Barros, Torres, & Bueno, 2010).

The larvae of *S. frugiperda* feed on maize leaves and ears, reducing the photosynthetic capacity of the plant and its production because of damage to the reproductive structures (Lima, Ohashi, Souza, & Gomes, 2006). This damage depends on the plant phenological state, infestation period and intensity of pest infestation (Cruz et al., 1999).

Spodoptera frugiperda control has been mainly performed with chemical insecticides (Yu, Nguyen, & Abo-Elghar, 2003). To reduce the use of chemical controls, alternative techniques are being studied. Among these alternatives, plant resistance to insects and the use of plant metabolites with insecticidal effects have shown promising results in integrated pest management programs (Pereira et al., 2002; Hoffmann-Campo, Ramos Neto, Oliveira, & Oliveira, 2006; Jesus et al., 2014).

Resistance can be express by morphological characteristics or the presence of chemical metabolites (allelochemicals), such as alkaloids, flavonoids, terpenoids, sterols etc., present in the plants (Piubelli, Campo, Moscardi, Miyakubo, & Oliveira, 2005; Hoffmann-Campo et al., 2006; Magarelli, Lima, Silva, Souza, & Castro, 2014).

The activity of plant chemical substances has been promising, and new components with insecticidal potential have been discovered, and they have the potential for use in pest management for crops of agricultural importance (Pereira et al., 2002; Deota & Upadhyay, 2005; Rajamma, Dubey, Sateesha, Tiwan, and Ghosh, 2011; Tavares, Pereira, Freitas, Serrão, & Zanuncio, 2014).

Rutin is a flavonoid that can be used in plant protection against insects, especially lepidopterans, because of its anti-nutritional effects (Harborne & Grayer, 1993; Salvador et al., 2010; Tavares et al., 2014). In soybean, the flavonoids rutin and genistin were identified in several parts of the plant (Piubelli et al., 2005; Hoffmann-Campo et al., 2006; Lucci & Mazzafera, 2009).

Rutin can prolong the life cycle of *Anticarsia gemmatalis* Hubner, 1818 (Lepidoptera: Noctuidae) and cause higher larval mortality when it is added to the insect diet (Gazzoni, Hulsmeyer, & Hoffmann-Campo, 1997; Hoffmann-Campo et al., 2006; Piubelli, Hoffmann-Campo, Moscardi, Miyakubo, & Oliveira, 2006). Larvae of *Manduca sexta* Linnaeus, 1763 (Lepidoptera: Sphingidae) fed a diet containing rutin showed negative growth performance (Stamp & Skrobola, 1993).

The addition of 2% rutin to the diet of *Trichoplusia ni* Hubner, 1803 (Lepidoptera: Noctuidae) negatively affected the survival, behavior, and physiology of the insect (Hoffmann-Campo, Harbone, & Mcaffery, 2001). In *S. frugiperda,* the metabolite astilbin from the *Dimorphandra mollis* (Fabaceae - Mimosoideae) plant reduced the larval weight and prolonged the larval and pupal stages (Pereira et al., 2002).

Therefore, the current study aimed to evaluate whether diets containing the flavonoid rutin affected the biology of *S. frugiperda*.

Material and methods

Spodoptera frugiperda mass rearing

The methodology proposed by Cruz (2000) with adjustments (Jesus et al., 2014) was used to obtain the *S. frugiperda* larvae. Briefly, male and female moths were maintained in polyvinyl chloride (PVC) tubes (10 cm diameter; 21.5 cm height) that were internally coated with bond paper sheets for egg laying and covered with voile at the top to prevent insect escape.

Cotton pads soaked in 10% honey solution were placed on top of the cages for moth feeding and changed every two days. The bond paper sheets with egg masses were recovered daily, and the egg masses were cut from the sheets with the aid of scissors and then transferred to plastic containers (100 mL) containing 5 g of the artificial diet (Kasten Júnior, Precetti, & Parra, 1978). These containers were covered and maintained in a climate-controlled room (temperature 27 ± 2°C, 70 ± 10% relative humidity (RH) and 12 hours photophase).

The larvae were individually housed when they reached the second instar (3 days, approximately 4 mm) because of their cannibalistic habits. The individuals were placed in 50 mL plastic cups containing 5 g of artificial diet, and the cups were sealed with acrylic lids and maintained in a climate-controlled room until the pupae developed. Subsequently, the pupae were segregated using sexual dimorphism as a parameter to differentiate males and females (Luginbill, 1928), and seven pairs of male and female moths were maintained in each cage.

Purification and structural characterization of rutin

Technical grade rutin was purchased and purified in the laboratory by recrystallization using methanol as the solvent.

After recrystallization, the isolated crystals were characterized by nuclear magnetic resonance spectroscopy (NMR H^1, C^{13}) (Figure 1).

The spectra were obtained at the Laboratory of Nuclear Magnetic Resonance (Laboratório de Ressonância Magnética Nuclear - RMN) of the Chemistry Institute (Instituto de Química - IQ) of the Federal University of Goiás (Universidade Federal de Goiás - UFG).

Once isolated, the rutin crystals were characterized by NMR H^1 (Figure 2a) and C^{13} spectra (Figure 2b) and compared with data from the literature to verify the purity of the rutin.

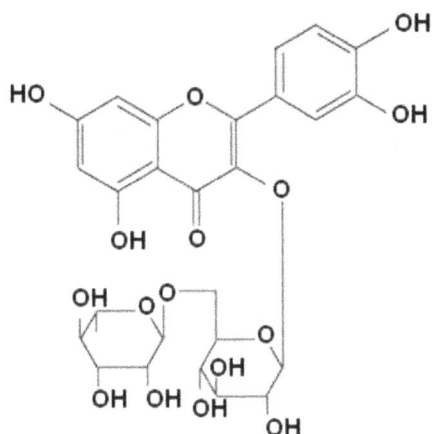

Figure 1. Chemical structure of rutin (quercetin 3 - O rhamnosyl glucoside).

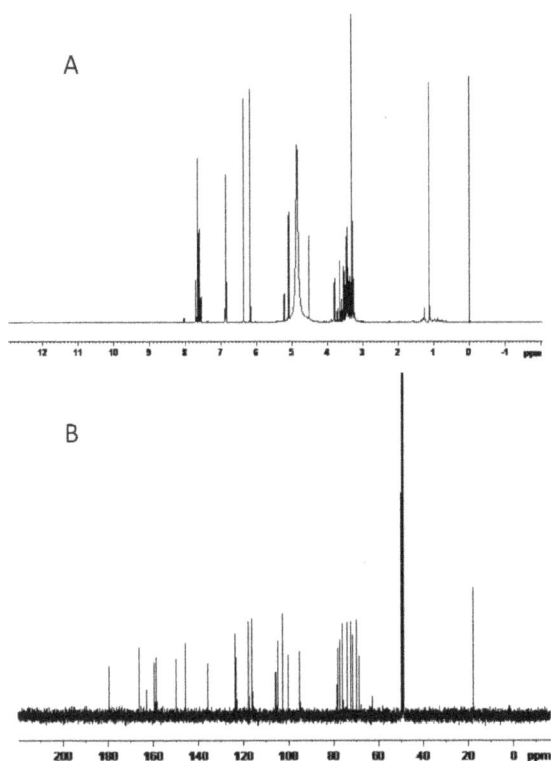

Figure 2. Spectra of rutin H^1 (A) and rutin C^{13} (B).

Biology of *Spodoptera frugiperda* fed rutin

The experimental design used in the current study was completely randomized with 4 treatments and 25 replicates, and each plot consisted of a 50 mL plastic cup. Five grams of the diet containing the flavonoid rutin at concentrations of 0 (control), 1, 2 and 3 mg g^{-1} and one newly hatched caterpillar were placed in each of the cups to monitor the pest life cycle.

The following biological parameters were evaluated: larvae and pupae development time and viability, weight of 10 day old larvae, weight of 24 hours old pupae, longevity and adult total life cycle.

Statistical analysis

The data were subjected to an analysis of deviance (Anodev), and the regression models were fitted with linear predictors containing polynomial effects of the first, second and third degrees for the rutin concentration. A Poisson distribution (Poisson regression) was assumed for the parameters larvae and pupae development time, total life cycle and adult longevity. For larval and pupal viability, a binomial distribution (binomial regression) was assumed. The weight of the larvae and pupae fit a Gaussian model. The nominal level of significance was 5%. The statistical analyses were performed using the software R version 3.0.3 (R Core Team, 2014).

Results and discussion

The development time of larvae and pupae of *S. frugiperda* can be observed in Figure 3. As shown, these parameters were affected by the different concentrations of rutin and there was a linear effect (p < 0.001) of rutin concentration on the larval development time, with higher concentrations of rutin inducing longer larval development time. A concentration of 3.0 mg g^{-1} of rutin prolonged the larval development time by 8 days on average compared with the control group (0 mg g^{-1}).

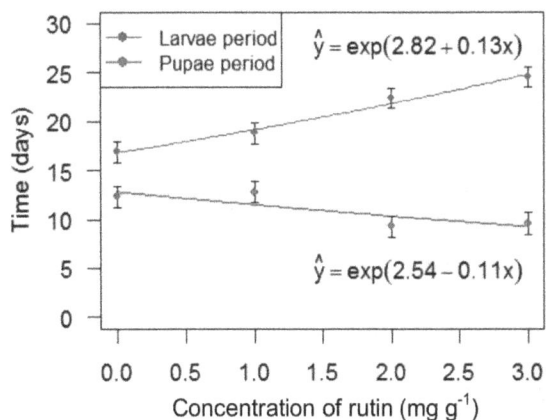

Figure 3. Development time (days) of the larvae and pupae of *Spodoptera frugiperda* (Lepidoptera: Noctuidae) fed an artificial diet containing different concentrations of rutin.

A linear effect (p = 0.0059) of rutin concentration on pupal development time was also observed; however, the effect was opposite to that of the larvae, with lower concentrations of rutin prolonging the developmental time of *S. frugiperda* pupae.

The shorter development time and lower weight of *S. frugiperda* larvae fed diets containing rutin may be associated with lower food intake by the caterpillars. This phenomenon may be related to the allelochemicals acting as feeding deterrents and digestion inhibitors and forming free radicals (Pereira et al., 2002; Salvador et al., 2010).

The weight of the larvae and pupae of *S. frugiperda* fed artificial diets containing different concentration of rutin can be observed in Figure 4. A linear effect (p < 0.001) was observed for the rutin concentrations on the larval and pupal weight.

Higher rutin concentrations correlated with lower weights in the *S. frugiperda* larvae and pupae (Figure 4). In the control group (0 mg g^{-1} rutin), the mean larval and pupal weights were 0.37 and 0.29 g, respectively, whereas at a rutin concentration of 3 mg g^{-1}, the mean larval weight drastically decreased to 0.15 g and mean pupal weight decreased to 0.23 g.

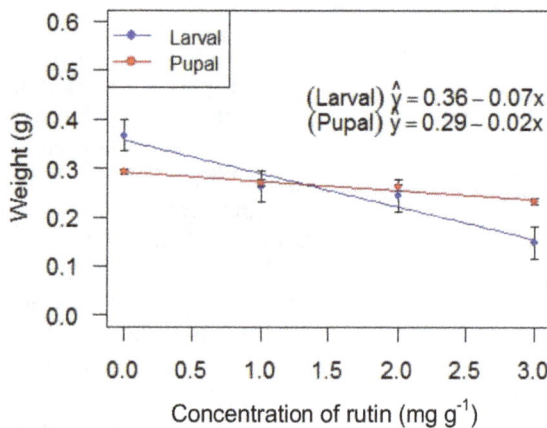

Figure 4. Weight (g) of the larvae and pupae of *Spodoptera frugiperda* (Lepidoptera: Noctuidae) fed an artificial diet containing different concentrations of rutin.

Many flavonoids can act as feeding deterrents for phytophagous insects, even at relatively low concentrations (Harborne & Grayer, 1993; Pereira et al., 2002). The deterrent effect is related to astringency, which is caused by phenolic compounds precipitating proteins (Appel & Maines, 1995). Reduced food intake and its consequent deleterious effects were also observed in *A. gemmatalis* fed artificial diets containing rutin (Hoffmann-Campo et al., 2006; Piubelli et al., 2006).

The nutritional quality of the diet without the flavonoid rutin was likely suitable for *S. frugiperda* larval development. Therefore, the adverse effects observed in the larvae fed diets containing rutin most likely occurred because of the presence of toxins in the artificial diets (Harborne & Grayer,

1993; Hoffmann-Campo et al., 2006; Piubelli et al., 2006; Salvador et al., 2010). These toxins may have produced damage in the digestive tract of *S. frugiperda* that was similar to the damage in the midgut of *A. gemmatalis* fed soybean genotypes containing the flavonoid rutin (Salvador et al., 2010).

The biological parameters of *S. frugiperda* pupae were affected by rutin concentrations in the diet, and the negative effects on the pupal stage were most likely related to the larval-pupal metamorphosis during ecdysis (Gazzoni et al., 1997). This suggests that rutin had a negative effect on the activity of enzymes and hormones, blocked biochemical pathways and reduced the assimilation of essential substances or the formation of reserves in the insect (Salvador et al., 2010; Tavares et al., 2014).

A significant effect on the larval viability was not observed (p > 0.05) for any of the rutin concentrations (Figure 5); however, an effect was observed for pupal viability (p < 0.001), although only the cubic effect was significant (p < 0.001). All of the pupae that were fed diets containing 2 mg g^{-1} rutin developed to adults, whereas insects fed diets containing 1 and 3 mg g^{-1} rutin had the lowest pupal viability (57.14%).

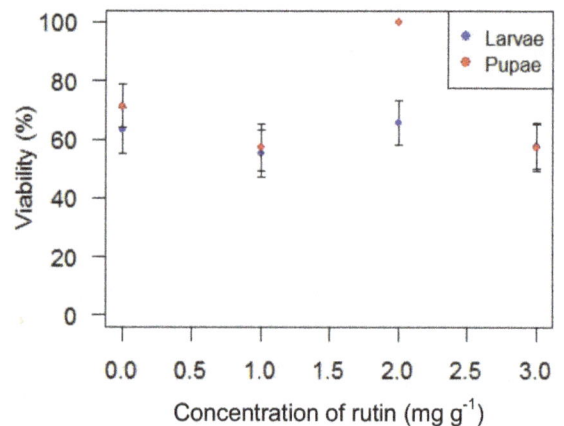

Figure 5. Viability (%) of the larvae and pupae of *Spodoptera frugiperda* (Lepidoptera: Noctuidae) fed an artificial diet containing different concentrations of rutin.

A linear effect (p < 0.001) was observed for the total life cycle of *S. frugiperda* fed diets with different rutin concentrations, with concentrations of 3 mg g^{-1} prolonging the life cycle by ten days on average compared with insects from the control group (0 mg g^{-1}) (Figure 6). The adult longevity was not affected (p > 0.05) by the different rutin concentrations.

These data corroborate those obtained by Hoffmann-Campo, Ramos Neto, Oliveira, and

Oliveira (2006), Piubelli, Hoffmann-Campo, Moscardi, Miyakubo, and Oliveira (2006) and Salvador et al. (2010) who studied the biological parameters of *A. gemmatalis* fed diets containing rutin. These authors observed a negative effect of rutin on the feeding and biology of this insect at the larval stage and pupal stage, including reduced weight of pupae fed diets containing the flavonoid (Hoffmann-Campo et al., 2006; Salvador et al., 2010).

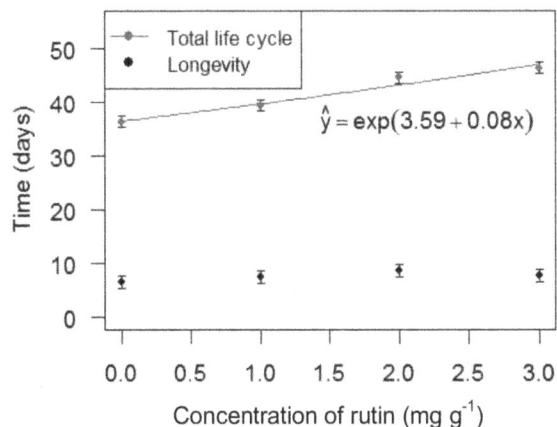

Figure 6. Total life cycle (days) and longevity (days) of *Spodoptera frugiperda* (Lepidoptera: Noctuidae) fed an artificial diet containing different concentrations of rutin.

In general, a negative effect on the biology of *S. frugiperda* was observed at the highest concentrations of the flavonoid rutin. The total life cycle of the insect was prolonged, which is a method used by insects to overcome dietary deficiencies and store more lipids for direct use by the adults (Panizzi, 1991).

Conclusion

The flavonoid rutin negatively affected the biology of *S. frugiperda* by prolonging the larval development time, reducing the larval and pupal weight and decreasing the pupal viability.

The addition of different concentrations of rutin prolonged the life cycle of *S. frugiperda*; therefore, the use of rutin is indicated in future studies evaluating the control of *S. frugiperda*.

Acknowledgements

The authors thank to Instituto Federal Goiano for financial support and American Journal Experts for revised and edited this manuscript.

References

Appel, H. M., Maines L. W. (1995). The influence of host plant on gut of *Gypsy Moth* (Lymantria dispar) Caterpillars. *Journal of Insect Physiology, 41*(3), 241-246.

Barros, E. M., Torres, J. B., & Bueno, A. F. (2010). Oviposição, desenvolvimento e reprodução de *Spodoptera frugiperda* (J.E. Smith) (Lepidoptera: Noctuidae) em diferentes hospedeiros de importância econômica. *Neotropical Entomology, 39*(6), 996-1001.

Boiça Junior, A. L., Souza, B. H. S., Neves, E. C., Ribeiro, Z. A., & Stout, M. J. (2015). Factors influencing expression of antixenosis in soybean to *Anticarsia gemmatalis* and *Spodoptera frugiperda* (Lepidoptera: Noctuidae). *Journal of Economic Entomology, 108*(1), 317-325.

Busato, G. R., Grutzmacher, A. D., Garcia, M. S., Giolo, F. P., Stefanello Junior, G. J., & Zotti, M. J. (2004). Preferencia para alimentação de biótipos de *Spodoptera frugiperda* (J.E Smith, 1797) (Lepidoptera: Noctuidae) por milho, sorgo, arroz e capim arroz. *Revista Brasileira de Agrociências, 10*(2), 215-218.

Campos, Z. R., Boiça-Júnior, A. L., Valério Filho, W. V., Campos, O. R., & Campos, A. R. (2012). The feeding preferences of *Spodoptera frugiperda* (J. E. Smith) (Lepidoptera: Noctuidae) on cotton plant varieties. *Acta Scientiarum. Agronomy, 34*(2), 125-130.

Cruz, I. (2000). Métodos de criação de agentes entomófagos de *Spodoptera frugiperda* (J.E. Smith). In V. H. P. Bueno (Ed.), *Controle biológico de pragas: produção massal e controle de qualidade* (p. 112-114). Lavras, MG: UFLA.

Cruz, I., Figueiredo, M. L. C., Oliveira, A. C., & Vasconcelos, C. A. (1999). Damage of *Spodoptera frugiperda* (Smith) in different maize genotypes cultivated in soil under three levels of aluminum saturation. *International Journal of Pest Management, 45*(4), 293-296.

Deota, P. T., & Upadhyay, P. R. (2005). Biological studies of azadirachtin and its derivatives against polyphagous pest, *Spodoptera litura*. *Journal of Scientific and Industrial Research, 19*(5), 529-539.

Gazzoni, D. L., Hulsmeyer, A., & Hoffmann-Campo, C. B. (1997). Efeito de diferentes doses de rutina e quercitina na biologia de *Anticarsia gemmatalis*. *Pesquisa Agropecuária Brasileira, 32*(7), 673-681.

Harborne, J. B., & Grayer, R. J. (1993). Flavonoids and insects. In J. B. Harborne (Ed.), *The Flavonoids: advances in research since 1986* (p. 589-618). London, UK: Chapman & Hall.

Hoffmann-Campo, C. B., Harbone, J. B., & Mcaffery, A. R. (2001). Pre-ingestive and post-ingestive effects of soya bean extracts and rutin on *Trichoplusia ni* growth. *Entomologia Experimentalis et Applicata, 98*(2), 181-194.

Hoffmann-Campo, C. B., Ramos Neto, J. A., Oliveira, M. C., & Oliveira, L. J. (2006). Detrimental effect of rutina on *Anticarsia gemmatalis*. *Pesquisa Agropecuária Brasileira, 41*(10), 1453-1459.

Jesus, F. G., Boiça Junior, A. L., Alves, G. C. S., & Zanuncio, J. C. (2014). Behavior, development, and predation of *Podisus nigrispinus* (Hemiptera: Pentatomidae) on *Spodoptera frugiperda* (Lepidoptera: Noctuidae) fed transgenic and conventional cotton cultivars. *Annals of the Entomological Society of America*, *107*(3), 601-606.

Kasten Júnior, P., Precetti, A. A. C. M., & Parra, J. R. P. (1978). Dados biológicos comparativos de *Spodoptera frugiperda* (J.E. Smith, 1797) em duas dietas artificiais e substrato natural. *Agricultura, 53*(1), 69-78.

Lima, F. W. N., Ohashi, O. S., Souza, F. R. S., & Gomes, F. S. (2006). Avaliação de acesso de milho para resistência a *Spodoptera frugiperda* (Lepidoptera: Noctuidae) em laboratório. *Acta Amazônica, 36*(2), 147-150.

Lucci, N., & Mazzafera, P. (2009). Distribution of rutin in fava d'anta (*Dimorphandra mollis*) seedlings under stress. *Journal of Plant Interactions, 4*(3), 203-208.

Luginbill, P. (1928). The fall armyworm. Washington: united states department of agriculture. *Technical Bulletin, 34*(1), 90-91.

Magarelli, G., Lima, L. H. C., Silva, J. G., Souza, J. R., & Castro, C. S. P. (2014). Rutin and total isoflavone determination in soybean at different growth stages by using voltammetric methods. *Microchemical Journal, 117*(2), 149-155.

Panizzi, A. R. (1991). Ecologia nutricional de insetos sugadores de sementes. In A. R. Panizzi & J. R. P. Parra (Eds.), *Ecologia nutricional de insetos e suas implicações no manejo integrado de pragas* (p. 253-278). São Paulo, SP: Manole.

Pereira, L. G. B., Petacci, F., Fernandes, J. B., Corrêa, A. G., Vieira, P. C., Silva, M. F., & Malaspina, O. (2002). Biological activity of astilbin from *Dimorphandra mollis* Bent. against *Anticarsia gemmatalis* Hubner and *Spodoptera frugiperda* Smith. *Pest Management Science, 58*(5), 503-507.

Piubelli, G. C., Campo, C. B. H., Moscardi, F., Miyakubo, S. H., & Oliveira, M. C. N. (2005). Are chemical compounds important for soybean resistance to *Anticarsia gemmatalis*? *Journal of Chemical Ecology, 31*(7), 1509-1525.

Piubelli, G. C., Hoffmann-Campo, C. B., Moscardi, F.,

Miyakubo, S. H., & Oliveira, M. C. N. (2006). Baculovirus resistant *Anticarsia gemmatalis* responds differently to dietary rutin. *Entomologia Experimentalis et Applicata, 119*(1), 53-60.

R Core Team. (2014). *R: A language and environment for statistical computing*. Vienna, AT: R Foundation for Statistical Computing. Retrieved from http://www.R-project.org/

Rajamma, A. J., Dubey, S., Sateesha, S. B., Tiwan, S. N., & Ghosh, S. K. (2011). Comparative larvicidal activity of different species of *Ocimum* against *Culex Quinquefasciatus*. *Natural Product Research, 25*(20), 1916-1922.

Sá, V. G. M., Fonseca, B. V. C., Boregas, K. G. B., & Waquil, J. M. (2009). Sobrevivência e desenvolvimento larval de *Spodoptera frugiperda* (J E Smith) (Lepidoptera: Noctuidae) em hospedeiros alternativos. *Neotropical Entomology, 38*(1), 108-115.

Salvador, M. C., Boiça Júnior, A. L., Oliveira, M. C. N., Graça, J. P., Silva, D. M., & Hoffmann-Campo, C. B. (2010). Do different casein concentrations increase the adverse effect of rutin on the biology of *Anticarsia gemmatalis* Hübner (Lepidoptera: Noctuidae)? *Neotropical Entomology, 39*(5), 774-783.

Stamp, N. E., & Skrobola, C. M. (1993). Failure to avoid rutin diets results in altered food utilization and reduced growth rate of *Manduca sexta* larvae. *Entomologia Experimentalis et Applicata, 68*(2), 127-142.

Tavares, W. S., Pereira, A. I. A. P., Freitas, S. S., Serrão, J. E., & Zanuncio, J. C. (2014). The chemical exploration of *Dimorphandra mollis* (Fabaceae) in Brazil, with emphasis on insecticidal response: A rewil. *Journal of Scientific and Industrial Research, 73*(3), 465-468.

Yu, S. J., Nguyen, S. N., & Abo-Elghar, G. E. (2003). Biochemical characteristics of insecticide resistance in the fall armyworm, *Spodoptera frugiperda* (J.E. Smith). *Pesticide Biochemistray and Physiology, 77*(1), 1-11.

8

Toxicity of extracts of *Cyperus rotundus* on *Diabrotica speciosa*

Flávia Silva Barbosa[*], Germano Leão Demolin Leite, Marney Aparecida de Oliveira Paulino, Denilson de Oliveira Guilherme, Janini Tatiane Lima Souza Maia and Rodrigo Carvalho Fernandes

*Instituto de Ciências Agrárias, Universidade Federal de Minas Gerais, Av. Universitária, 1000, Cx Postal 135, 39404-006, Montes Claros, Minas Gerais, Brazil. *Author for correspondence. E-mail: barbosasilva_f@yahoo.com.br*

ABSTRACT. The objective of this work was to evaluate the insecticidal effect or repellency of *Cyperus rotundus*, an important weed plant, through alternative methods of extraction, on *Diabrotica speciosa*, a pest that affects several plant species. The experimental design was completely casual, and consisted of five repetitions. The *C. rotundus* extracts were prepared using leaves and roots by alcoholic extraction, aqueous (hot water) extraction and aqueous (cold water) extraction and diluted to four different concentrations (0, 5, 10, and 15% of the volume of each extract). These dilutions were then tested and compared with a control. The higher mortality of *D. speciosa* adults as well as a smaller leaf consumption area were observed after treatments with increasing dosages of different *C. rotundus* extracts generated by alcoholic extraction (55% of mortality and 28% leaf consumption). Therefore, the alcoholic extract of the *C. rotundus* foliage is an option for the control of *D. speciosa* in agroecologic systems.

Keywords: rootworm, nutgrass, Chrysomelidae, alternative control.

Introduction

Cyperus rotundus L. (Cyperaceae) is an herbaceous perennial plant. Its stem is herbaceous and rhizome ramified (ARANTES et al., 2005). It is considered one of the most persistent species in the world (ARRUDA et al., 2005) and is the main weed species in cultivated soils in tropical areas (JAKELAITIS et al., 2003a and b). According to Costa (2000), *C. rotundus* is rich in alkaloids, anthraquinone, coumarins, steroids, triterpenes, flavonoids, saponins, tannins, and resins. In addition, it has possible insecticidal properties and could be a repellent against arthropods.

Control of polyphagous pests is difficult because of their easy adaptation to different cultures. *Diabrotica speciosa* (Germar) (Coleoptera: Chrysomelidae) affects diverse cultures in Brazil (GALLO et al., 2002) and usually occurs in large population densities (LEITE et al., 2008; PICANÇO et al., 2004). Young pests feed on the root system, while the adults feed on leaves, new budburst, husks or fruits, thus reducing productivity or depreciating the product. The pest population has considerably increased in bean and soy crops, resulting in systematic application of organosynthetic insecticides (MARQUES et al., 1999).

Pest control in agriculture has usually been done using chemical insecticides, causing harm to the farmers who apply the chemicals and to the environment, and decimating populations of beneficial species. In addition to these negative factors, these products are expensive, burdening the family producers who make them (VENZON et al., 2006).

Therefore, there is a need to control pests without causing imbalance and harm to the environment. This can be accomplished by alternative methods, such as the use of plant extracts (BARBOSA et al., 2011; VIEGAS JÚNIOR, 2003) that favor their natural enemies. These compounds are required to establish the biological balance and reduce production costs. Thus, organic cultivation, since it uses environmentally safe technologies, constitutes a promising alternative (VENZON et al., 2006).

However, there are few studies using plant extracts for pest control, and even when they are used as pesticides, they are extracted with expensive and toxic substances such as methanol (BENEVIDES et al., 2001) and hexane. Other studies using plants for insect control use plant essential oils, which require more sophisticated extraction equipment, making their use incompatible with the reality of family producers.

The objective of this study was to determine the insecticidal effect of roots and leaves of C. rotundus on D. speciosa through alternative methods of extraction.

Material and methods

This study was carried out in the "Instituto de Ciências Agrárias da Universidade Federal de Minas Gerais" between December 2006 and February 2007. The experimental design was completely casual, with five repetitions. Each repetition consisted of one Petri dish (10 x 2 cm) with 10 D. speciosa adults (sex and age unknown) collected in the organic bean crop, incubated at 25°C, and evaluated for mortality after 24 and 48 hours.

Three extraction methods were tested for each part of the plant (leaves and roots): 1) For the alcoholic extraction, we used 25% of the fresh weight of leaves or roots (25 g), both picked separately, in commercial 100% hydrated ethyl alcohol (100 mL). This mixture was left for 15 days at room temperature in glass amber and was agitated twice a day. 2) For the aqueous extraction, we used 25% of the fresh weight of leaves or cut roots, which were softened separately (25 g), in 100% of distilled water (100 mL). The extract was left in glass amber for 24 hours at room temperature. 3) For the aqueous extraction for infusion (tea), we used 25% fresh weight of the leaves or separately picked roots (25 g) in 100% distilled water (100 mL). The mixture was boiled (100°C) then conditioned in glass amber until it was cool. All of the obtained extracts were filtered and conditioned in glass amber flasks until their use.

After each plant extract was obtained, four dosages of each extract were tested: 0, 5, 10 and 15%

of the volume. An apical bean leaflet (Phaseolus vulgare L.) was immersed for five seconds in each concentration of the appropriate extract. This leaflet was maintained in the shade and free air for two hours until all excess water had evaporated. After this, 10 D. speciosa adults were conditioned in Petri dishes (10 x 2 cm), and their mortality after 24 and 48h as well as their consumption of the foliage (visual inspection) was recorded. The control plant (dosage = 0) was immersed in the solvent used in the extraction process and dried similarly to the sample. The data were submitted to variance analysis and regression analysis, all to a 5% significance level.

Results and discussion

A higher mortality of D. speciosa adults as well as a smaller foliage consumption area was observed with increased dosage of the different extracts of leaves and roots of C. rotundus (Figures 1 and 2). Among the different extraction types, higher concentrations of alcohol extract caused the greatest mortality of D. speciosa. After 24 or 48h of treatment, the highest concentration of root extract caused 35 and 45% mortality, respectively, while the highest concentration of leaf extract caused 50 and 55% mortality, respectively (Figure 1). In addition, the alcohol extract of leaves had a higher straight line inclination than the alcohol root extract, indicating greater insecticidal capacity (Figure 1). This effect was similar when area of foliage consumption was evaluated; alcohol extracts were more effective at reducing this consumption area. After 24 or 48h of treatment, the greatest concentration of root extract decreased consumption by 13.75 and 46.25%, respectively, and the greatest concentration of leaf extract decreased consumption by 22.75 and 28%, respectively. The leaf extract again had a larger straight line inclination (Figure 2).

Sharma and Gupta (2007) observed that methanolic extract of C. rotundus tubers strongly inhibited the activity of acetylcholinesterases (AChE). These authors suggested that the AChE inhibitors in nutgrass could possibly act as the plants' defense against herbivores. Raja et al. (2001) reported that aqueous extracts from C. rotundus tubers effectively protected pulses stored post-harvest without any infestation by Callosobruchus maculatus (Coleoptera : Bruchidae) for up to 6 months. The insecticidal effect of C. rotundus is probably a result of compounds in the plant such as alkaloids, anthraquinones, coumarins, steroids and triterpenes, sesquiterpenoid, flavonoids, saponins, tannins, and resins (COSTA, 2000; JEONG et al., 2000; KILANI et al., 2005). These compounds have an insecticidal effect against other insects. Examples include the effect of flavonoids against *S. zeamais* Mots. and *Aedes aegypti* L. larvae (Diptera: Culicidae) (SILVA et al., 2005; TREVISAN et al., 2006), coumarins against *A aegypti* (CHAITHONG et al., 2006), or triterpenes against mites such as *Oligonychus ilicis* McGregor (Acari: Tetranychidae) and *Iphiseiodes zuluagai* Denmark & Muma (Acari: Phytoseiidae) (MARTINEZ, 2002; MOURÃO et al., 2004). Saponins, tannins and alkaloids are generally toxic and are deterrents for most herbivores (CAVALCANTE et al., 2006; MARTINS et al., 2005). *C. rotundus* extract is effective against wood, agricultural, hygiene, cereals, and domestic insect pests and is nontoxic to fish and animals due to alpha-cyperone, cyperotundone, and rotundone (KUMIAI CHEM. IND. CO. LTD, 2008). Their sesquiterpenes are effective against domestic and agricultural insect pests (e.g., flies, cockroaches, mosquitoes, slugs, wolf moths, and rice weevils) and are used to make conventional repellents as well as wide-spectrum and long-lasting insect repellents without fear of toxicity and environmental pollution (YUGAKI YAKUHIN KOGYO KK, 2008).

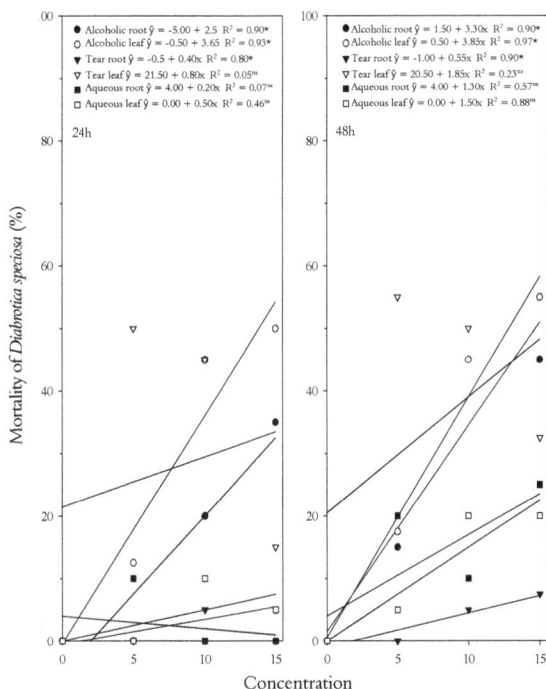

Figure 1. Mortality of adults of *Diabrotica speciosa* (%) under differents extracts of *Cyperus rotundus*. *Significative to 5% by F test. ns = no significative.

This plant has an allelopathic effect on other vegetable species, probably due to the presence of these secondary compounds, reducing the growth of the aerial part and root system of *Brassica oleraceae* L. and *B. rapa* L. (Brassicaceae) (YAMAGUSHI et al., 2007), as well as the germination of *B. oleraceae*,

B. rapa, *Zea mays* L. (Gramineae), *Phaseolus vulgaris* L. and *Glyicine max* (L.) (Fabaceae), and *Lactuca sativa* L. (Asteraceae) (MUNIZ et al., 2007; YAMAGUSHI et al., 2007).

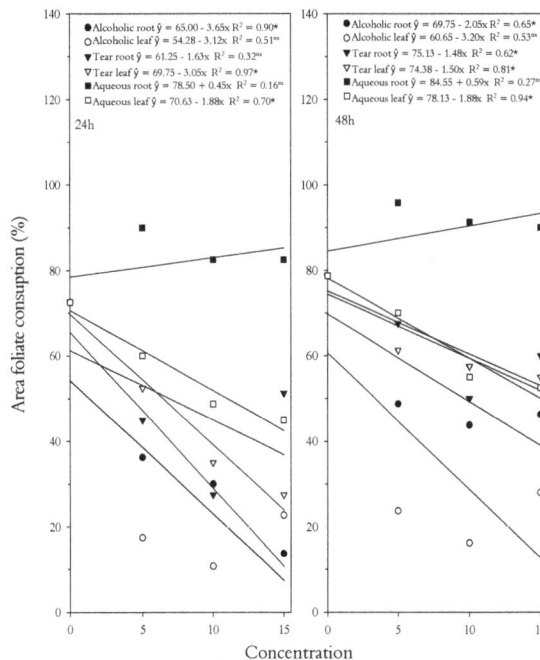

Figure 2. Area foliate consumpion (%) dor *Diabrotica speciosa* under differents extracts of *Cyperus rotundus*.*Significative to 5% by F test. ns = no significative.

The antibacterial effects of ethyl acetate extracts of *C. rotundus* observed by KILANI et al. (2008) and anti-malarial action of *C. rotundus* of the *Plasmodium falciparum* verified by Thebtaranonth et al. (1995) are strong indicators of their possible action on insects.

On the other hand, similar mortality values to those seen in this work were also observed in other natural plant extracts, such as piretro against *Argyrotaenia sphaleropa* (Meyrick) (Lepidoptera: Tortricidae) (MORANDI FILHO et al., 2006), *Ruta graveolens* L. (Rutaceae) against *Sitophilus* spp. (Coleoptera: Curculionidae), and *Ctenocephalides canis* (Curtis) (Siphonaptera: Pulicidae) (ALMEIDA et al., 1999; LEITE et al., 2006), *Prosopis juliflora* Sw. (D.C.) (Fabaceae) on *Bemisia tabaci* Genn. (Hemiptera: Aleyrodidae) (CAVALCANTE et al., 2006), as well as the synthetic organic insecticide thiamethoxam against *D. speciosa* (CALAFIORI; BARBIERI, 2001).

It is worth emphasizing that the studied plant is easily generated by treating a spontaneous plant, thus generating an alcoholic extract, a good option for use as an alternative for pest control in agroecologic systems.

Conclusion

We conclude that extracts of *C. rotundus* generated by alcoholic extraction have more insecticides effect on *D. speciosa* well as demonstrated a smaller area of leaf consumption that extracts prepared with aqueous (hot water) and aqueous extraction (cold water) with (55% of mortality and 28% leaf consumption).

Acknowledgements

We thank Capes, CNPq, and Fapemig for research support.

References

ALMEIDA, F. A. C.; GOLDFARB, A. C.; GOUVEIA, J. P. G. Avaliação de extratos vegetais e métodos de aplicação no controle de *Sitophilus* spp. **Revista Brasileira de Produtos Agroindustriais**, v. 1, n. 1, p. 13-20, 1999.

ARANTES, M. C. B.; OLIVEIRA, L. M. G.; FREITAS, M. R. F.; SILVA, L. N. M.; NOGUEIRA, J. C. M.; PAULA, J. R.; BARA, M. T. F. Estudo farmacognóstico do *Cyperus rotundus* L. **Revista Eletrônica de Farmácia Suplemento**, v. 2, n. 2, p. 17-20, 2005.

ARRUDA, F. P.; ANDRADE, A. P.; BELTRÃO, N. E. M.; PEREIRA, W. E.; LIMA, J. R. F. Viabilidade econômica de sistemas de preparo do solo e métodos de controle de Tiririca em algodoeiro. **Revista Brasileira Engenharia Agrícola e Ambiental**, v. 9, n. 4, p. 481-488, 2005.

BARBOSA, F. S.; LEITE, G. L. D.; ALVES, S. M.; NASCIMENTO, A. F.; D'AVILA, V. A.; COSTA, C. A. Insecticide effects of *Ruta graveolens*, *Copaifera langsdorffii* and *Chenopodium ambrosioides* against pests and natural enemies in commercial tomato plantation. **Acta Scientiarum. Agronomy**, v. 33, n. 1, p. 37-43, 2011.

BENEVIDES, P. J. C.; YOUNG, M. C. M.; GIESBRECHT, A. M.; ROQUE, N. F.; BOLZANI, V. S. Antifungal polysulphides from *Petiveria alliacea* L. **Phytochemistry**, v. 57, n. 5, p. 743-747, 2001.

CALAFIORI, M. H.; BARBIERI, A. A. Effects of seed treatment with insecticide on the germination, nutrients, nodulation, yield and pest control in bean (*Phaseolus vulgaris* L.) culture. **Revista Ecossistema**, v. 26, n. 1, p. 97-104, 2001.

CAVALCANTE, G. M.; MOREIRA, A. F. C.; VASCONCELOS, S. D. Potencialidade inseticida de extratos aquosos de essências florestais sobre mosca-branca. **Pesquisa Agropecuária Brasileira**, v. 41, n. 1, p. 9-14, 2006.

CHAITHONG, U.; CHOOCHOTE, W.; KAMSUK, K.; JITPAKDI, A.; TIPPAWANGKOSOL, P.; CHAIYASIT, D.; CHAMPAKAEW, D.; TUETUN, B.; PITASAWAT, B. Larvicidal effect of pepper plants on *Aedes aegypti* (L.) (Diptera: Culicidae). **Journal of Vector Ecology**, v. 31, n. 1, p. 138-144, 2006.

COSTA, A. F. **Farmacognosia**. 3. ed. Lisboa: Fundação Calouste Gulbenkian, 2000. v. III.

GALLO, D.; NAKANO, O.; SILVEIRA, NETO, S.; CARVALHO, R. P. L.; BAPTISTA, G. C.; BERTI FILHO, E.; PARRA, J. R. P.; ZUCCHI, R. A.; ALVES, S. B.; VENDRAMIM, J. D.; MARCHINI, L. C.; LOPES, J. R. S.; OMOTO, C. **Entomologia agrícola**. Piracicaba: Fealq, 2002.

JAKELAITIS, A.; FERREIRA, L. R.; SILVA, A. A.; AGNES, E. L.; MIRANDA, G. V.; MACHADO, A. F. L. Dinâmica populacional de plantas daninhas sob diferentes sistemas de manejo nas culturas de milho e feijão. **Planta Daninha**, v. 21, n. 1, p. 71-79, 2003a.

JAKELAITIS, A.; FERREIRA, L. R.; SILVA, A. A.; AGNES, E. L.; MIRANDA, G. V.; MACHADO, A. F. L. Efeitos de sistemas de manejo sobre a população de tiririca. **Planta Daninha**, v. 21, n. 1, p. 89-95, 2003b.

JEONG, S. J.; MIYAMOTO, T.; INAGAKI, M.; KIM, Y. C.; HIGUCHI, R. Rotundines A−C, Three novel sesquiterpene alkaloids from *Cyperus rotundus*. **Journal of Natural Products**, v. 63, n. 5, p. 673-675, 2000.

KILANI, S.; AMMAR, R. B.; BOUHLEL, I.; ABDELWAHED, A.; HAYDER, N.; MAHMOUD, A.; GHEDIRA, K.; CHEKIR-GHEDIRA, L. Investigation of extracts from (Tunisian) *Cyperus rotundus* as antimutagens and radical scavengers. **Environmental Toxicology and Pharmacology**, v. 20, n. 3, p. 478-484, 2005.

KILANI, S.; BEN SGHAIER, M.; LIMEM, I.; BOUHLEL, I.; BOUBAKER, J.; BHOURI, W.; SKANDRANI, I.; NEFFATTI, A.; BEN AMMAR, R.; DIJOUX-FRANCA, M. G.; GHEDIRA, K.; CHEKIR-GHEDIRA, L. In vitro evaluation of antibacterial, antioxidant, cytotoxic and apoptotic activities of the tubers infusion and extracts of *Cyperus rotundus*. **Bioresource Technology**, v. 99, n. 18, p. 9004-9008, 2008.

KUMIAI CHEM IND. CO. LTD. Insect pest controlling agent for domestic insects, etc - contains crude extract of *Cyperus rotundus* obtd. by extn. of raw or dried plants with steam distn. or organic solvent. **International Patent Classification**: A01N-065/00, 2008.

LEITE, G. L. D.; PIMENTA, M.; FERNANDES, P. L.; VELOSO, R. V. S.; MARTINS, E. R. Fatores que afetam artrópodes associados a cinco acessos de ginseng-brasileiro (*Pfaffia glomerata*) em Montes Claros, Estado de Minas Gerais. **Acta Scientiarum. Agronomy**, v. 30, n. 1, p. 7-11, 2008.

LEITE, G. L. D.; SANTOS, M. M. O.; GUANABENS, R. E. M.; SILVA, F. W. S.; REDOAN, A. C. M. Efeito de boldo chinês, do sabão de côco e da cipermetrina na mortalidade de pulgas em cachorro doméstico. **Revista Brasileira de Plantas Medicinais**, v. 8, n. 3, p. 96-98, 2006.

MARQUES, G. B. C.; ÁVILA, C. J.; PARRA, J. R. P. Danos causados por larvas e adultos de *Diabrotica speciosa* (Coleoptera: Chrysomelidae) em milho. **Pesquisa Agropecuária Brasileira**, v. 34, n. 11, p. 1983-1986, 1999.

MARTINEZ, S. S. **O NIM** *Azadirachta indica*: natureza, usos múltiplos, produção. Londrina: Iapar-Instituto Agronômico do Paraná, 2002.

MARTINS, A. G.; ROSÁRIO, D. L.; BARROS, M. N.; JARDIM, M. A. G. Levantamento etnobotânico de plantas medicinais, alimentares e tóxicas da Ilha do Combu, Município de Belém, Estado do Pará, Brasil. **Revista Brasileira de Farmácia**, v. 86, n. 1, p. 21-30, 2005.

MORANDI FILHO, W. J.; BOTTON, M.; GRÜTZMACHER, A. D.; GIOLO, F. P.; MANZONI, C. G. Ação de produtos naturais sobre a sobrevivência de *Argyrotaenia sphaleropa* (Meyrick) (Lepidoptera: Tortricidae) e seletividade de inseticidas utilizados na produção orgânica de videira sobre *Trichogramma pretiosum* Riley (Hymenoptera: Trichogrammatidae). **Ciência Rural**, v. 36, n. 4, p. 1072-1078, 2006.

MOURÃO, S. A.; SILVA, J. C. T.; GUEDES, R. N. C.; VENZON, M.; JHAM, G. N. Selectivity of neem extracts (*Azadirachta indica* A. Juss.) to the predatory Mite *Iphiseiodes zuluagai* (Denmark & Muma) (Acari: Phytoseiidae). **Neotropical Entomology**, v. 33, n. 5, p. 613-617, 2004.

MUNIZ, F. R.; CARDOSO, M. G.; VON PINHO, E. V. R.; VILELA, M. Qualidade fisiológica de sementes de milho, feijão, soja e alface na presença de extrato de tiririca. **Revista Brasileira de Sementes**, v. 29, n. 2, p. 195-204, 2007.

PICANÇO, M. C.; SEMEÃO, A. A.; GALVÃO, J. C. C.; SILVA, E. M.; BARROS, E. C. Fatores de perdas em cultivares de milho safrinha. **Acta Scientiarum. Agronomy**, v. 26, n. 2, p. 161-167, 2004.

RAJA, N.; BABU, A.; DORN, S.; IGNACIMUTHU, S. Potential of plants for protecting stored pulses from *Callosobruchus maculatus* (Coleoptera: Bruchidae) infestation. **Biological Agriculture and Horticulture**, v. 19, n. 1, p. 19-27, 2001.

SHARMA, R.; GUPTA, R. *Cyperus rotundus* extract inhibits acetylcholinesterase activity from animal and plants as well as inhibits germination and seedling growth in wheat and tomato. **Life Sciences**, v. 80, n. 24-25, p. 2389-2392, 2007.

SILVA, G.; ORREGO, O.; HEPP, R.; TAPIA, M. Búsqueda de plantas con propiedades insecticidas para el controlde *Sitophilus zeamais* en maíz almacenado. **Pesquisa Agropecuária Brasileira**, v. 40, n. 1, p. 11-17, 2005.

THEBTARANONTH, C.; THEBTARANONTH, Y.; WANAUPPAPHAMKUL, C.; YUTHAVONG, Y. Antimalarial sesquiterpenes from tubers of *Cyperus rotundus*. Structure of 10,12 peroxycalamenene, Asesquiterpenes endoperoxide. **Phytochemistry**, v. 40, n. 1, p. 125-128, 1995.

TREVISAN, M. T. S.; BEZERRA, M. Z. B.; SANTIAGO, G. M. P.; FEITOSA, C. M. F. Atividades larvicida e anticolinesterásica de plantas do gênero *Kalanchoe*. **Química Nova**, v. 29, n. 3, p. 415-418, 2006.

VENZON, M.; TUELHER, E. S.; BONOMO, I. S.; TINOCO, R. S.; FONSECA, M. C. M.; PALLINI, A. Potencial de defensivos alternativos para o controle de pragas do cafeeiro. In: VENZON, M.; PAULA JÚNIOR, T.J.; PALLINI, A. (Ed.). **Tecnologias alternativas para o controle de pragas e doenças**. Viçosa: Epamig/CTZM, 2006. p. 117-136.

VIEGAS JÚNIOR, C. Terpenos com atividade inseticida: uma alternativa para o controle químico de insetos. **Química Nova**, v. 26, n. 3, p. 390-400, 2003.

YAMAGUSHI, M. Q.; ANDRADE, H. M.; PARDÓCIMO, E. M.; PORTILHO, G. P.; BITTENCOURT, A. H. C.; VESTENA, S. Efeito de extratos aquosos de tiririca (*Cyperus rotundus* L.) sobre a germinação e crescimento de repolho (*Brassica oleracea* L.) e de nabo (*Brassica rapa* L.). **Revista Científica da FAMINAS**, v. 3, n. 1, p. 262, 2007.

YUGAKI YAKUHIN KOGYO KK. Broad spectrum insect pest repellent - contg. extract of Cyperaceae plants. **International Patent Classification:** A01N-065/00, 2008.

9

Fungicide and insecticide residues in rice grains

Gustavo Mack Teló[1*], Enio Marchesan[1], Renato Zanella[2], Sandra Cadore Peixoto[2], Osmar Damian Prestes[2] and Maurício Limberger de Oliveira[1]

[1]Grupo de Pesquisa em Arroz Irrigado, Departamento de Fitotecnia, Universidade Federal de Santa Maria, 97105-900, Santa Maria, Rio Grande do Sul, Brazil. [2]Laboratório de Análise de Resíduos de Pesticidas, Departamento de Química, Universidade Federal de Santa Maria, Santa Maria, Rio Grande do Sul, Brazil. *Author for correspondence. E-mail: gustavo.telo@yahoo.com.br

ABSTRACT. The objective of this study was to analyse residues of fungicides and insecticides in rice grains that were subjected to different forms of processing. Field work was conducted during three crop seasons, and fungicides and insecticides were applied at different crop growth stages on the aerial portion of the rice plants. Azoxystrobin, difenoconazole, propiconazole, tebuconazole, and trifloxystrobin fungicides were sprayed only once at the R2 growth stage or twice at the R2 and R4 growth stages; cypermethrin, lambda-cyhalothrin, permethrin, and thiamethoxam insecticides were sprayed at the R2 growth stage; and permethrin was sprayed at 5-day intervals from the R4 growth stage up to one day prior to harvest. Pesticide residues were analysed in uncooked, cooked, parboiled, polished and brown rice grains as well as rice hulls during the three crop seasons, for a total of 1,458 samples. The samples were analysed by gas chromatography with electron capture detection (GC-ECD) using modified QuEChERS as the extraction method. No fungicide or insecticide residues were detected in rice grain samples; however, azoxystrobin and cypermethrin residues were detected in rice hull samples.

Keywords: *Oryza sativa* L., food, gas chromatography, pesticides, grains quality.

Introduction

The use of pesticides in irrigated rice has intensified in recent years due to the higher incidence of foliar diseases and insect pests. In many cases, damage from these pests occurs close to harvest, causing economic loss. Thus, the use of pesticides has been determined to be an essential management practice to ensure optimal agricultural yield and food quality. However, the presence of pesticide residues in food is a major public health concern, and identifying the presence of such residues in all types of food (both fresh and industrialized) is important to guarantee food safety (Wang, Wu, & Zhang, 2012; Hou, Han, Dai, Yang, & Yi, 2013).

Several studies have analysed the presence of pesticide residues in rice grains. In India, a study involving the application of thiamethoxam and lambda-cyhalothrin insecticides at both the recommended rate and twice the recommended rate to the aerial parts of rice plants concluded that no residues were detected in the grains (Barik, Ganguly, Kunda, Kole, & Bhattacharyya, 2010). However, Zhang, Chai, and Wu (2012) detected residues in brown rice grains

(0.027 mg kg^{-1}) following the application of chlorantraniliprole insecticide both during and at the end of the rice crop cycle in China. The presence of insecticide residues in rice grain may vary depending on the part of the grain evaluated (hull, bran or polished grain) and its chemical group, as an insecticide from a certain group may be more likely to associate with a certain part of the grain (Teló et al., 2015a).

Studies of fungicide residues have also shown varying results. In China, some studies have found no residues of difenoconazole fungicide in rice grains, and such studies concluded that the 30-day period between the application of the fungicide and the harvest of rice, a period established by the country's legislation, is safe for grain consumption. Even when rates of 90 and 135 g ai ha^{-1} were applied, no residues were detected in rice grains (Wang et al., 2012). The temperature during rice grain processing may influence the persistence of pesticide residues because high temperatures may reduce up to 38% of the concentration of difeconazole found in the rice hull (Teló et al., 2015b). Furthermore, due to the distinct characteristics of the rice grain, the grain's external structures may present higher residue concentrations because they serve as physical barriers. Pareja et al. (2012) assessed the concentrations of azoxystrobin, epoxiconazole, difenoconazole, lambda-cyhalothrin, tebuconazole, thiamethoxam and tricyclazole residues in the rice bran and in both polished grains and whole grain rice. The results show that the highest detected concentrations are associated with rice bran and whole grain rice. Other rice processing techniques such as parboiling reduced up to 11% of the detected residue of tebuconazole in the hull, bran and polished rice grain compared to non-parboiled grains. The highest residue concentration was detected in the rice hull, with a value that was 71% higher than the value detected in the bran and 99% higher than the value found in the non-parboiled polished rice grain (Dors, Primel, Fagundes, Mariot, & Badiale-Furlong, 2011).

Notably, several other studies from different countries have reported both the presence and absence of pesticide residues in rice. Thus, the objective of this study was to analyse the presence or absence of azoxystrobin, difenoconazole, propiconazole, tebuconazole and trifloxystrobin fungicide residues and of cypermethrin, lambda-cyhalothrin, permethrin and thiamethoxam insecticide residues, which are commonly used in rice crops in Brazil, in rice grains and hulls when these pesticides are applied to the aerial parts of the plants.

Material and methods

This study was carried out in two stages: the first stage was performed in the field; the second stage, in the laboratory. The first stage was conducted during the 2007/08, 2008/09, and 2009/10 crop seasons in the lowland area belonging to the Department of Plant Science at the Universidade Federal de Santa Maria-UFSM (Federal University of Santa Maria) in southern Brazil. In all crop seasons, rice was planted in the second week of October, and levees were placed around the plots. Each plot consisted of nine 7-m long seed rows that were spaced 0.17 m from each other, where the IRGA 417 rice cultivar was planted at a rate of 95 kg ha^{-1} in three replications. Management practices were conducted according to technical recommendations (Sociedade Sul-brasileira de Arroz Irrigado [SOSBAI], 2010).

Fungicides and insecticides were applied at different crop growth stages to the aerial parts of the rice plants as described in Table 1. In one treatment, the selected fungicides were applied at the R_2 growth stage, and in the other treatment, they were applied at both the R_2 and R_4 growth stages (phenology based on Counce, Keisling, & Mitchell, 2000). The selected insecticides were applied at the R_2 stage in one treatment, and in another treatment, only the permethrin insecticide was applied to the different plots at 5-day intervals from the R_4 growth stage up to one day before harvest. The applications were performed according to the labelled rates for each pesticide using a CO_2 pressurized back sprayer (pressure of 276 kPa) attached to a boom spray with four hollow cone nozzles (JA-2).

Table 1. Active ingredient, chemical group, rates applied to the aerial part of the plants, safety interval from application to harvest set by the Brazilian National Agency for Sanitary Surveillance (ANVISA) and octanol-water partition coefficient (log K_{ow}).

	Active ingredients	Chemical group	Rate (g ai ha^{-1})	Safety interval (days)[1]	Log K_{ow}[2]
Fungicides	Azoxystrobin	Strobilurin	100.0	30	2.5
	Trifloxystrobin[3]	Strobilurin	75.0	15	4.5
	Trifloxystrobin[4]	Strobilurin	93.7	15	4.5
	Difenoconazole	Triazole	75.0	45	4.3
	Propiconazole[3]	Triazole	93.7	45	1.3
	Tebuconazole[4]	Triazole	150.0	35	3.7
Insecticides	Cypermethrin	Pyrethroid	25.0	10	5.3
	Lambda-cyhalothrin	Pyrethroid	21.2	21	6.9
	Permethrin	Pyrethroid	25.0	20	6.1
	Thiamethoxam	Neonicotinoid	28.2	21	-0.1

[1]Agência Nacional de Vigilância Sanitária (ANVISA, 2016). [2]International Union of Pure and Applied Chemistry (IUPAC, 2015). [3]Commercial product with two active ingredients applied in the same plot. [4]Commercial product with two active ingredients applied in the same plot.

In each crop season, 54 rice plots were planted (18 treatments with 3 replications). For the residue analyses, grains were harvested from a 4.76 m^2 (4.0×1.19 m) area in the centre of each plot when the average moisture content reached 22%. After the grain was threshed, a 3 kg sample of homogeneous rice was collected from the previously mentioned area in each plot, for a total of 162 plots for the entire study period. Afterwards, samples were cleaned and dried by forced air ventilation at 38±2°C until a rice grain moisture of 13% was reached. The samples were then stored at -20°C.

To analyse fungicide and insecticide residues, each of the 3 kg samples harvested was subdivided according to processing method into 486 subsamples per crop season and 1458 total subsamples for the entire study period. The different treatments used in the analyses were the rice hulls from rice processing on a testing rice mill (Zaccaria PAZ-1-DTA), polished rice grains obtained by a grain polishing process, brown rice grains (processed grains with no polishing process), cooked polished rice grains (polished and processed grains that were cooked in a 1:2.5 grain mass:water ratio with ultrapure water heated at 45°C until the sample was completely dried), cooked brown rice grains (unpolished processed grain, cooked as described above), parboiled polished rice grains (raw grains [with hull] that were soaked [1:1.5 grain mass:water ratio] in water heated up to 65±1°C for 300 min. and autoclaved at 110±1°C [pressure of 0.6±0.05 kPa] for 10 min.). After the subsamples were divided into treatment groups, each grain sample was dried with forced air ventilation until the moisture reached 13% and was then processed to obtain parboiled polished rice grains, parboiled brown rice grains, cooked parboiled polished rice grains and cooked parboiled brown rice grains as previously described.

In the second stage of the study, the analysis of pesticide residues in the rice grains was conducted at the Laboratory of Pesticide Residue Analysis (LARP) at UFSM. Analytical standards of the pesticides were obtained from Dr. Ehrenstorfer (Augsburg, Germany) with purities above 95%. A full-scan analysis revealed no contamination. In total, 9 pesticides were analysed; these pesticides are listed in Table 1.

Prior to chromatographic analysis, the samples were subjected to an extraction process using the modified QuEChERS method (Prestes, Friggi, Adaime, & Zanella, 2009), which involves an initial extraction step in a 50-mL tube with 10 g of slurry prepared as described by Kolberg, Prestes, Adaime, and Zanella (2010) with a mixture 1:1 (w/w) rice:water and 1:3 (w/w) hull:water, into which 10 mL of acetonitrile containing 1% (v/v) of acetic acid was added. The tube was shaken vigorously by hand for 1 min., followed by a partition step after 3 g of anhydrous magnesium sulphate and 1.7 g of sodium acetate were added. Afterwards, the tube was centrifuged at 3,300 rpm for 8

min. Cleaning was done by the addition of 500 mg of the sorbent C_{18} and 600 mg of anhydrous magnesium sulphate in a 15-mL tube containing 4 mL of extract. The tube was capped, shaken vigorously by hand for 1 min and centrifuged at 3,300 rpm for 8 min. The extract was then ready for analysis.

Analyses were done by gas chromatography with electron capture detection (GC-ECD) using a CP 3800 gas chromatograph from Varian (USA) with a CP 8410 autosampler and using a DB-5 fused silica capillary 30-meter column with an internal diameter of 0.25 mm and film thickness of 0.25 μm. Helium was used as carrier gas with a constant flow rate of 1.3 mL min.$^{-1}$. The injector used programmed temperature vaporization as follows: initial temperature of 80°C (0.1 min), increasing at a rate of 200°C min.$^{-1}$ until 250°C was reached. The injection volume was 2 μL in the splitless mode. The temperature of the column was 80°C (0.1 min.), increasing at a rate of 25°C min.$^{-1}$ until a temperature of 215°C was reached (6 min) and then at a rate of 5°C min.$^{-1}$ until 250°C was reached. The detector temperature was maintained at 330°C.

The method was validated by determining the limits of detection and the quantification (LOD and LOQ), linearity, precision and accuracy, in terms of recovery. The pesticides lindane and methyl parathion were used as the internal and surrogate standars, respectively.

Results and discussion

Method validation

The method validation confirmed that the sample preparation by the modified QuEChERS method and the analysis by GC-ECD were adequate for the determination of residues of the selected pesticides in different parts of the rice grains subjected to different processing procedures.

The calibration curves obtained from the matrix matched standards of rice grains and rice hulls extracts presented good linearity with coefficients of determination greater than 0.995 for all studied pesticides. The method presented recovery values between 80.7 and 119.7% with good precision in terms of relative standard deviation (RSD), with values ranging from 0.4 to 19.4% for the different processed grains. For the rice hull, recovery ranged from 80.1 to 111.8%, with RSD values from 1.2 to 18.7%. No interferences were observed.

Pesticide residue analysis in rice grains

Residues of azoxystrobin and cypermethrin were detected in the rice hulls from the 2007/08 crop season (Table 2), and cypermethrin was detected at a higher concentration than azoxystrobin. Residues of fungicides and insecticides in the rice grains were not detected by the method of analysis used in this study, regardless of the form of rice used during the analysis.

Table 2. Limit of detection (LOD) and concentration of pesticide residues detected in rice hull, uncooked grains, and cooked grains (brown and polished grains) and in uncooked and cooked parboiled grains (brown and polished grains).

| Active ingredients | LOD (mg kg⁻¹) | | Pesticide concentration (mg kg⁻¹) | | | | |
| | | | Rice hull | Cooked and uncooked grains | | Parboiled grains Cooked and uncooked | |
	Grains	Hull		Brown	Polished	Brown	Polished
Azoxystrobin[1]	0.005	0.01	0.02±0.005*	n.d.	n.d.	n.d.	n.d.
Azoxystrobin[2]	0.005	0.01	0.03±0.008*	n.d.	n.d.	n.d.	n.d.
Difenoconazole[1,2]	0.02	0.04	n.d.	n.d.	n.d.	n.d.	n.d.
Propiconazole[1,2]	0.02	0.04	n.d.	n.d.	n.d.	n.d.	n.d.
Tebuconazole[1,2]	0.02	0.04	n.d.	n.d.	n.d.	n.d.	n.d.
Trifloxystrobin[1,2]	0.002	0.004	n.d.	n.d.	n.d.	n.d.	n.d.
Cypermethrin[1]	0.01	0.02	0.12±0.021*	n.d.	n.d.	n.d.	n.d.
Lambda-cyhalothrin[1]	0.002	0.004	n.d.	n.d.	n.d.	n.d.	n.d.
Permethrin[1]	0.01	0.02	n.d.	n.d.	n.d.	n.d.	n.d.
Thiamethoxam[1]	0.01	0.02	n.d.	n.d.	n.d.	n.d.	n.d.
Permethrin (30)[3]	0.01	0.02	n.d.	n.d.	n.d.	n.d.	n.d.
Permethrin (25)	0.01	0.02	n.d.	n.d.	n.d.	n.d.	n.d.
Permethrin (20)	0.01	0.02	n.d.	n.d.	n.d.	n.d.	n.d.
Permethrin (15)	0.01	0.02	n.d.	n.d.	n.d.	n.d.	n.d.
Permethrin (10)	0.01	0.02	n.d.	n.d.	n.d.	n.d.	n.d.
Permethrin (5)	0.01	0.02	n.d.	n.d.	n.d.	n.d.	n.d.
Permethrin (1)	0.01	0.02	n.d.	n.d.	n.d.	n.d.	n.d.

[1]Application at the R_2 growth stage. [2]Application at the R_2+R_4 growth stages. [3]Application before grain harvest (days). *Residues detected only in the 2007/08 crop season. n.d.=not detected.

The literature presents several hypotheses regarding what factors may influence the concentrations of pesticide residues in irrigated rice crops, and many of these hypotheses have not yet been fully tested because of the numerous variables involved. Some of the primary factors that can influence the persistence of pesticides are weather conditions, characteristics of the pesticide (Komárek, Cadková, Bollinger, Bordas, & Chrastný, 2010), level of plant development at the moment of application (Navarro, Vela, & Navarro, 2011), solar radiation, temperature (Peña, Rodríguez-Liébana, & Mingorance, 2011), and the persistence of the pesticides in the plant (Macedo, Araujo, & Castro, 2013).

The results found in this study regarding the non-detection of fungicide residues in rice grains may be related to the time of application, which occurred 40 days before grain harvest. During this period, degradation and/or metabolism of the fungicide molecules by the plant may have occurred. This hypothesis could explain such results. In a study conducted with the objective of evaluating the persistence of the fungicide tricyclazole in rice plants (Phong, Nhung, Yamazaki, Takagi, & Watanabe, 2009), residues were detected until the 13th day after application in the first year and until 14th day after application in the second year. In this case, the pesticide molecules were metabolized within a short period; thus, their translocation might not have occurred in the original structure of tricyclazole during grain filling. Similar results were observed for azoxystrobin applied at different rates (125, 250, and 500 g ai ha⁻¹) on the aerial part of rice plants in which residues were not detected in the rice grains

from harvest, which supports the hypothesis of metabolization of the original molecule by plants (Sundravadana, Kutalam, & Samiyappan, 2007).

Metabolism of the original structure of the molecule of propiconazole was studied using a ¹⁴C-labeled molecule to assess its behaviour in rice plants (Kim, Beaudette, Shim, & Trevors, 2002). The authors analysed the metabolites after harvest and detected only 0.0004% of the initial concentration of 1.7% of the radioactive molecule that had been applied to the plant; they attributed this result to the fragmentation of the original molecular structure. These results highlight the need for more studies to identify and detect metabolites in rice grains. Thus, triazole group fungicides such as propiconazole, difenoconazole, and tebuconazole are metabolized within the plant and may undergo oxidation, reduction, hydrolysis, or the formation of other compounds, or even total degradation into simpler compounds (Zambolin, Picanco, Silva, Ferreira, & Ferreira, 2008).

The hypothesis related to the degradation of fungicides in plants is also relevant for insecticides. Studies related to the dynamics and persistence of imidacloprid, from the neonicotinoid chemical group, did not detect insecticide residues in rice grains (Kanrar et al., 2006). The authors linked the results to the degradation of imidacloprid molecules in the plant. A reduction of 80% of the initial concentration was measured in plants 30 days after the application of the insecticide; it has a half-life of 11 days. Regarding the dissipation and degradation of thiamethoxam and lambda-cyhalothrin in rice plants, there was an 88% reduction of the initial concentration measured in the plants on the 15th day after the application of thiamethoxam and a reduction of more than 50% of the initial concentration measured after the first day for lambda-cyhalothrin; residues were not detected in rice grains (Barik et al., 2010).

It is a well-known fact that several factors may influence the persistence of pesticides in plants by interfering with the absorption and metabolization of pesticides by the plant (Santos, Areas, & Reyes, 2007; Macedo et al., 2013). Thus, residues of the fungicides azoxystrobin and difenoconazole and of the insecticide lambda-cyhalothrin were assessed in the irrigated rice plants after 40 days (the period from the moment of the pesticide application up to the rice harvest); however, according to a previous study by Teló et al. (2015b), no residues from the previously mentioned pesticides could be found in rice grains.

For the sequential applications of permethrin, which was applied up to 1 day before harvest, residue was also not detected in rice grains. Several processes, such as photodegradation, volatilization, chemical degradation, and biological degradation (Linders, Mensink, Stephenson, Wauchope, & Racke, 2000), may be associated with the non-detection of residues in rice grains. The pyrethroid chemical group has a high rate of degradation mainly due to solar radiation, which is characteristic of the molecules in this group (Laskowski, 2002), and the permethrin insecticides have a half-life of 1 day under aqueous photolysis (IUPAC, 2015). In this context, it is noteworthy that the insecticides were applied during the months of the highest solar radiation incidence. The results of this study could be explained by the hypothesis that molecules are degraded by solar radiation. It is also notable that insecticides of the pyrethroid chemical group have low absorption and translocation rates in plants (Garrido, Martínez, Martínez, & López-López, 2005).

Another important point refers to pesticide applications performed in the period close to harvest, which may be associated with low or almost zero mobility of photoassimilates in the plant due to natural senescence. The reduction of respiration in the caryopsis, which acts as a barrier to absorption of the insecticide by the plant and thereby contributes to the degradation of the molecules by exposure to solar radiation, must also be considered.

Additionally, analyses were carried out with limits set to 20 times lower than the Maximum Residue Limit (MRL) allowed and regulated for rice grains by the Brazilian National Agency for Sanitary Surveillance (ANVISA), as shown in Table 3.

The MRL of rice grains among different countries varies according to the legislation of each administrative organization, and some pesticides used in this study were not registered for rice in certain countries. The MRL in Brazil for the pesticides used in this study are below the limits set by the *Codex Alimentarius*, which is the United Nations' international reference from the Food and Agriculture Organization (FAO), an international authority for solving disputes over food safety and consumer protection. This may be a positive point for Brazilian rice exports because of the existing legislation being very stringent regarding the quality of rice produced. It should be noted that the limit of quantification for the method used is always set equal to or below the regulated MRLs presented by the different agencies responsible for food quality.

It is important that management practices respect the minimum time between pesticide application and grain harvest, which is regulated by ANVISA in Brazil. Overall, if the pesticides are applied according to Good Agricultural Practices (GAP), MRLs are not exceeded; however, the misuse of these compounds is worrying and may result in significant amounts of residues found in both food and the environment (Thurman, Ferrer, & Fernández-Alba, 2006; Jardim, Andrade, & Queiroz, 2009).

Table 3. Comparisons among the maximum residue limits (mg kg^{-1}) for rice grains set by the Codex Alimentarius, Brazil, the United States (U.S.), the European Union (E.U.), Canada, India, Japan, China and Korea.

Pesticide	Codex[1]	Brazil[2]	U.S.[1]	E.U.[1]	Canada[1]	India[1]	Japan[1]	China[1]	Korea[1]
Azoxystrobin	5.0	0.1	5.0	5.0	5.0	-*	0.2	-	1.0
Difenoconazole	-	1.0	-	0.05	0.01	-	-	-	0.2
Propiconazole	-	0.1	1.0	0.05	0.05	-	0.1	-	0.1
Tebuconazole	1.0	0.1	1.0	0.02	-	-	0.05	-	0.05
Trifloxystrobin	5.0	0.2	3.5	0.02	3.5	-	1.6	-	-
Cypermethrin	2.0	0.05	-	2.0	-	-	0.9	-	1.0
Lambda-cyhalothrin	1.0	1.0	1.0	0.02	0.01	-	-	-	-
Permethrin	-	0.1	-	0.05	-	-	2.0	-	-
Thiamethoxam	-	1.0	0.02	0.05	0.02	0.02	0.3	0.1	0.1

[1]United States Department of Agriculture (USDA, 2016). [2]ANVISA (2016). *no registration for rice crop.

Usually, fungicides and insecticides are applied to the aerial part of rice plants, so the determination of residues in rice grains is fundamental as a means of ensuring the quality of the grain. However, pesticides may also affect human life due to pesticide build-up throughout the food chain, which may lead the ecological phenomenon of biomagnification, or bioaccumulation, of a pesticide in an ecological food chain by residue transfer from the food into a person's body tissues. However, there are no consolidated studies on pesticide residue concentrations detected in food items, especially those which exceed the maximum limits for human consumption; the information regarding the effects of accumulation of these compounds in humans is still very restricted (Mostafalou & Abdollahi, 2013) but indicates that such effects may be associated with health problems linked to the nervous system (Costa, Giordano, Guizzetti, & Vitalone, 2008), reproductive system (Saadi, & Abdollahi, 2012) and diseases such as cancer (Alavanja & Bonner, 2012).

The results show that the method of analysis used in this study did not detect any pesticide residues in rice hulls. Thus, the identification of a method that allows the detection of residues at lower concentrations in grains is a goal that should be pursued to ensure optimal food quality.

Conclusion

Residues of azoxystrobin (0.02 and 0.03 mg kg^{-1}) and cypermethrin (0.12 mg kg^{-1}) were detected in rice hulls only in the 2007/08 crop season.

Residues of azoxystrobin, difenoconazole, propiconazole, tebuconazole, and trifloxystrobin fungicides and of cypermethrin, lambda-cyhalothrin, permethrin, and thiamethoxam insecticides were not found in rice grains, regardless of the processing method used.

Acknowledgements

The CNPq for granting the funds, in addition to the assistance in this project provided by the edict MCT/CNPq 14/2008 - Universal. The Laboratory of Post-Harvest, Processing and Quality of Grain at Universidade Federal de Pelotas for parboiling the rice grains.

References

Alavanja, M. C., & Bonner, M. R. (2012). Occupational pesticide exposures and cancer risk: a review. *Journal of Toxicology and Environmental Health Part B, 15*(4), 238-263. doi: 10.1080/10937404.2012.632358

Agência Nacional de Vigilância Sanitária [ANVISA]. (2016). *Monografia de agrotóxicos 2015*. Retrieved on January 11, 2016 from www.portal.anvisa.gov.br/wps/portal/ anvisa/ anvisa/home

Barik, S. F., Ganguly, P., Kunda, S. K., Kole, R. K., & Bhattacharyya, A. (2010). Persistence behaviour of thiamethoxam and lambda cyhalothrin in transplanted paddy. *Bulletin of Environmental Contamination and Toxicology, 85*(4), 419-422. doi: 10.1007/s00128-010-0101-2

Costa, L. G., Giordano, G., Guizzetti, M., & Vitalone, A. (2008). Neurotoxicity of pesticides: a brief review. *Frontiers Bioscience, 13*(4), 1240-1249. doi: 10.2741/2758

Counce, P. A., Keisling, T. C., & Mitchell, A. J. (2000). A uniform, objective, and adaptive system for expressing rice development. *Crop Science, 40*(2), 436-443. doi: 10.2135/cropsci2000.402436x

Dors, G. C., Primel, E. G., Fagundes, C. A. A., Mariot, C. H. P., & Badiale-Furlong, E. (2011). Distribution of pesticide residues in rice grain and in its coproduces. *Journal of the Brazilian Chemical Society, 22*(10), 1921-1930. doi: 10.1590/S0103-50532011001000013

Garrido, F. A., Martínez, S. I., Martínez, V. J. L., & López-López, T. (2005). Determination of multiclass pesticides in food commodities by pressurized liquid extraction using GC-MS/MS and LC-MS/MS. *Analytical and Bioanalytical Chemistry, 383*(7-8), 1106-1118. doi: 10.1007/s00216- 005-0139-x

Hou, X., Han, M., Dai, X., Yang, X., & Yi, S. (2013). A multi-residue method for the determination of 124 pesticides in rice by modified QuEChERS extraction and gas chromatography-tandem mass spectrometry. *Food Chemistry, 138*(2-3), 1198-1205. doi: 10.1016/j. foodchem.2012.11.089

International Union of Pure and Applied Chemistry [IUPAC]. (2015). *Pesticide Properties Database*. Retrieved on January 11, 2016 from www.sitem.herts.ac.uk/aeru/iupac/index.htm.

Jardim, I. C. S. F., Andrade, J. A., & Queiroz, S. C. N. (2009). Pesticide residues in food: global environmental preoccupation. *Química Nova, 32*(4), 996-1012. doi: 10.1590/S0100-40422009000400031

Kanrar, B. G., Ghosh, T., Pramanik, S. K., Dutta, S., Bhattacharyya, A., & Dhuri, A. V. (2006). Degradation dynamicand persistence of imidacloprid in a rice ecosystem under west bengal climatic conditions. *Bulletin of Environmental Contamination and Toxicology, 77*(5), 631-637. doi: 10.1007/s00128-006-1109-5

Kim, I. S., Beaudette, L. L., Shim, J. H., & Trevors, J. T. (2002). Environmental fate of the triazole fungicide propiconazole in a rice-paddy-soilly lysimeter. *Plant and Soil, 238*(2), 321-331. doi: 10.1023/A:1015000328350

Kolberg, D. I., Prestes, O. D., Adaime, M. B., & Zanella, R. (2010). A new gas chromatography/mass spectrometry (GC-MS) method for the multiresidue analysis of pesticides in bread. *Journal of the Brazilian Chemical Society, 21*(6), 1065-1070. doi: 10.1590/S0103-50532010000600016

Komárek, M., Cadková, E., Chrastný, V., Bordas, F., & Bollinger, J. (2010). Contamination of vineyard soils with fungicides: A review of environmental and toxicological aspects. *Environment International, 36*(1), 138-151. doi:10.1016/j.envint.2009.10.005

Laskowski, D. A. (2002). Physical and chemical properties of pyrethroids. *Environmental Contamination Toxicology*, *174*, 49-170. doi: 10.1007/978-1-4757-4260-2_3

Linders, J., Mensink, H., Stephenson, G., Wauchope, D., & Racke, K. (2000). Foliar interception and retention values after pesticide application. A proposal for standardized values for environmental risk assessment. *Pure and Applied Chemistry*, *72*(11), 2199-2218. doi: 10.1002/ps.448

Macedo, W. R., Araujo, D. K., & Castro, P. R. C. (2013). Unravelling the physiologic and metabolic action of thiamethoxam on rice plants. *Pesticide Biochemistry and Physiology*, *107*(2), 244-249. doi: 10.1016/j.pestbp.2013.08.001

Mostafalou, S., & Abdollahi, M. (2013). Pesticides and human chronic diseases: Evidences, mechanisms and perspective. *Toxicology and Applied Pharmacology*, *268*(2), 157-177. doi: 10.1016/j.taap.2013.01.025

Navarro, S., Vela, N., & Navarro, G. (2011). Fate of triazole fungicide residues during malting, mashing and boiling stages of beermaking. *Food Chemistry*. *124*(1), 278-284. doi: 10.1016/j.foodchem.2010.06.033

Pareja, L., Colazzo, M., Perez-Parada, A., Bbesil, N., Heinzen, H., Bocking, B., ... Fernandez-Alba, A. R. (2012). Occurrence and distribution study of residues from pesticides applied under controlled conditions in the field during rice processing. *Journal of Agricultural and Food Chemistry*, *60*(18), 4440-4448. doi: 10.1021/jf205293j

Peña, A., Rodríguez-Liébana, J. A., & Mingorance, M. D. (2011). Persistence of two neonicotinoid insecticides in wastewater, and in aqueous solutions of surfactants and dissolved organic matter. *Chemosphere*, *84*(4), 464-470. doi: 10.1016/j.chemosphere.2011.03.039

Phong, T. K., Nhung, D. T., Yamazaki, K., Takagi, K., & Watanabe, H. (2009). Behavior of sprayed tricyclazole in rice paddy lysimeters. *Chemosphere*, *74*(8), 1085-1089. doi: 10.1016/j.chemosphere.2008.10.050

Prestes, O. S., Friggi, C. A., Adaime, M. B., & Zanella, R. (2009). QuEChERS-Um método moderno de preparo de amostra para determinação multirresíduo de pesticidas em alimentos por métodos cromatográficos acoplados à espectrometria de massas. *Química Nova*, *32*(6), 1620-1634. doi:10.1590/S0100-40422009000600046

Saadi, H. S., & Abdollahi, M. (2012). Is there a link between human infertilities and exposure to pesticides? *International Journal of Pharmacology*, *8*(8), 708-710. doi: 10.3923/ijp.2012.708.710

Santos, M. A. T., Areas, M. A., & Reyes, F. G. (2007). Piretróides - uma visão geral. *Alimentos e Nutrição*, *18*(3), 339-349.

Sociedade Sul-brasileira de Arroz Irrigado. Arroz Irrigado [SOSBAI] (2010). *Recomendações Técnicas da Pesquisa para o Sul do Brasil* (188p). Bento Gonçalves, RS: Palotti.

Sundravadana, S., Kutalam, S., & Samiyappan, R. (2007). Azoxystrobin activity on *Rhizoctonia solani* and its efficacy against rice sheath blight. *Journal of Plant Protection*, *2*(2), 79-84.

Teló, G. M., Senseman, S. A., Marchesan, E., Camargo, E. R., Jones, T., & McCauley, G. (2015a). Residues of thiamethoxam and chlorantraniliprole in rice grain. *Journal of Agricultural and Food Chemistry*, *63*(8), 2119-2126. doi: 10.1021/jf5042504

Teló, G. M., Marchesan, E., Zanella, R., Oliveira, M. L., Coelho, L. L., & Martins, M. L. (2015b). Residues of fungicides and insecticides in rice field. *Agronomy Journal*, *107*(3), 851-863. doi: 10.2134/agronj14.0475

Thurman, E. M., Ferrer, I., & Fernández-Alba, A. R. (2006). Feasibility of LC/TOFMS and elemental database searching as a spectral library for pesticides in food. *Food Additives & Contaminants*, *23*(11), 1169-1178. doi: 10.1080/ 02652030600838241

United States Department of Agriculture [USDA]. (2015). *Foreign agricultural service: pesticide MRL database*. Retrieved on January 12, 2016 from www.fas.usda.gov

Wang, K., Wu, J. X., & Zhang, H. Y. (2012). Dissipation of difenoconazole in rice, paddy soil, and paddy water under field conditions. *Ecotoxicology and Environmental Safety*, *86*(1), 111-115. doi: 10.1016/j.ecoenv. 2012.08.026

Zambolin, L., Picanco, M. C., Silva, A. A., Ferreira, L. R., & Ferreira, F.A. (2008). *Penetração e translocação de fungicidas sistêmicos nos tecidos das plantas. Produtos fitossanitários (fungicidas, inseticidas e herbicidas)* (652p). Viçosa, MG: Produção Independente.

Zhang, J. M., Chai, W. G., & Wu, Y. L. (2012). Residues of chlorantraniliprole in rice field ecosystem. *Chemosphere*, *87*(2), 132-136. doi: 10.1016/j. chemosphere.2011.11.076

10

Parasitism capacity of *Trichogramma pretiosum* on eggs of *Trichoplusia ni* at different temperatures

José Romário de Carvalho[1], Dirceu Pratissoli[1], Leandro Pin Dalvi[1], Marcos Américo Silva[1], Regiane Cristina Oliveira de Freitas Bueno[2] and Adeney de Freitas Bueno[3*]

[1]Centro de Ciências Agrárias, Departamento de Produção Vegetal, Núcleo de Desenvolvimento Científico e Tecnológico em Manejo Fitossanitário, Setor de Entomologia, Universidade Federal do Espírito Santo, Alegre, Espírito Santo, Brazil. [2]Departamento de Produção Vegetal Defesa Fitossanitária, Faculdade de Ciências Agrônomicas, Botucatu, São Paulo, Brazil. [3]Embrapa Soja, Rodovia Carlos João Strass, 86001-970, Londrina, Paraná, Brazil. *Author for correspondence. E-mail: adeney.bueno@embrapa.br

ABSTRACT. *Trichogramma* spp. are egg parasitoids of various pest species of Lepidoptera including *Trichoplusia ni*, an important pest of plants in the genus *Brassica*. Of the climatic conditions that can impair *Trichogramma* spp. parasitism capacity, the temperature is critical. Thus, the objective of this research was to evaluate the parasitism capacity of *Trichogramma pretiosum* on eggs of *T. ni* at 18, 21, 24, 27, 30, and 33°C; 70±10% RH; and 12/12 hours photophase (L/D). Fresh eggs of the host moth were offered to *T. pretiosum* daily. The parasitism rate varied between 8 and 11.4 eggs/female at the temperatures evaluated for the first 24 hours. The highest number of parasitized eggs per female occurred at 24°C (53.0 parasitized eggs/female). The period of parasitism and the mean longevity of females were inversely related to the temperature. Temperature heavily influences the parasitism rate of *T. pretiosum* on eggs of *T. ni*, and the best overall performance of the parasitoid occurs from 24 to 27°C.

Keywords: insecta, biological control, horticultural plants.

Introduction

The cabbage looper, *Trichoplusia ni* Hübner (Lepidoptera: Noctuidae), is a pest with a wide range of hosts including plants of the families Brassicaceae, Solanaceae, and Curcubitaceae, such as cotton, soybean, and different weeds (GRECCO et al., 2010; MILANEZ et al., 2009). The ability of *T. ni* to feed simultaneously on a large variety of host species is crucial to its success. Furthermore, *T. ni* causes large amounts of crop damage, especially among leafy green vegetables (LANDOLT, 1993), which can impair commercialization due to the physical damage caused by the feeding caterpillars.

Trichoplusia ni is most commonly controlled by the use of chemicals. However, the abusive use of insecticides with high levels of biological activity and persistence has caused high economic and environmental costs (BRITO et al., 2004; GRECCO et al., 2010). An alternative measure to mitigate these costs might be biological control. Biological control agents, such as egg parasitoids, constitute an important tool in the development of Integrated Pest Management (IPM) programs, aiming to reduce the use of insecticides and to manage insect resistance to insecticides in a more sustainable manner (PRATISSOLI et al., 2008). In this regard, Godin and Boivin (1998) emphasized that the utilization of egg parasitoids on brassicas is capable of promoting the regulation of the population of pest insects to a level below the economic threshold, which underscores the importance of these biological control agents.

Among egg parasitoids, the genus *Trichogramma* is notable because of its wide geographical distribution and high parasitism capacity on eggs of different species, primarily those in the order Lepidoptera (PRATISSOLI et al., 2002). The success of *Trichogramma* spp. releases, however, depends on knowledge of the ecological characteristics of the parasitoid and on its interaction with the targeted host. Therefore, before using *Trichogramma* species to control insect pests in commercial releases, laboratory studies investigating the parasitism capacity of the selected *Trichogramma* species are needed (PARRA et al., 2002). Thus, Milanez et al. (2009) selected the strain 'Tspd' of *Trichogramma pretiosum* Riley, 1879 (Hymenoptera: Trichogrammatidae) from several different species and strains of *Trichogramma* as the most suitable to parasitize eggs of *T. ni*. Consequently, tests to evaluate the parasitoid in relation to environmental factors should be conducted because the potential that is observed under optimal conditions may be impaired under adverse conditions.

Among abiotic factors, temperature is the most influential, altering life cycle duration, parasitism rate, sex ratio, and longevity of the parasitoids (HOFFMANN; HEWA-KAPUGE, 2000; MOLINA et al., 2005). Therefore, the present work aims to evaluate the parasitism capacity of *T. pretiosum* on eggs of *T. ni* at different temperatures in the laboratory, with the objective of eventually using these data to manage *T. ni* in the field.

Material and methods

The experiment was carried out at the Núcleo de Desenvolvimento Científico e Tecnológico em Manejo Fitossanitário (Nudemafi) located at the Agrarian Sciences Center of the Federal University of Espírito Santo in Alegre, Espírito Santo, Brazil.

Trichoplusia ni rearing and maintenance

Trichoplusia ni pupae were sexed and placed into plastic pots containing moistened filter paper at the bottom. After emergence, the adults were transferred to 60 x 50 x 50 cm wooden framed cages containing a leaf of cabbage for oviposition. The adults were fed a 10% honey solution, which was placed inside 20 mL flasks containing cotton pads that were in contact with the honey solution. Food was renewed every 48 hours. The cabbage leaf was replaced daily, and those containing the host eggs were transferred to lidded plastic containers. After

eclosion, the caterpillars were transferred to 8.5 x 2.5 cm glass tubes containing the artificial diet proposed by Greene et al. (1976) until the pupal phase.

Trichogramma pretiosum rearing and maintenance

Parasitoid rearing and multiplication was performed on eggs of the factitious host, *Anagasta kuehniella* Zeller (Lepidoptera: Pyralidae), according to the methodology described by Parra et al. (2002). Eggs of *A. kuehniella* were glued onto 8.0 x 2.5 cm Bristol board cards with the aid of 30% (w/v) gum arabic and subsequently exposed to ultraviolet light for 45 min. for sterilization. Next, the cards were transferred into 8.5 x 2.5 cm glass tubes containing honey droplets, into which parasitoid females were introduced in sequence. The rearing procedure was performed inside climatic chambers set at 25±1°C, 70±10% RH, and 14/10 hours photophase (L/D).

Trichogramma pretiosum parasitism capacity on eggs of T. ni at different temperatures

Based on the results obtained by Milanez et al. (2009), the 'Tspd' strain of *T. pretiosum* was selected for this experiment. Fifteen newly emerged *T. pretiosum* females were separated into 8.5 x 2.5 cm glass tubes containing a droplet of honey for food. New small Bristol board cards containing 20 *T. ni* eggs (< 24 hours old) were introduced daily into the tubes to allow parasitism by *T. pretiosum*. The tubes were maintained in climatic chambers set at 70±10% RH, 14/10h photophase (L/D), and temperatures of 18°±1°C, 21°±1°C, 24°±1°C, 27°±1°C, 30°±1°C, and 33±1°C. The cards containing the parasitized eggs were removed from the tubes daily and transferred to 23.0 x 4.0 cm plastic bags, which were then sealed and maintained at the same climatic conditions until offspring emergence.

The characteristics evaluated were as follows: daily parasitism, cumulative parasitism, lifetime number of parasitized eggs per female, time span of parasitism, parental adult female longevity, sex ratio, number of parasitoid individuals per egg, and parasitism viability (% parasitoid emergence). A completely randomized experimental design with six treatments (different temperatures) and 15 replications was used.

Statistical analysis

The data were subjected to ANOVA, and the means were compared by Tukey's test at 5% probability (p ≤ 0.05).

Results and discussion

The rate of parasitism during the first 24 hours varied between 8 and 11.4 eggs per *T. pretiosum* female between 18 and 33°C (Figure 1). After the first day, the parasitism rate decreased at all temperatures, with the highest figures observed in the first three days of parasitism activity. These results demonstrate that the parasitism rate is not constant at all temperatures and may depend on the intrinsic characteristics of the species and/or strain of the parasitoid and host. The rate of parasitism at different temperatures is a biological characteristic specific to each parasitoid strain or species reared on each host (PRATISSOLI; PARRA, 2000, 2001; PRATISSOLI et al., 2004). Similar results were reported by Pratissoli et al. (2004) and Bueno et al. (2010), who recorded that the rate of T. pretiosum parasitism varied with temperature on Plutella xylostella (L., 1758) (Lepidoptera: Plutellidae) and Spodoptera frugiperda Smith (Lepidoptera: Noctuidae) eggs. Other authors have reported high parasitism during the first 24 hours when studying different species or strains of Trichogramma and host species (INOUE; PARRA, 1998). This behavior might be associated with the lower longevity of the parasitoids when reared under higher temperatures compared to lower temperatures (INOUE; PARRA, 1998). Under higher temperatures the metabolic expenses are higher so it is advantageous for the parasitoid to conduct the parasitism in the first hours (GERLING, 1972). Having the majority of parasitism concentrated on the first day is a positive feature for mass releases in the field as this might guarantee quick pest control and allow growers to apply herbicides or fungicides shortly after the parasitoid release if necessary. Therefore, when choosing a release strategy it is important to consider whether parasitism is concentrated in the first days of life or evenly distributed throughout adulthood and to consider that this might vary due to differences in temperature (REZNIK; VAGHINA, 2006), hosts (REZNIK et al., 2001), or parasitoid species/strain (PIZZOL et al., 2010). These factors can influence the success of biological control programs using egg parasitoids of the genus Trichogramma (SMITH, 1996).

Moreover, in T. pretiosum these characteristics should also be considered when choosing the most suitable parasitoid species or strain to be used in the field. It is better if the parasitoid reaches 80% of its lifetime parasitism soon after release. A longer amount of time between release and parasitism increases the chances of being influenced by biotic and/or abiotic factors that may impair parasitism. In this context, the higher rate of parasitism in the first 24 hours observed in this study is a positive characteristic of the parasitoid strain because they are less vulnerable to the side effects of any pesticide spraying that might take place after the field release. After releasing parasitoids into the field, it is not unusual to use a fungicide or herbicide that might impair parasitism. Egg parasitoids are typically tiny little wasps that may be more susceptible to

chemicals used in agriculture, including herbicides and fungicides, than their hosts (CARMO et al., 2010). Furthermore, parasitoids differ from herbivorous insects by their inability to synthesize lipids as adults, and this makes them more vulnerable to temperature increases than most pest species (DENIS et al., 2011). Similar results have been observed by several authors using different species of parasitoids and hosts. Pratissoli et al. (2004) studied T. pretiosum parasitism on eggs of P. xylostella and verified a higher rate of parasitism on the first day, with the number of parasitized eggs decreasing with time. The decrease in the performance of the females may be directly related to the age of the insects (PASTORI et al., 2007; SÁ, PARRA, 1994; ZAGO et al., 2007). The highest parasitism rate of T. exiguum Pinto and Platner, 1978 on eggs of P. xylostella occurred at 25°C (PEREIRA et al., 2007), and for T. pratissolii Querino and Zucchi, on eggs of Corcyra cephalonica Stainton, 1865 (Lepidoptera: Pyralidae) and A. kuehniella, the highest parasitism rates observed were between 24 and 30°C (ZAGO et al., 2007).

The index of 80% cumulative parasitism, which represents an estimate of the efficiency of the parasitoid in the field (due to abiotic factors, this index will never reach 100%), also varied with the temperature (Figure 1). The observed variation was approximately four days. At the lowest temperatures (18 and 21°C), the time required was six and eight days, respectively; at the median temperatures (24 and 27°C), it was six and five days, respectively; and at the highest temperatures (30 and 33°C), it was five and three days, respectively. The variation observed may be associated with the characteristics of the parasitoid species and/or strain, as well as with the choice of the host studied because the parasitoids used in this experiment were reared on eggs of A. kuehniella.

Pereira et al. (2007) reported similar results to those found here. Studying different species of parasitoids under distinct thermal conditions, those authors attributed the variation of the parasitism of Trichogramma sp. to temperature differences. Pastori et al. (2007), studying the performance of T. pretiosum on eggs of Bonagota salubricola Meyrick (Lepidoptera: Tortricidae), verified that the strain collected on eggs of B. salubricola, and even those reared on eggs of Sitotroga cerealella (Oliv., 1819) (Lepidoptera: Gelechiidae), did not show reduction of potential, being influenced solely by the thermal regime to which they had been submitted. Another possibility is that the intrinsic characteristics of the host species might influence the parasitoid performance, as verified by Pratissoli et al. (2004)

and Zago et al. (2007) in studies performed with *T. pretiosum* and *T. pratissolii*, respectively, on eggs of different hosts. Hoffmann et al. (2001), however, reported that the development of *T. ostriniae* collected on *Ostrinia nubilalis* (Lepidoptera: Crambidae), on different hosts and during several generations of the parasitoid did not affect their performance on the original host. These differences in results emphasize the importance of studying these biological characteristics for each parasitoid strain and target host.

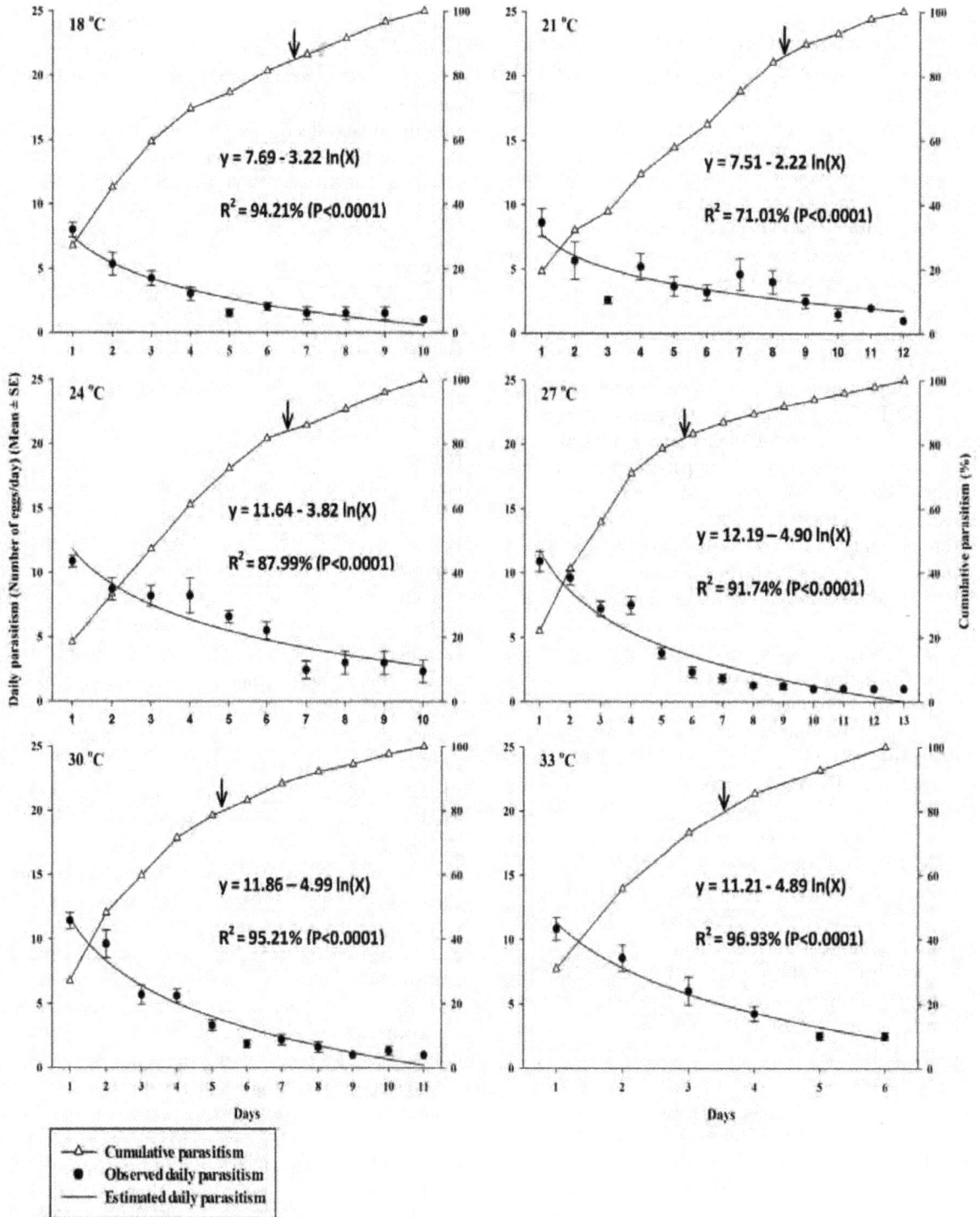

Figure 1. Daily and accumulated parasitism rates and parasitism time span of the egg parasitoid *Trichogramma pretiosum* on eggs of the cabbage looper *Trichoplusia ni* at different temperatures (18°±1°C, 21°±1°C, 24°±1°C, 27°±1°C, 30°±1°C, and 33±1°C); 70±10% RH; and 14/10 hours photophase (L/D) (↓ 80% parasitism).

The lifetime number of parasitized eggs varied with temperature (Table 1). However, there were no statistically significant differences at 18, 21, 30, and 33°C. The highest number of parasitized eggs occurred at 24°C (53.0 parasitized eggs/female) and the lowest number was observed at 33°C (23.2 parasitized eggs/female). The figures obtained in this study were higher than those reported for different species of *Trichogramma* and different hosts at the same temperatures, as observed by Pratissoli et al. (2004), Pereira et al. (2007), Zago et al. (2007) and Pratissoli et al. (2008), demonstrating that *T. pretiosum* displays a real potential for use in biological control programs of the caterpillar *T. ni*. Contrary to the results presented in this study, Pastori et al. (2007) found that *T. pretiosum*, strain bonagota, is better adapted to 18°C, a fact that is linked to adaptation of the parasitoid strains to specific climatic conditions, according to the authors. This illustrates the importance of this type of study for each parasitoid strain to successfully develop a biological control program.

The period of parasitism varied between six and 13 days at the temperatures studied, with the smallest value observed at 33°C (Figure 1). This might be directly correlated with the intrinsic characteristics of the *Trichogramma* strain studied, which displays differentiated responses as a function of thermal adaptability. Reinforcing such results, Zago et al. (2007), studying *T. pratissolii*, verified an inverse correlation of the parasitism period with the temperature, observing figures between two and 10 days from 18 to 33°C. Pratissoli et al. (2009), in studies with *T. acacioi* on eggs of *A. kuehniella* and *S. cerealella*, found that between 20 and 35°C, there was variation in the period of parasitism between five and 16 days. Pereira et al. (2007) verified an inferior variation on the parasitism period in the same temperature range (nine to 13 days).

The mean longevity of parental adult *T. pretiosum* females reared on eggs of *T. ni* presented statistically significant variation at different temperatures, demonstrating an inverse behavior to increases in temperature (Table 1). The longevity of the female parasitoids were statistically similar at 18, 21, 24, and 27°C (11.5, 10.9, 10.1, and 9.7 days, respectively), but significantly different at 30 and 33°C. At 27 and 30°C,

this parameter was not significantly different, displaying values of 9.7 and 7.7 days, respectively. The lowest mean longevity of females occurred at 33°C (4.9 days) illustrating that the increase in temperature places more stress on the parasitoid than the lower temperatures. Other authors have found similar results from different species of the same parasitoid genus on the same host (PASTORI et al., 2007; PEREIRA et al., 2007; PRATISSOLI et al., 2004; ZAGO et al., 2007). According to Pastori et al. (2007), the reduction of the temperature promotes an increase in parasitoid longevity due to a reduction of metabolic rate.

Temperature is not the only factor responsible for variations of longevity. Other factors, such as photoperiod, relative humidity, interspecific and intraspecific competition (PRATISSOLI; PARRA, 2001), and the presence of the host (CAÑETE; FOERSTER, 2003), can interfere with the biological characteristics of an insect. Nevertheless, in the present study, temperature was the primary factor responsible for the variation of parasitoid longevity.

The sex ratio of *T. pretiosum* descendants reared on eggs of the moth *T. ni* was statistically different at 24 and 33°C, varying between 0.57 and 0.80 (Table 1). At 18, 21, 27, and 30°C, the sex ratios were statistically similar. This may have occurred due to the intrinsic characteristics of the parasitoid species and/or strain. However, the results are within the range found by other authors (BUENO et al., 2009; DIAS et al., 2008; PEREIRA et al., 2007; PRATISSOLI et al., 2010), but environmental factors such as temperature and relative humidity, as well as the host used, can influence this parameter. In their study with *T. pretiosum*, Pastori et al. (2007) observed a variation in the sex ratio between 18 and 32°C, detecting a higher number of females at 32°C. In contrast, Pratissoli and Parra (2000) studying *T. pretiosum* on eggs of *Tuta absoluta* (Lepidoptera: Gelechiidae) and *Phthorimaea operculella* Zeller (Lepidoptera: Gelechiidae) found no variation in sex ratio of the descendants from the eggs of *T. absoluta* in the same thermal range (18 to 32°C), while the highest sex ratio figures were found between 20 and 32°C in the descendants reared on eggs of *P. operculella*.

Table 1. Biological characteristics of the egg parasitoid *Trichogramma pretiosum* reared on eggs of the cabbage looper *Trichoplusia ni* at different temperatures (18°±1°C, 21°±1°C, 24°±1°C, 27°±1°C, 30°±1°C, and 33°±1°C); 70±10% RH; and 14/10 hours photophase (L/D).

Temperature (°C)	Lifetime number of parasitized eggs/female (± SE)[1]	Parental adult female longevity (days ± SE)[1]	Sex ratio[1]	Number of parasitoids/egg[1]	Parasitism viability (%)[1]
18	26.6±2.18 c	11.5±0.51 a	0.65±0.04 bc	1.9±0.08 ab	100.0±0.00 a
21	29.2±4.24bc	10.9±0.78 a	0.74±0.02 ab	1.9±0.04 ab	100.0±0.00 a
24	53.0±3.11a	10.1±0.61 a	0.57±0.01 c	1,3±0.03 c	100.0±0.00 a
27	40.2±2.82b	9.7±0.56 ab	0.66±0.03 bc	1.7±0.16 ab	95.4±2.39 ab
30	38.4±2.54bc	7.7±0.49 b	0.74±0.02 ab	1.5±0.14 b	88.2±5.16 b
33	23.2±2.29 c	4.9±0.46 c	0.80±0.04 a	2.0±0.12 a	93.5±3.41 ab

[1]Means (Mean±Standard Error) followed by the same letter in the column are not significantly different from each other by Tukey's test at 5% probability.

The number of descendants that emerged per egg significantly differed between 24, 30, and 33°C (Table 1). Despite the variation, the number of individuals emerged per host egg was always higher than one. The highest mean number of parasitoids emerging per egg occurred at 33°C (2.0 individuals/egg), while the smallest figures were obtained at 24 and 30°C (1.3 and 1.5 individuals/egg, respectively). At the remaining temperatures studied, the figures varied from 1.7 and 1.9 individuals/egg. The temperature can cause alterations in the physiology and development of the insects, and the suitability and the behavioral and functional responses of the insects to certain environmental conditions clearly vary between individuals (BOIVIN, 2010; CHOWN; NICHOLSON, 2004; GULLAN; CRANSTON, 2010; SPEIGHT et al., 2008).

The host directly influences the development of the parasitoids because it is the source of food as well as shelter (ÖZDER; KARA, 2010). Therefore, the size of the host egg not only influences the number of eggs deposited by the parasitoid female but also the size of the adult *Trichogramma*, which will depend of the nutritional resources available for the development of the larva (VINSON, 1997). Molina and Parra (2006) have reported similar results obtained with eggs of *Gymnandrosoma aurantianum* Lima (Lepidoptera: Tortricidae) in which they found a variation between 1.4 and 1.8 individuals/egg at 25°C. The values found in this study, however, are higher than the figures determined for other species of pest lepidopterans such as *Bonagota cranaodes* Meyrick (Lepidoptera: Tortricidae) and *Chrysodeixis includens* Walker (Lepidoptera: Noctuidae) (BUENO et al., 2009; FONSECA et al., 2005). These authors reported values of 1.3 and 1.0 individuals/egg, respectively, at 25°C. Similarly, Pratissoli and Parra (2000) verified that on two different hosts, the highest number of parasitoids per host egg was obtained at 25°C. Another study by Pastori et al. (2008) reported a higher number of descendants per egg at 22 and 30°C.

There were statistically significant differences in the viability parameter at different temperatures (Table 1). At all temperatures, however, the values observed were above 88%. Several other authors have reported similar values (MELO et al., 2007; MILANEZ et al., 2009; PRATISSOLI; PARRA, 2000), demonstrating that the cabbage looper *T. ni* is a potentially suitable host for the egg parasitoid *T. pretiosum*.

Conclusion

Temperature heavily influences the parasitism rate of *T. pretiosum* reared on eggs of *T. ni*.

The egg parasitoid *T. pretiosum* has suitable characteristics for the control of the cabbage looper *T. ni* within a thermal range between 24 and 27°C.

Acknowledgements

The authors would like to thank *Embrapa Soja*, *Coordenação de Aperfeiçoamento de Pessoal de Nível Superior (CAPES)* and *Conselho Nacional de Desenvolvimento Científico e Tecnológico (CNPq)* for the support provided.

References

BOIVIN, G. Reproduction and immature development of egg parasitoids. In: CÔNSOLI, F. L.; PARRA, J. R. P. ZUCCHI, R. A. (Ed.). **Egg parasitoids in agroecosystems with emphasis on *Trichogramma*, progress in biological control**. New York: Springer, 2010. v. 9, p. 1-23.

BRITO, G. G.; COSTA, E. C.; MAZIERO, H.; BRITOY, A. B.; DÖRR, F. A. Preferência da broca-das-cucurbitáceas [*Diaphania nitidalis* Cramer, 1782 (Lepidoptera: Pyralidae)] por cultivares de pepineiro em ambiente protegido. **Ciência Rural**, v. 34, n. 2, p. 577-579, 2004.

BUENO, R. C. O. F.; BUENO, A. F.; PARRA, J. R. P.; VIEIRA, S. S.; OLIVEIRA L. J. Biological characteristicis and parasitism capacity of Trichogramma pretiosum Riley (Hymenoptera, Trichogrammatidae) on eggs of Spodoptera frugiperda (J. E. Smith) (Lepidoptera, Noctuidae). **Revista Brasileira de Entomologia**, v. 54, n. 2, p. 322-327, 2010.

BUENO, R. C. O. F.; PARRA, J. R. P.; BUENO, A. F.; HADDAD, M. Desempenho de Tricogramatídeos como potenciais agentes de controle de Pseudoplusia includens Walker (Lepidoptera: Noctuidae). **Neotropical Entomology**, v. 38, n. 3, p. 389-394, 2009.

CAÑETE, C. L.; FOERSTER, L. A. Incidência natural e biologia de Trichogramma atopovirilia Oatman and Platner, 1983 (Hymenoptera: Trichogrammatidae) em ovos de Anticarsia gemmatalis Hubner, 1818 (Lepidopera, Noctuidae). **Revista Brasileira de Entomologia**, v. 47, n. 2, p. 201-204, 2003.

CARMO, E. L.; BUENO, A. F.; BUENO, R. C. O. F. Pesticide selectivity for the insect egg parasitoid Telenomus remus. **BioControl**, v. 55, n. 4, p. 455-464, 2010.

CHOWN, S. L.; NICHOLSON, S. W. **Letal temperature limits**. In: CHOWN, S. L.; NICHOLSON, S. W. (Ed.). Insect physiological ecology: mechanisms and patterns. Oxford: Oxford University Press, 2004. p. 115-153.

DENIS, D.; PIERRE, J. S.; VAN BAAREN, J.; VAN ALPHEN, J. J. M. How temperature and habitat quality affect parasitoid lifetime reproductive success – A simulation study. **Ecological Modelling**, v. 222, n. 9, p. 1604-1613, 2011.

DIAS, N. S.; PARRA, J. R. P.; LIMA, T. C. C. Seleção de hospedeiro alternativo para três espécies de tricogramatídeos neotropicais. **Pesquisa Agropecuária Brasileira**, v. 43, n. 11, p. 1467-1473, 2008.

FONSECA, F. L.; KOVALESKI, A.; FORESTI, J.; RINGENBERG, R. Desenvolvimento e exigências térmicas de Trichogramma pretiosum Riley (Hymenoptera: Trichogrammatidae) em ovos de Bonagota cranaodes (Meyrick) (Lepidoptera: Tortricidae). **Neotropropical Entomology**, v. 34, n. 6, p. 945-949, 2005.

GERLING, D. The developmental biology of *Telenomus remus* Nixon (Hymenoptera: Scelionidae). **Bulletin of Entomological Research**, v. 61, n. 3, p. 385-388, 1972.

GODIN, C.; BOIVIN, G. Lepidopterous pests of Brassica crops and their parasitoids in southwestern Quebec. **Environmental Entomology**, v. 27, n. 5, p. 1157-1165, 1998.

GRECCO, E. D.; POLANCZYK, R. A.; PRATISSOLI, D. Seleção e caracterização molecular de *Bacillus thuringiensis* Berliner com atividade tóxica para *Trichoplusia ni* Hübner (Lepidoptera: Noctuidae). **Arquivos do Instituto Biológico**, v. 77, n. 4, p. 685-692, 2010.

GREENE, G. L.; LEPPLA, N. C.; DICKERSON, W. A. Velvetbean caterpillar: a rearing procedure and artificial medium. **Journal of Economic Entomology**, v. 69, n. 4, p. 487-488, 1976.

GULLAN, P. J.; CRANSTON, P. S. **Insects**: An outline of entomology. 4th ed. Hoboken: Blackwell Science, 2010.

HOFFMANN, A. A.; HEWA-KAPUGE, S. Acclimation for heat resistance in *Trichogramma* nr. *brassicae*: can it occur without costs? **Functional Ecology**, v. 14, n. 1, p. 55-60, 2000.

HOFFMANN, M. P.; ODE, P. R.; WALKER, D. L.; GARDNER, J.; VAN NOUHUYS, S.; SHELTON, A. M. Performance of *Trichogramma ostriniae* (Hymenoptera: Trichogrammatidae) reared on factitious hosts, including the target host, *Ostrinia nubilalis* (Lepidoptera: Crambidae). **Biological Control**, v. 21, n. 1, p. 1-10, 2001.

INOUE, M. S. R.; PARRA, J. R. P. Efeito da temperatura no parasitismo de *Trichogramma pretiosum* Riley, 1879 sobre ovos de *Sitrotroga cerealella* (Oliv., 1819). **Scientia Agricola** v. 55, n. 2, p. 222-226, 1998.

LANDOLT, P. J. Effects of host leaf damage on cabbage looper moth attraction and oviposition. **Entomologia Experimentalis Et Applicata**, v. 67, n. 1, p. 79-85, 1993.

MELO, R. L.; PRATISSOLI, D.; POLANCZYK, R. A.; MELO, D. F.; BARROS, R.; MILANEZ, A. M. Biologia e Exigências Térmicas de *Trichogramma atopovirilia* Oatman and Platner (Hymenoptera: trichogrammatidae) em Ovos de *Diaphania hyalinata* L. (Lepidoptera: Pyralidae). **Neotropical Entomology**, v. 36, n. 3, p. 431-435, 2007.

MILANEZ, A. M.; PRATISSOLI, D.; POLANCZYK, R. A.; BUENO, A. F.; TUFIK, C. B. A. Avaliação de *Trichogramma* spp. para o controle de *Trichoplusia ni*. **Pesquisa Agropecuária Brasileira**, v. 44, n. 10, p. 1219-1224, 2009.

MOLINA, R. M. S.; FRONZA, V.; PARRA, J. R. P. Seleção de *Trichogramma* spp., para o controle de *Ecdytolopha aurantiana*, com base na biologia e exigências térmicas. **Revista Brasileira de Entomologia**, v. 49, n. 1, p. 152-158, 2005.

MOLINA, R. M. S.; PARRA, J. R. P. Seleção de linhagens de *Trichogramma* (Hymenoptera, Trichogrammatidae) e determinação do número de parasitoides a ser liberado para o controle de *Gymnandrosoma aurantianum* Lima (Lepidoptera, Tortricidae). **Revista Brasileira de Entomologia**, v. 50, n. 4, p. 534-539, 2006.

ÖZDER, N.; KARA, G. Comparative biology and life tables of *Trichogramma cacoeciae, T. brassicae* and *T. evanescens* (Hymenoptera: Trichogrammatidae) with *Ephestia kuehniella* and *Cadra cautella* (Lepidoptera: Pyralidae) as hosts at three constant temperatures. **Biocontrol Science and Technology**, v. 20, n. 3, p. 245-255, 2010.

PARRA, J. R. P.; BOTELHO, P. S. M.; FERREIRA, C.; BENTO, J. M. S. Controle biológico: uma visão inter e multidisciplinar. In: PARRA, J. R. P.; BOTELHO, P. S. M.; FERREIRA, C.; BENTO, J. M. S. (Ed.). **Controle biológico no Brasil**: parasitóides e predadores. São Paulo: Manole, 2002. cap. 8, p. 125-137.

PASTORI, P. L.; MONTEIRO, L. B.; BOTTON, M. Biologia e exigências térmicas de *Trichogramma pretiosum* Riley (Hymenoptera, Trichogrammatidae) 'linhagem bonagota' criado em ovos de *Bonagota salubricola* (Meyrick) (Lepidoptera, Tortricidae). **Revista Brasileira de Entomologia**, v. 52, n. 3, p. 472-476, 2008.

PASTORI, P. L.; MONTEIRO, L. B.; BOTTON, M.; PRATISSOLI, D. Capacidade de parasitismo de *Trichogramma pretiosum* Riley (Hymenoptera: Trichogrammatidae) em ovos de *Bonagota salubricola* (Meyrick) (Lepidoptera: Tortricidae) sob diferentes temperaturas. **Neotropical Entomology**, v. 36, n. 6, p. 926-931, 2007.

PEREIRA, F. F.; BARROS, R.; PRATISSOLI, D.; PEREIRA, C. L. T.; VIANNA, U. R.; ZANUNCIO, J. C. Capacidade de parasitismo de Trichogramma exiguum Pinto and Platner, 1978 (Hymenoptera: Trichogrammatidae) em ovos de Plutella xylostella (L., 1758) (Lepidoptera: Plutellidae) em diferentes temperaturas. **Ciência Rural**, v. 37, n. 2, p. 297-303, 2007.

PIZZOL, J., PINTUREAU, B.; KHOUALDIA, O.; DESNEUX, N. Temperature-dependent differences in biological traits between two strains of Trichogramma cacoeciae (Hymenoptera: Trichogrammatidae). **Journal of Pest Science**, v. 83, n. 4, p. 447-452, 2010.

PRATISSOLI, D.; DALVI, L. P.; POLANCZYK, R. A.; ANDRADE, G. S.; HOLTZ, A. M.; NICOLINE, H. O. Características biológicas de Trichogramma exiguum em ovos de Anagasta kuehniella e Sitotroga cerealella. **Idesia**, v. 28, n. 1, p. 39-42, 2010.

PRATISSOLI, D.; FORNAZIER, M. J.; HOLTZ, A. M.; GONÇALVES, J. R.; CHIORAMITAL, A. B.; ZAGO, H. Ocorrência de *Trichogramma pretiosum* em áreas comerciais de tomate, no Espírito Santo, em regiões de diferentes altitudes. **Horticultura Brasileira**, v. 21, n. 1, p. 73-76, 2002.

PRATISSOLI, D.; OLIVEIRA, H. N.; POLANCZYK, R. A.; HOLTZ, A. M.; BUENO, R. C. O. F.; BUENO, A. F.; GONÇALVEZ, J. R. Adult feeding and mating effects on the biological potential and parasitism of *Trichogramma pretiosum* and *T. acacioi* (Hymenoptera: Trichogrammatidae). **Brazilian Archives of Biology and Technology**, v. 52, n. 5, p. 1057-1062, 2009.

PRATISSOLI, D.; PARRA, J. R. P. Seleção de linhagens de *Trichogramma pretiosum* Riley (Hymenoptera:

Trichogrammatidae) para o controle das traças *Tuta absoluta* (Meyrich) e *Phthorimaea operculella* (Zeller) (Lepidoptera: Gelechiidae). **Neotropical Entomology**, v. 30, n. 2, p. 277-282, 2001.

PRATISSOLI, D.; PARRA, J. R. P. Desenvolvimento e exigências térmicas de *Trichogramma pretiosum* Riley, criados em duas traças do tomateiro. **Pesquisa Agropecuária Brasileira**, v. 35, n. 7, p. 1281-1288, 2000.

PRATISSOLI, D.; PEREIRA, F. F.; BARROS, R.; PARRA, J. R. P.; PEREIRA, C. L. T. Parasitismo de *Trichogramma pretiosum* em ovos da traça-das-crucíferas sob diferentes temperaturas. **Horticultura Brasileira**, v. 22, n. 4, p. 754-757, 2004.

PRATISSOLI, D.; POLANCZYK, R. A.; HOLTZ, A. M.; DALVI, L .P.; SILVA, A. F.; SILVA, L. N. Selection of *Trichogramma* species for controlling the Diamondback moth. **Horticultura Brasileira**, v. 26, n. 2, p. 259-261, 2008.

REZNIK, S. YA., VAGHINA, N. P. Temperature effects on induction of parasitization by females of *Trichogramma principium* (Hymenoptera, Trichogrammatidae). **Entomological Review**, v. 86, n. 2, p. 133-138, 2006.

REZNIK, S. YA.; UMAROVA, T. YA.; VOINOVICH, N. D. Long-term egg retention and parasitization in *Trichogramma principium* (Hymenoptera, Trichogrammatidae).

Journal of Applied Entomology, v. 125, n. 4, p. 169-175, 2001.

SÁ, L. A. N.; PARRA, J. R. P. Biology and parasitism of *Trichogramma pretiosum* Riley (Hym.: Trichogrammatidae) on *Ephestia kuehniella* (Zeller) (Lep.: Pyralidae) and *Heliothis zea* (Boddie) (Lep.: Noctuidae) egg. **Journal of Applied Entomology**, v. 118, n. 1-5, p. 38-43, 1994.

SMITH, S. M. Biological control with *Trichogramma*: Advances, successes, and potencial of their use. **Annual Review of Entomology**, v. 41, n. 1, p. 375-406, 1996.

SPEIGHT, M. R.; HUNTER, M. D.; WATT, A. D. **Ecology of insects**: Concepts and applications. Hoboken: Wiley-Blackwell, 2008.

VINSON, S. B. Comportamento de seleção hospedeira de parasitoides de ovos, com ênfase na família Trichogrammatidae. In: PARRA, J. R. P.; ZUCCHI, R. A. (Ed.). *Trichogramma* **e o controle biológico aplicado**. Piracicaba: Fealq, 1997. p. 67-120.

ZAGO, H. B.; PRATISSOLI, D.; BARROS, R.; GONDIM JR., M. G. C.; SANTOS JR., H. J. G. Capacidade de parasitismo de *Trichogramma pratissolii* Querino and Zucchi (Hymenoptera: Trichogrammatidae) em hospedeiros alternativos, sob diferentes temperaturas. **Neotropical Entomology**, v. 36, n. 1, p. 84-89, 2007.

11

Assessment of *Trichogramma* species (Hymenoptera: Trichogrammatidae) for biological control in cassava (*Manihot esculenta* Crantz)

Marcus Alvarenga Soares[1*], Germano Leão Demolin Leite[2], José Cola Zanuncio[3], Cleidson Soares Ferreira[2], Silma Leite Rocha[3] and Veríssimo Gibran Mendes de Sá[4]

[1]*Programa de Pós-graduação em Produção Vegetal, Universidade Federal dos Vales do Jequitinhonha e Mucuri, Rodovia MGT-367, Km 583, 5000, 39100-000, Diamantina, Minas Gerais, Brazil.* [2]*Insetário George Washington Gomez de Moraes, Instituto de Ciências Agrárias, Universidade Federal de Minas Gerais, Montes Claros, Minas Gerais, Brazil.* [3]*Departamento de Biologia Animal, Universidade Federal de Viçosa, Viçosa, Minas Gerais, Brazil.* [4]*Faculdade de Engenharia, Universidade do Estado de Minas Gerais, João Monlevade, Minas Gerais, Brazil. *Author for correspondence. E-mail: marcusasoares@yahoo.com.br*

ABSTRACT. Cassava is the sixth most important crop in the world, and it is attacked by many pests, such as *Erinnyis ello* (L.) (Lepidoptera: Sphingidae). This lepidopteran pest has natural enemies that can efficiently control its population, such as *Trichogramma* spp. (Hymenoptera: Trichogrammatidae). The objective of this research was to assess the flight capacity, parasitism and emergence of *Trichogramma pretiosum*, *T. marandobai* and *T. demoraesi* and to select the most efficient species among them for biological control programs. The flight capacity of these species was assessed in test units consisting of a plastic PVC cylinder with a rigid, transparent plastic circle on the upper portion of the cylinder and an extruded polystyrene disk to close the bottom of the cylinder. A tube was placed in each test unit containing a card with 300 *Anagasta kuehniella* (Zeller) (Lepidoptera: Pyralidae) eggs that had been parasitised by *Trichogramma*. These cards were later assessed to determine the parasitism rate and adult emergence of these natural enemies. *Trichogramma pretiosum* presented the highest flight capacity (68 ± 5%), parasitism (74 ± 2%) and percentage of adults emerged (91 ± 3%) in the laboratory, making this species suitable for mass rearing and release in biological control programs.

Keywords: Lepidoptera, egg parasitoid, *Erinnyis ello*.

Introduction

Cassava (*Manihot esculenta* Crantz) is cultivated on tropical plains between 30° N and 30° S of the Equator in areas where the average annual temperature is greater than 18°C (COURSEY; HAYNES, 1970; NUWAMANYA et al., 2012). Cassava is considered the sixth most important crop in the world behind rice, wheat, corn, potatoes, and sweet potatoes (FAO, 2012). Africa accounts for more than half of the world's cassava production (51%), and Asia and Latin America produce 29% and 19%, respectively. In addition, Nigeria and Brazil produce approximately one third of the global output (NASSAR; ORTIZ, 2007).

The cassava crop is attacked by several pests on the African continent, especially scale insects (Sternorrhyncha: Pseudococcidae) and drill roots *Stictococcus vayssierei* (Richard) (Hemiptera: Stictococcidae) (OERKE, 2006; TINDO et al., 2009). In Brazil, the main pest is *Erinnyis ello* (L.) (Lepidoptera: Sphingidae); others, such as the cassava fly (Diptera: Lonchaeidae), mites (Acari), thrips (Thysanoptera: Thripidae) and the white fly (Hemiptera: Aleyrodidae), may occur but with secondary importance (BARRIGOSSI et al., 2002; BELLOTTI et al., 1999; GISLOTI; PRADO, 2011; LEITE et al., 2002, 2003).

Erinnyis ello infestations can result in the complete defoliation of plants, the destruction of thinner branches and the spread of bacterial diseases, reducing cassava production by 50% (BARRIGOSSI et al., 2002). However, natural enemies can efficiently control the *E. ello* population, especially the egg parasitoids *Trichogramma pretiosum* (Riley) and *T. marandobai* (Brun, Moraes; Soares) (Hymenoptera: Trichogrammatidae) (BRUN et al., 1986; OLIVEIRA et al., 2010). Moreover, *T. demoraesi* (Nagaraja) (Hymenoptera: Trichogrammatidae), which was originally described as a parasitizing forest species in Minas Gerais State (NAGARAJA, 1983), also parasitizes the eggs of *E. ello* in the Amazonas State (RONCHI-TELES; QUERINO, 2005).

Trichogramma spp. can be easily multiplied in the laboratory with high efficiency and at low cost for biological control programs (ÖZTEMIZ; KORNOSOR, 2007). However, *Trichogramma* species are typically released onto crops after several generations of rearing in the laboratory, which may reduce their efficiency due to inbreeding and genetic erosion (PRATISSOLI et al., 2005).

Thus, quality control of *Trichogramma* production in the laboratory is important to ensure efficiency in the field (SOARES et al., 2012). The methodology used by several authors for the evaluation of natural enemies includes tests of parasitism, emergence and flight for different species or strains of *Trichogramma* (DUTTON; BIGLER, 1995; PRASAD et al., 1999; PRATISSOLI et al., 2008; PREZOTTI et al., 2002; SOARES et al., 2007; SOARES et al., 2012). These biological parameters can assist in choosing the individuals with the best characteristics for use in biological control programs of pests such as *E. ello*.

The objective of this research was to assess the capacity of parasitism, emergence and flight of *T. pretiosum*, *T. marandobai* and *T. demoraesi* and select the most efficient species for biological control programs.

Material and methods

The experiment was carried out in the Entomology Laboratory and the George Washington Gomez de Morais Insectarium at the Agrarian Science Institute of the Federal University of Minas Gerais in the Municipality of Montes Claros, Minas Gerais State, Brazil. The parasitoids were reared and multiplied at $25 \pm 2°C$ with a 12-hours light period.

Eggs from the alternative host, *Anagasta kuehniella* (Zeller) (Lepidoptera: Pyralidae), which were obtained using the technique of Soares et al. (2012), were fixed with gum arabic diluted to 30% on 7.20 x 0.7 cm pieces of white card and placed under an ultraviolet lamp that was 25 cm high for 60 min.; these eggs were used to maintain the parasitoids in the laboratory (SOARES et al., 2012). Each card, with 300 eggs of this alternative host, was placed in a test tube (12.0 x 2.0 cm) with 30 newly emerged thelytokous females of one of the *Trichogramma* species (STOUTHAMER et al., 1993) at a ratio of one parasitoid for every 10 eggs, and parasitism was permitted for five hours (PRATISSOLI; PARRA, 2000; SOARES et al., 2007). A drop of honey was placed in each test tube to feed the parasitoids (LUNDGREN; HEIMPEL, 2003).

The flight capacity of the *Trichogramma* species was assessed in test units (Figure 1) (SOARES et al., 2007) consisting of a PVC cylinder that was 18 cm long, 11 cm in diameter and painted with black latex paint on the inside. The bottom of the cylinder was closed with black flexible plastic (larger in size than the tube diameter) held firmly in place by an extruded polystyrene disk that was approximately 1 cm thick with the same diameter as the tube. The excess plastic that overlapped the edges of the tube after fitting the disk were secured to the tube with elastic to create a better seal and prevent the escape of the parasitoids.

Each card, with 300 *A. kuehniella* eggs parasitized by *Trichogramma* and near the adult emergence of these parasitoids (twenty days after parasitism), was placed inside a test tube (7.5 x 1.0 cm) and affixed to the center of the bottom portion of the test unit (the extruded polystyrene disk) with adhesive tape (PREZOTTI et al., 2002). A ring of gum was applied to the inside wall of the cylinder 4 cm from the lower end as a barrier to prevent the parasitoids from walking up the sides of the tube (SOARES et al., 2007; SOARES et al., 2012). A circle of rigid, transparent plastic of a larger diameter than the PVC was brushed with glue 24 hours prior to the beginning of the experiment and placed on top of the cylinder as a trap for flying parasitoids.

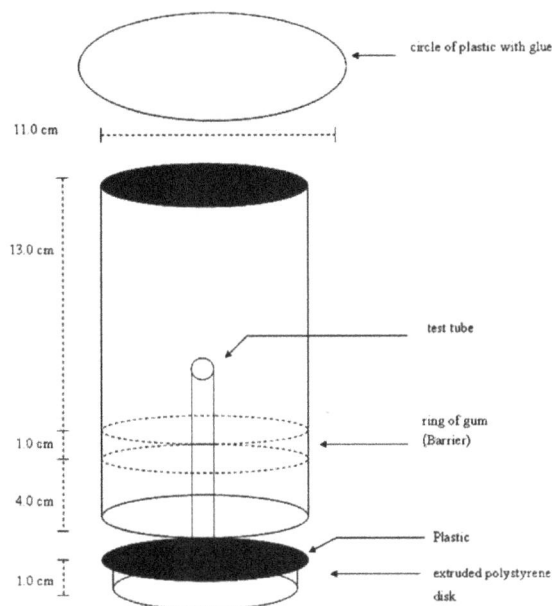

Figure 1. Schematic drawing of the test unit to assess *Trichogramma* flight capacity in the laboratory.

Four replicate experiments were conducted for each of the *Trichogramma* species at 25 ± 2°C with a 24-hours light period. The plots were randomly distributed on a counter under a continuous light source from the date of mounting because *Trichogramma* species are phototropic positive (SOARES et al., 2007; SOARES et al., 2012). The parasitoids were kept in the test unit for three days after the start of their emergence and subsequently frozen.

The numbers of parasitoids caught in the glue ring, called 'walkers', on the plastic circle, called 'flyers', and on the bottom, called 'non-flyers', were recorded by direct counting with a handheld magnifying glass to determine the percentage of each group in relation to the total adults emerged. The cards with parasitized *A. kuehniella* eggs were removed from the test tube after the death of the *Trichogramma* adults and assessed under a magnifying glass (40x) to determine the number of eggs that had been parasitized (SOARES et al., 2007; SOARES et al., 2012) by counting the number of blackened *A. kuehniella* eggs. The emergence rate was calculated by counting the number of eggs with *Trichogramma* emergence, observing the opening on the corium of the host *A. kuehniella* eggs.

The data were subjected to tests of the assumptions of the mathematical model (normality and homogeneity of variances). Subsequently, data were transformed to the arcsine of x, analysis of variance was performed, and the means were compared with the Scott-Knott test (SCOTT;

KNOTT, 1974) at 5% probability using the SAEG 9.1 statistics program (UFV) (GOMES, 1985). The data are presented as percentages.

Results and discussion

Trichogramma pretiosum showed a higher parasitism capacity than *T. marandobai*, followed by *T. demoraesi*, in the eggs of *A. kuehniella* (Figure 2). The parasitism rate can vary with physical and chemical barriers and by the type and characteristics of host eggs such as their size, hardness, scales and kairomones or differences in fertility among species of parasitoids (BESERRA; PARRA, 2004; SOARES et al., 2006; SOARES et al., 2009; SOARES et al., 2012). Temperature can also affect the parasitism capacity, with many *Trichogramma* species presenting maximum fertility between 21 and 30°C that decreases at extreme temperatures (DA FONSECA et al., 2005; PRASAD et al., 1999; ZAGO et al., 2007). Life tables for *T. pretiosum* and *Trichogramma atopovirilia* (Oatman; Platner) (Hymenoptera: Trichogrammatidae) reared with *Helicoverpa zea* (Boddie) (Lepidoptera: Noctuidae) eggs showed higher specific fertilities for the first species with all temperatures tested (NAVARRO; MARCANO, 2000). The ideal temperature to rear *T. pretiosum*, which had a higher number of females produced per female for this parasitoid (Σm x= 104.48), was 28°C. However, *T. atopovirilia* showed better development at 23°C (Σm x= 42.16) and a lower tolerance to high temperatures, with no development at 33°C (NAVARRO; MARCANO, 2000). The absence of food (carbohydrates) could also affect parasitism, as was reported for *T. pretiosum* (BAI et al., 1992) and *Trichogramma maxacalii* (Voegelé; Pointel) (Hymenoptera: Trichogrammatidae) (OLIVEIRA et al., 2003).

Trichogramma species may differ in biological characteristics, such as parasitism, as a result of being more aggressive or the adaptation of the species. Each species of parasitoid possesses an evolutionary history that shapes its particular adaptation route. Thus, great variability in biological characteristics can be observed among *Trichogramma* species due to their genotypic plasticity. Differences in the biological characteristics of strains of the same species are also observed due to phenotypic changes. When these changes occur, a genotype can produce different phenotypes, altering their physiology, morphology or development in response to changes in the environment (COLINET et al., 2007; PIGLIUCCI, 2005).

Trichogramma pretiosum and *T. demoraesi* showed a higher emergence of adults than *T. marandobai* (Figure 2). The emergence of these species is similar to that reported for *T. pretiosum* and *Trichogramma exiguum* (Pinter; Platner) (Hymenoptera: Trichogrammatidae), both with 86% emergence at 28°C in the eggs of *Plutella xylostella* (Linnaeus) (Lepidoptera: Plutellidae) (PEREIRA et al., 2004), indicating that emergence is usually high in this genus. *Trichogramma pretiosum* had rates of emergence of 100%, 100% and 99.8% at 25°C in the eggs of hosts *Anagasta kuehniella* (Zeller) (Lepidoptera: Pyralidae), *Spodoptera frugiperda* (J.E. Smith) (Lepidoptera: Noctuidae) and *P. xylostella*, respectively (VOLPE et al., 2006). The emergence rate of *Trichogramma* adults can also vary with the size and quality of the host egg, the number of parasitoids that develop per egg, the development period in the host eggs and the temperature (DOYON; BOIVIN, 2005; PRATISSOLI et al., 2005), but generally approaches 100%, as observed in this study.

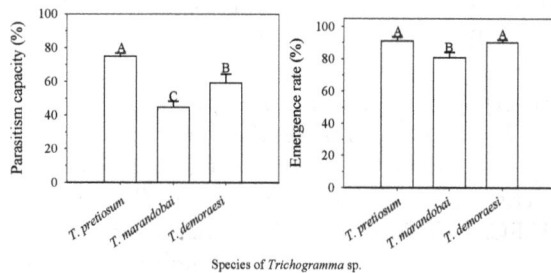

Figure 2. Mean standard +/- error of the percentage of eggs parasitized and of emergence rate of *Trichogramma pretiosum*, *T. marandobai* and *T. demoraesi* Municipality of Montes Claros, Minas Gerais State, Brasil. Bars followed by the same letter do not differ by the Scott-Knott test at 5% probability.

Trichogramma pretiosum showed a higher flight capacity than *T. demoraesi* and *T. marandobai* (Figure 3). For *Trichogramma acacioi* (Brun), *T. bruni* (Nagaraja), *T. demoraesi*, *T. maxacalii* and *T. soaresi* (Nagaraja) (Hymenoptera: Trichogrammatidae), 39%, 50%, 45%, 57% and 46%, respectively, were flying individuals when reared with *A. kuehniella* eggs (SOARES et al., 2007). Two strains of *T. pretiosum*, in different geographical regions, had a similar flight capacity (69% and 73%) but different parasitism capacity (78% and 50%) (SOARES et al., 2012). The geographical origin of the parasitoid may have a strong influence on the biological parameters of the strain because the expression of several characteristics changes with the environment in which the organism is raised. With *T. pretiosum*, reared on *A. kuehniella* eggs, 81.1 ± 4.20% of individuals showed flight capacity (PREZOTTI et al., 2002), while two *T. brassicae* lineages had 72% and 61% of individuals with flight capacity with the same host (DUTTON; BIGLER, 1995). Furthermore, the environmental temperature can affect *Trichogramma* flight capacity, as reported

for *Trichogramma sibericum* (Sorkina) (Hymenoptera: Trichogrammatidae) that presented variation in the proportion of adults, with the highest flight capacity (51.74 ± 2.3%) at 26°C and the lowest (1.57 ± 0.5%) at 16°C (PRASAD et al., 1999). Comparing the flight ability of various species of *Trichogramma* in the unit tests, it can be concluded that species with higher flight capacity are better able to disperse. The dispersal of *Trichogramma* species in the field is usually limited, but once in flight (flight capacity), the species can disperse more easily with the help of the wind. However, species with a high percentage of non-flying insects are less effective in the field because the parasitoid walking toward the bottom of the PVC is going against several intrinsic behavioral characteristics of species (negative geotropism and positive phototropism) and certainly has a reduced propensity for flight or develops deformed wings (PREZOTTI et al., 2002; SOARES et al., 2007).

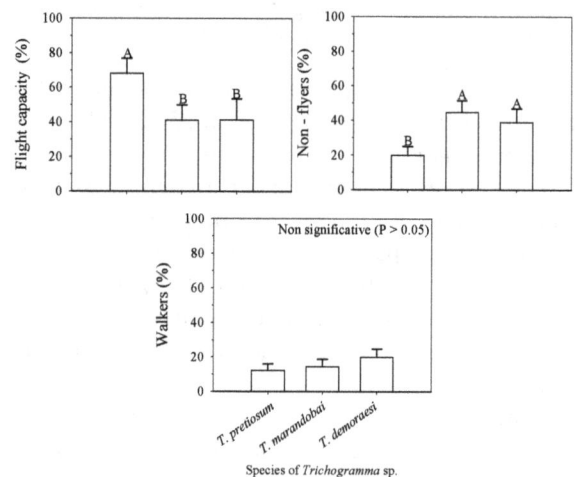

Figure 3. Mean standard +/- error of the percentage of flyers, not-flyers and walkers adults of *Trichogramma pretiosum*, *T. marandobai* and *T. demoraesi* Municipality of Montes Claros, Minas Gerais State, Brasil. Bars followed by the same letter do not differ by the Scott-Knott test at 5% probability.

The results indicate that *T. pretiosum* is the most efficient species for biological control of *E. ello*, with a high capacity for parasitism, emergence and flight in adults. These results corroborate other studies that have reported improved biological characteristics for this species of parasitoid (PARRA; ZUCCHI, 2004). Furthermore, *T. pretiosum* is widely distributed throughout South America and is often found in studies that assess natural occurrence (BARBOSA et al., 2011; PARRA; ZUCCHI, 2004), which confirms these observations. *Trichogramma pretiosum* may be more likely to increase its population in the field in a shorter period of time and still disperse more easily, making the species a better option for biological control programs of *E. ello* in cassava plantations.

Conclusion

Trichogramma pretiosum showed biological characteristics superior to the other species tested, being more efficient and suitable for mass rearing and release for biological control programs on cassava plantations.

Acknowledgements

To Dr. George W. G. de Morais for the donation of the *Anagasta kuehniella* and *Trichogramma* species, and laboratory technicians Aurélio Gomes dos Santos (*in memoriam*) and Cézar Guimarães. To the Brazilian agencies 'Conselho Nacional de Desenvolvimento Científico e Tecnológico (CNPq)' and 'Fundação de Amparo à Pesquisa do Estado de Minas Gerais (FAPEMIG)'.

References

BAI, B.; LUCK, R. F.; FORSTER, L.; STEPHENS, B.; JANSSEN, J. A. M. The effect of host size on quality attributes of the egg parasitoid, *Trichogramma pretiosum*. **Entomologia Experimentalis et Applicata**, v. 64, n. 1, p. 37-48, 1992.

BARBOSA, F. S.; LEITE, G. L. D.; ALVES, S. M.; NASCIMENTO, A. F.; DAVILA, V. A.; COSTA, C. A. Insecticide effects of *Ruta graveolens*, *Copaifera langsdorffii* and *Chenopodium ambrosioides* against pests and natural enemies in commercial tomato plantation. **Acta Scientiarum. Agronomy**, v. 33, n. 1, p. 37-43, 2011.

BARRIGOSSI, J. A. F.; ZIMMERMANN, F. J. P.; LIMA, P. S. C. Consumption rates and performance of *Erinnyis ello* L. on four cassava varieties. **Neotropical Entomology**, v. 31, n. 3, p. 429-433, 2002.

BELLOTTI, A. C.; SMITH, L.; LAPOINTE, L. S. Recent advances in cassava pest management. **Annual Review of Entomology**, v. 44, p. 343-370, 1999.

BESERRA, E. B.; PARRA, J. R. P. Biologia e parasitismo de *Trichogramma atopovirilia* Oatman and Platner e *Trichogramma pretiosum* Riley (Hymenoptera, Trichogrammatidae) em ovos de *Spodoptera frugiperda* (J.E. Smith) (Lepidoptera, Noctuidae). **Revista Brasileira de Entomologia**, v. 48, n. 1, p. 119-126, 2004.

BRUN, P. G.; MORAES, G. W. G.; SOARES, L. A. *Trichogramma marandobai* sp. n. (Hymenoptera: Trichogrammatidae) parasitóide de *Erinnyis ello* (Lepidoptera: Sphingidae) desfolhador da mandioca. **Pequisa Agropecuária Brasileira**, v. 21, n. 12, p. 1245-1248, 1986.

COLINET, H.; BOIVIN, G.; HANCE, T. Manipulation of parasitoid size using the temperature-size rule: fitness consequences. **Oecologia**, v. 152, n. 3, p. 425-433, 2007.

COURSEY, D. G.; HAYNES, P. H. Root crops and their potential as food in the tropics. **World Crops**, v. 22, n. 2, p. 261-265, 1970.

DA FONSECA, F. L.; KOVALESKI, A.; FORESTI, J.; RINGENBERG, R. Desenvolvimento e exigências térmicas de *Trichogramma pretiosum* Riley (Hymenoptera:

Trichogrammatidae) em ovos de *Bonagota cranaodes* (Meyrick) (Lepidoptera: Tortricidae). **Neotropical Entomology**, v. 34, n. 6, p. 945-949, 2005.

DOYON, J.; BOIVIN, G. The effect of development time on the fitness of female *Trichogramma evanescens*. **Journal of Insect Science**, v. 5, n. 4, p. 1-5, 2005.

DUTTON, A.; BIGLER, F. Flight activity assessment of the egg parasitoid *Trichogramma brassicae* (Hym.: Trichogrammatidae) in laboratory and field conditions. **Entomophaga**, v. 40, n. 1, p. 223-233, 1995.

FAO-Food and Agriculture Organization. **Statistical databases – Agriculture**. Rome: FAO. Available from: <http:www.//apps.fao.org>. Access on: 4 Jun. 2012.

GISLOTI, L.; PRADO, A. P. Cassava shoot infestation by larvae of *Neosilba perezi* (Romero and Ruppell) (Diptera: Lonchaeidae) in Sao Paulo state, Brazil. **Neotropical Entomology**, v. 40, n. 3, p. 312-315, 2011.

GOMES, J. M. **SAEG 3.0**: Sistema de análises estatísticas e genéticas. Viçosa: Imprensa Universitária, 1985.

LEITE, G. L. D.; PICANÇO, M.; JHAM, G. N.; GUSMÃO, M. R. Effects of leaf compounds, climate and natural enemies on the incidence of thrips in cassava. **Pesquisa Agropecuária Brasileira**, v. 37, n. 11, p. 1657-1662, 2002.

LEITE, G. L. D.; PICANÇO, M.; ZANUNCIO, J. C.; GUSMÃO, M. R. Natural factors affecting the whitefly infestation on cassava. **Acta Scientiarum. Agronomy**, v. 25, n. 2, p. 291-297, 2003.

LUNDGREN, J. G.; HEIMPEL, G. E. Quality assessment of three species of commercially produced *Trichogramma* and the first report of thelytoky in commercially produced *Trichogramma*. **Biological Control**, v. 26, n. 1, p. 68-73, 2003.

NAGARAJA, H. Descriptions of new *Trichogramma* (Hymenoptera: Trichogrammatidae). **Oriental Insects**, v. 7, p. 275-290, 1983.

NASSAR, N. M. A.; ORTIZ, R. Cassava improvement: challenges and impacts. **Journal of Agricultural Science**, v. 145, p. 163-171, 2007.

NAVARRO, R. V.; MARCANO, R. Tablas de vida de *Trichogramma pretiosum* Riley y *T. atopovirilia* Oatman y Platner en el laboratorio. **Agronomia Tropical**, v. 50, n. 1, p. 123-134, 2000.

NUWAMANYA, E.; CHIWONA-KARLTUN, L.; KAWUKI, R. S.; BAGUMA, Y. Bio-ethanol production from non-food parts of cassava (*Manihot esculenta* Crantz). **Ambio**, v. 41, n. 3, p. 262-270, 2012.

OERKE, E. C. Crop losses to pests. **Journal of Agricultural Science**, v. 144, n. 1, p. 31-43, 2006.

OLIVEIRA, H. N.; GOMEZ, S. A.; ROHDEN, V. S.; ARCE, C. C. M.; DUARTE, M. M. Record of *Trichogramma* (Hymenoptera: Trichogrammatidae) species on *Erinnyis ello* Linnaeus (Lepidoptera: Sphingidae) eggs in Mato Grosso do Sul State, Brazil. **Pesquisa Agropecuaria Tropical**, v. 40, n. 3, p. 378-379, 2010.

OLIVEIRA, H. N.; ZANUNCIO, J. C.; PRATISSOLI, D.; PICANÇO, M. C. Biological characteristics of *Trichogramma maxacalii* (Hymenoptera: Trichogrammatidae) on eggs of

Anagasta kuehniella (Lepidoptera: Pyralidae). **Brazilian Journal of Biology**, v. 63, n. 4, p. 647-653, 2003.

ÕZTEMIZ, S.; KORNOSOR, S. The effects of different irrigation systems on the inundative release of *Trichogramma evanescens* westwood (Hymenoptera: Trichogrammatidae) against *Ostrinia nubilalis* Hubner (Lepidoptera: Pyralidae) in the second crop maize. **Turkish Journal of Agriculture and Forestry**, v. 31, n. 1, p. 23-30, 2007.

PARRA, J. R. P.; ZUCCHI, R. A. *Trichogramma* in Brazil: feasibility of use after twenty years of research. **Neotropical Entomology**, v. 33, n. 3, p. 271-281, 2004.

PEREIRA, F. F.; BARROS, R.; PRATISSOLI, D.; PARRA, J. R. P. Biologia e exigências térmicas de *Trichogramma pretiosum* Riley e *T. exiguum* Pinto and Platner (Hymenoptera: Trichogrammatidae) criados em ovos de *Plutella xylostella* (L.) (Lepidoptera: Plutellidae). **Neotropical Entomology**, v. 33, n. 2, p. 231-236, 2004.

PIGLIUCCI, M. Evolution of phenotypic plasticity: where are we going now? **Trends in Ecology and Evolution**, v. 20, n. 9, p. 481-486, 2005.

PRASAD, R. P.; ROITBERG, B. D.; HENDERSON, D. The effect of rearing temperature on flight initiation of *Trichogramma sibericum* Sorkina at ambient temperatures. **Biological Control**, v. 16, n. 3, p. 291-298, 1999.

PRATISSOLI, D.; PARRA, J. R. P. Desenvolvimento e exigências térmicas de *Trichogramma pretiosum* Riley, criados em duas traças do tomateiro. **Pesquisa Agropecuária Brasileira**, v. 35, n. 7, p. 1281-1288, 2000.

PRATISSOLI, D.; ZANUNCIO, J. C.; VIANNA, U. R.; ANDRADE, J. S.; ZANOTTI, L. C. M.; SILVA, A. F. Biological characteristics of *Trichogramma pretiosum* and *Trichogramma acacioi* (Hym: Trichogrammatidae), parasitoids of the avocado defoliator *Nipteria panacea* (Lep.: Geometridae), on eggs of *Anagasta kuehniella* (Lep.: Pyralidae). **Brazilian Archives of Biology and Technology**, v. 48, n. 1, p. 7-13, 2005.

PRATISSOLI, D.; ZANUNCIO, J. C.; VIANNA, U. R.; ANDRADE, J. S.; ZINGER, F. D.; ALENCAR, J. R. C. C.; LEITE, G. L. D. Parasitism capacity of *Trichogramma pretiosum* and *Trichogramma acacioi* (Hym.: Trichogrammatidae) on eggs of *Sitotroga cerealella* (Lep.: Gelechiidae). **Brazilian Archives of Biology and Technology**, v. 51, n. 6, p. 1249-1254, 2008.

PREZOTTI, L.; PARRA, J. R. P.; VENCOVSKY, R.; DIAS, C. T. S.; CRUZ, I.; CHAGAS, M. C. M. Teste de vôo como critério de avaliação da qualidade de *Trichogramma pretiosum* Riley (Hymenoptera: Trichogrammatidae): Adaptação de metodologia. **Neotropical Entomology**, v. 31, n. 3, p. 411-417, 2002.

RONCHI-TELES, B.; QUERINO, R. B. Registro de *Trichogramma demoraesi* Nagaraja (Hymenoptera: Trichogrammatidae) parasitando ovos de *Erynnyis ello* (Lepidoptera: Sphingidae) na Amazônia Central.

Neotropical Entomology, v. 34, n. 3, p. 515, 2005.

SCOTT, A. J.; KNOTT, M. A. A cluster analysis method for grouping means in the analysis of variance. **Biometrics**, v. 30, n. 3, p. 507-512, 1974.

SOARES, M. A.; TORRES-GUTIERREZ, C.; ZANUNCIO, J. C.; BELLINI, L. L.; PREZOTTO, F.; SERRAO, J. E. *Pachysomoides* sp. (Hymenoptera: Ichneumonidae: Cryptinae) and *Megaselia scalaris* (Diptera: Phoridae) parasitoids of *Mischocyttarus cassununga* (Hymenoptera: Vespidae) in Viçosa, Minas Gerais state, Brazil. **Sociobiology**, v. 48, n. 3, p. 1-8, 2006.

SOARES, M. A.; LEITE, G. L. D.; ZANUNCIO, J. C.; ROCHA, S. L.; SÁ, V. G. M.; SERRÃO, J. E. Flight capacity, parasitism and emergence of five *Trichogramma* (Hymenoptera: Trichogrammatidae) species from forest areas in Brazil. **Phytoparasitica**, v. 35, n. 3, p. 314-318, 2007.

SOARES, M. A.; TORRES-GUTIERREZ, C.; ZANUNCIO, J. C.; PEDROSA, A. R. P.; LORENZON, A. S. Superparasitismo de *Palmistichus elaeisis* (Hymenoptera: Eulophidae) y comportamiento de defensa de dos hospederos. **Revista Colombiana de Entomología**, v. 35, n. 1, p. 62-65, 2009.

SOARES, M. A.; LEITE, G. L. D.; ZANUNCIO, J. C.; SÁ, V. G. M.; FERREIRA, C. S.; ROCHA, S. L.; PIRES, E. M.; SERRÃO, J. E. Quality Control of *Trichogramma atopovirilia* and *Trichogramma pretiosum* (Hym.: Trichogrammatidae) adults reared under laboratory conditions. **Brazilian Archives of Biology and Technology**, v. 55, n. 2, p. 305-311, 2012.

STOUTHAMER, R.; BREEUWER, J. A. J.; LUCK, R. F.; WERREN, J. H. Molecular identification of microorganisms associated with parthenogenesis. **Nature**, v. 361, n. 6407, p. 66-68, 1993.

TINDO, M.; HANNA, R.; GOERGEN, G.; ZAPFACK, L.; TATA-HANGY, K.; ATTEY, A. Host plants of *Stictococcus vayssierei* Richard (Stictococcidae) in non-crop vegetation in the Congo Basin and implications for developing scale management options. **International Journal of Pest Management**, v. 55, n. 4, p. 339-345, 2009.

VOLPE, H. X. L.; DE BORTOLI, S. A.; THULER, R. T.; VIANA, C. L. T. P.; GOULART, R. M. Avaliação de características biológicas de *Trichogramma pretiosum* Riley (Hymenoptera: Trichogrammatidae) criado em três hospedeiros. **Arquivos do Instituto Biológico**, v. 73, n. 3, p. 311-315, 2006.

ZAGO, H. B.; PRATISSOLI, D.; BARROS, R.; GONDIM JR., M. G. C.; SANTOS JR., H. J. G. Capacidade de parasitismo de *Trichogramma pratissolii* Querino and Zucchi (Hymenoptera: Trichogrammatidae) em hospedeiros alternativos, sob diferentes temperaturas. **Neotropical Entomology**, v. 36, n. 1, p. 48-89, 2007.

12

Mutation of Trp-574-Leu ALS gene confers resistance of radish biotypes to iodosulfuron and imazethapyr herbicides

Joanei Cechin[1*], Leandro Vargas[2], Dirceu Agostinetto[1], Fabiane Pinto Lamego[3], Franciele Mariani[4] and Taísa Dal Magro[5]

[1]Faculdade de Agronomia Eliseu Maciel, Universidade Federal de Pelotas, Campus Universitário s/n., Cx. Postal 354, 96010-900, Pelotas, Rio Grande do Sul, Brazil. [2]Empresa Brasileira de Pesquisa Agropecuária, Passo Fundo, Rio Grande do Sul, Brazil. [3]Empresa Brasileira de Pesquisa Agropecuária, Embrapa Pecuária Sul, Bagé, Rio Grande do Sul, Brazil. [4]Instituto Federal de Educação, Ciência e Tecnologia, Campus Sertão, Sertão, Rio Grande do Sul, Brazil. [5]Universidade de Caxias do Sul, Vacaria, Rio Grande do Sul, Brazil. *Author for correspondence. E-mail: joaneicechin@yahoo.com.br

ABSTRACT. Acetolactate synthase inhibitors are the main group of herbicides used in winter crops in Southern Brazil where their intensive use has selected for herbicide-resistant biotypes of radish. The resistance affects the efficacy of herbicides, and identifying the resistance mechanism involved is important for defining management strategies. The aim of this study was to elucidate the resistance mechanism of radish biotypes by quantifying the enzyme activity, ALS gene sequencing and evaluating the response of biotypes to iodosulfuron and imazethapyr herbicide application after treatment with a cytochrome P_{450} monooxygenase inhibitor. The susceptible (B_1) and resistant (B_4 and B_{13}) biotypes were from wheat fields in the Northwest of Rio Grande do Sul State. The results demonstrated that the enzyme affinity for the substrate (K_M) was not affected in biotypes B_4 and B_{13} but that the V_{max} of the resistant biotypes was higher than that of biotype B_1. The resistant biotypes showed no differential metabolic response to iodosulfuron and imazethapyr herbicides when inhibited by malathion and piperonyl butoxide. However, gene sequencing of ALS showed a mutation at position 574, with an amino acid substitution of tryptophan for leucine (Trp-574-Leu) in resistant biotypes.

Keywords: *Raphanus sativus*, mechanism of resistence, ALS enzyme activity, gene mutation, metabolism.

Introduction

Raphanus sativus L. (radish) is a dicotyledonous weed species that is found in wheat, barley and canola fields of southern Brazil, where it causes yield reduction (Rigoli, Agostinetto, Schaedler, Dal Magro, & Tironi, 2008). Acetolactate synthase (ALS) is the most common enzyme in the branched-chain amino acid biosynthetic pathway and produces leucine, isoleucine and valine (McCourt & Duggleby, 2006). ALS inhibitor herbicides are essential for a variety of crops due to their selectivity, low effective dosage, reduced toxicity to animals and high potential for inhibiting the ALS enzyme (Yu, Han, & Vila-Aiub, 2010; Endo, Shimizu, Fujimori,

Yanagisawa, & Toki, 2013). Iodosulfuron and imazethapyr are the main herbicides used in wheat and canola where their intensive use has favored the selection of herbicide-resistant radish biotypes (Pandolfo, Presotto, Poverene, & Cantamutto, 2013).

The survival of biotypes can occur due to factors that may be related to the herbicide target site or non-target-site (Yuan, Tranel, & Stewart, 2007). The occurrence of DNA mutations in the gene sequence and the overexpression of the ALS enzyme are possible factors resulting in reduced sensitivity to the herbicide due to insufficient or excessive levels of biosynthetic product (Duggleby, McCourt, & Guddat, 2008; Han et al., 2012). Mutations in the ALS gene can affect enzyme structure and function, thereby reducing the enzyme activity and herbicide affinity with the target site (Han et al., 2012). In *Raphanus raphanistrum* L., ALS gene mutations affecting the efficacy of ALS inhibitor herbicides were identified for proline (Pro_{197}), aspartate (Asp_{376}), tryptophan (Trp_{574}) and alanine (Ala_{122}); (Tan & Medd, 2002; Yu, Hashem, Walsh, & Powles, 2003; Yu, Han, Purba, Walsh, & Powles, 2012; Han et al., 2012).

Non-target-site herbicide resistance mechanisms can occur due to increased metabolism and compartmentalization, a reduction in absorption or the differential translocation of the herbicide molecule (Powles & Yu, 2010; Délye, Jasieniuk, & Le Corre, 2013). These mechanisms of resistance are also characterized by higher rates of herbicide detoxification due to increased glutathione-s-transferase, cytochrome P_{450} monooxygenase or glycosyltransferase activities (Délye et al., 2013). Nevertheless, both mechanisms affect herbicide efficacy and should be evaluated due to the possibility of their coexistence (Ahmad-Hamdani et al., 2013, Brosnan et al., 2016).

Therefore, elucidating the resistance mechanism in radish biotypes is important for determining alternative weed management strategies and for reducing the herbicide-resistance selective pressure. The aim of this study was to elucidate the resistance mechanism of radish biotypes by quantifying ALS enzyme activity, ALS gene sequencing and evaluating the response of biotypes to iodosulfuron and imazethapyr herbicide application after treatment with a cytochrome P_{450} monooxygenase inhibitor.

Material and methods

Seeds from radish plants that survived application of iodosulfuron herbicide were collected in wheat fields in the Northwest region of Rio Grande do Sul State. To screen for resistant biotypes, plants grown from these seeds were sprayed with 5 g a.i. ha^{-1} of iodosulfuron and 106 g. i.a. ha^{-1} of imazethapyr; and dose response studies were then carried out for biotypes B_1, B_4 and B_{13}. The B_4 and B_{13} biotypes were from Três de Maio and Boa Vista do Cadeado, Rio Grande do Sul State municipalities, respectively, and demonstrated cross resistance to these herbicides and a high level of resistance to the iodosulfuron herbicide. Biotype B_1, from Três de Maio, Rio Grande do Sul State, was susceptible to the iodosulfuron and imazethapyr herbicides (Cechin et al., 2016).

ALS enzyme activity and in vitro assays with the herbicides

The enzymatic extraction method was adapted from methods described by Singh, Stidham, and Shaner (1988). Seven grams of young plant leaves was collected, frozen in liquid nitrogen (N_2) and ground to a fine powder. Then, 70 mL (1:10 p/v) of 100 mM phosphate extraction buffer (pH = 7.5) containing 0.5 mM magnesium chloride ($MgCl_2$), 10 mM sodium pyruvate, 0.5 mM thiamine pyrophosphate (TPP), 10 μM flavin adenine dinucleotide (FAD), 10% glycerol, 1 mM dithiothreitol and 5% polyvinylpolypyrrolidone (PVPP) was added. The material was homogenized for 20 minutes at 4°C, and the mixture was filtered to remove solid sediments. The liquid portion was centrifuged at 12,000 rpm for 20 minutes, the supernatant was collected and the solid residue was discarded.

The methodology used for the *in vitro* biossay with the herbicide was adapted from the method of Gerwick (Gerwick, Mireles, & Eilers, 1993). The assay was performed in test tubes using three replicates with a factorial treatment design, where factor A was the different biotypes (B_1, B_4 and B_{13}) and factor B consisted of different concentrations of iodosulfuron or imazethapyr (zero, 0.001, 0.01, 0.1, 1.0, 10, 100, and 1,000 μM). Each tube received 600 μL of enzyme solution, 100 μL of herbicide solution and 300 μL of 80 μM phosphate reaction buffer (pH = 7.0) containing 20 mM de magnesium chloride, 200 mM sodium piruvate, 2 mM thiamine pyrophosphate and 20 μM flavin adenine dinucleotide. The assay had two standard treatments without herbicide to measure zero and 100% enzyme activity. The zero activity standard received 50 μL sulfuric acid (H_2SO_4 - 3 M) at the start of assay to prevent enzyme activity, and the 100% activity standard received 100 μL milli-Q water instead of the herbicide solution. The absorbance values for the zero activity control were subtracted from the values read in other treatments. After preparation of the reaction, samples were incubated for 60 minutes at 30°C. The reactions were stopped with 50 μL of 3 M sulfuric acid for all treatments other than the zero activity control treatment, where the reaction had been stopped initially. Next, the tubes were incubated for 15 minutes at 60°C to create acetoin from the reaction of sulfuric acid with acetolactate. Then, 0.5% creatin (1,000 μL) and 0.5% 1-naphtol (1,000 μL) were prepared in 2.5 M sodium hydroxide (NaOH) and added to produce a colored complex. The final reaction was incubated for 15 minutes at 60°C, and the absorbance at 530 nm was read in a spectrophotometer. The ALS enzyme activity (μM acetoin min.$^{-1}$ mL^{-1}) was determined by the amount of acetoin produced. For the standard curve, three repetitions were used; each tube receveid 1000 μL with different concentrations of solution acetoin (0, 10, 20, 40, 60, 80, 100, 200, and 400 μM). The colored complex was obtained by adding creatine, 1-naphtol and NaOH and incubated as described above.

The kinetic parameters of enzyme activity (K_M and V_{max}) were obtained using ten pyruvate concentrations (zero, 0.5, 1, 2, 4, 8, 16, 32, 64, and 100 mM). The substrate concentrations (pyruvate) were obtained by diluting phosphate reaction buffer (80 μM, pH = 7.0) with 100 mM pyruvate solution. The buffer contained 20 mM magnesium chloride, 2 mM thiamine pyrophosphate and 20 μM flavin adenine dinucleotide. The values of K_M and V_{max} were determined using the Michaelis-Menten equation: $y = V_{max} \star X / K_M + X$ (Nelson & Cox, 2008), where y = ALS enzyme activity (μmol min.$^{-1}$ mL^{-1}); V_{max} = maximum reaction velocity; X = substrate concentration (pyruvate); and K_M = substrate concentration, where the initial velocity is equal to half of the maximum reaction velocity. The data obtained were analyzed for normality (Shapiro-Wilk test) and submitted to analysis of variance (p ≤ 0.05), where the K_M and V_{max} values of biotypes were compared by a Duncan Test (p ≤ 0.05).

The absorbance values were corrected with the zero standard, and I_{50} (amount of herbicide to inhibit 50% enzyme activity) was calculated using the logistic non-linear regression model: $y = a / [1 + (x / x_{I50})]^b$ (Seefeldt, Jensen, & Fuerst, 1995), where y = ALS enzyme activity (%); a = maximum point; x = dose of iodosulfuron or imazethapyr (μM); x_{I50} = dose of iodosulfuron or imazethapyr that corresponds to a 50% inhibition of the ALS enzyme; and b = curve declivity. The resistance factor (RF) was calculated dividing the I_{50} of the resistant biotype by values from the susceptible biotype (Hall, Stromme, & Horsman, 1998). The

level of total protein was obtained using the Bradford method (Bradford, 1976).

ALS gene sequencing

RNA was extracted from leaf tissue (100 mg) in biotypes (B_1, B_4, and B_{13}) using 500 μL of extraction buffer (reagent Kit PureLink™ Plant RNA) according to the manufacturer's instructions. The quality and quantity of RNA were verified by electrophoresis gel and spectrophotometry, respectively. cDNA was synthesized from 2 μg RNA using the SuperScript™ III First-Strand Synthesis System for RT-PCR (Invitrogen™ - USA) according to the manufacturer's instructions. Polymerase chain reaction (PCR) products were obtained using the *primers* WR122F, WR653R, WR205R, WR376R, and WR574F (Han et al., 2012) and the ALS gene sequence of *R. raphanistrum* L. (AJ344986) that was deposited in GenBank of National Center for Biotechnology Information (NCBI).

PCR was conducted in a final volume of 25 μL containing 1.25 μL cDNA, 0.25 μM each of the *forward* (F) and *reverse* (R) *primers*, 12.5 μL 2x GoTaq™ Green Master Mix (Promega™) and nuclease-free water. cDNA denaturation was conducted at 94°C for 4 minutes, with 40 cycles of 94°C for 30 seconds. The annealing of *primers* occurred at 55°C for 30 seconds, and sequence extension was performed at 72°C for 30 to 120 seconds (Han et al., 2012). *Amplicons* were purified using the PCR Purification Combo Kit (Invitrogen™ - USA) and sequenced using the ABI-PRISM 3100 Genetic Analyzer (Applied Biosystems). Sequences from the various biotypes were aligned using the Bioedit version 7.2.5 software and compared with the sequence of *R. raphanistrum* L. (AJ344986) deposited at GenBank (www.ncbi.nlm.nih.gov/genbank).

Metabolization

The metabolization of radish biotypes was analyzed in two additional experiments that were performed in a greenhouse with a completely randomized design and four replications. Seeds were sown in plastic trays, and two days after emergence, each seedling was transplanted to plastic pots with volume capacities of 0.75 L that contained soil and PlantMax substrate at a 2:1 ratio.

The treatments were arranged in a factorial design, where factor A was the radish biotypes (B_1, B_4 and B_{13}) and factor B were the cytochrome P_{450} monooxygenase inhibitors malathion and piperonyl butoxide. Spraying occurred when plants were at the three to four leaf stage by using a CO_2 backpack sprayer calibrated to deliver 120 L ha^{-1}. The metabolism inhibitors were sprayed 30 minutes before herbicide application at a dose of 500 g a.i. ha^{-1} for malathion and 525 g a.i. ha^{-1} for piperonyl butoxide (PBO). The doses of iodosulfuron and imazethapyr herbicides sprayed were 3.5 and 106 g a.i. ha^{-1}, respectively.

The control and shoot dry matter (SDM) were evaluated at 28 days after application (DAA). A percentage scale for control was adopted, in which zero (0) and one hundred (100) corresponded to the absence of damage and complete death of the plants, respectively (Frans & Crowley, 1986). The SDM was determined by drying the vegetable material in kiln with circulated forced air at 60°C for 72 hours and expressed as grams per plant.

The obtained data were analyzed for normality (Shapiro-Wilk test), which did not require data transformation, and were then submitted to analysis of variance ($p \leq 0.05$). When statistical significance was observed, the data were submitted to the Duncan test ($p \leq 0.05$).

Results and discussion

ALS enzyme activity and in vitro assay with the herbicides

The K_M values (pyruvate) of resistant biotypes (B_4 and B_{13}) did not show statistically significant differences in enzyme affinity for the substrate, and their V_{max} was higher than that of the the susceptible biotype (Figure 1). However, the K_M values of the B_4 and B_{13} biotypes were 20 and 25% higher, respectively, than that of the susceptible biotype (Table 1). Further, the V_{max} values were 7.78, 5.73, and 1.87 μM acetoin mg^{-1} protein h^{-1} for B_4, B_{13} and B_1 biotypes, respectively (Table 1). Other studies have demonstrated that biotypes that are resistant to ALS inhibitor herbicides do not present changes in the kinetic parameters (K_M and V_{max}) compared to susceptible biotypes (Ashigh & Tardif, 2007; Dal Magro et al., 2010), which suggests that resistance does not affect pyruvate binding. Similar results were found in *Cyperus difformis* L. biotypes resistant to pirazosulfuron-ethyl herbicide, in which the affinity of enzyme was not affected and the V_{max} was 52% greater (Dal Magro et al., 2010).

The ALS inhibitor herbicides act by blocking the channel that leads to the active site of the enzyme; therefore, any changes in this channel can impede the binding of the herbicide while maintaining the conformation and function of the enzyme (McCourt, Panf, King-Scott, Guddat, & Duggleby, 2006). In *Lolium rigidum* (Gaudin) biotypes resistant to sulfometuron and imazapyr herbicides, a mutation of tryptophan to leucine in position 574

(Trp$_{574}$-Leu) did not change the K_M, and V_{max} was 287% greater in resistant biotypes (Yu et al., 2010). The results obtained for in vitro assays with herbicide treatment demonstrated that 0.043 μM iodosulfuron and 3.2 μM imazethapyr inhibited 50% of the ALS enzyme activity (I_{50}) in the susceptible biotype. For the resistant biotypes, the I_{50} was 0.65 and 0.82 μM for iodosulfuron and 718 and 425 μM for imazethapyr in B_4 and B_{13} biotypes, respectively (Figure 2, Table 2).

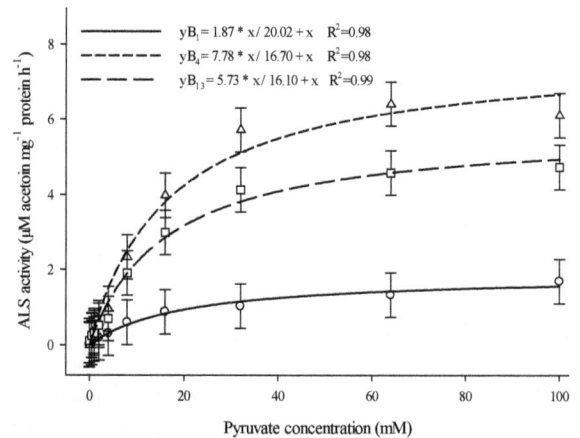

Figure 1. ALS activity (μM acetoin mg^{-1} protein h^{-1}) of susceptible (B_1) and resistant (B_4 and B_{13}) radish biotypes to iodosulfuron and imazethapyr herbicides subjected to differential pyruvate concentrations (mM). Points represent the mean values and bars represent least significant difference ($p < 0.05$).

Table 1. Kinetic parameters K_M (mM) and V_{max} (μM acetoin mg^{-1} protein h^{-1}) of susceptible (B_1) and resistant (B_4 and B_{13}) radish biotypes in response to iodosulfuron and imazethapyr herbicide treament.

Biotypes	K_M (mM)	V_{max} (μmol mg^{-1} protein h^{-1})
B_1 (Susceptible)	20.02^{ns}	1.87 C
B_4 (Resistant)	16.70	7.78 A
B_{13} (Resistant)	16.10	5.73 B
V.C. (%)		6.44

*means followed by the same uppercase letter (column) do not differ by Duncan's test ($p \leq 0.05$). ^{ns} = not significant ($p > 0.05$).

These values of I_{50} for biotypes B_4 and B_{13} resulted in a resistance factor (RF) of 15 and 19 to iodosulfuron herbicide and a RF of 224 and 133 to imazethapyr herbicide, respectively (Figure 2, Table 2). High enzyme inhibition was observed in *R. raphanistrum* L. where the I_{50} value in resistant and susceptible biotypes to clorosulfuron herbicide was 1.55 and 0.009 μM, respectively (Yu et al., 2012). The results demonstrated that B_4 and B_{13} biotypes present high levels of resistance to the herbicide imazethpayr (Table 2). Similar results were reported for radish biotypes resistant to ALS inhibitor herbicides, for which the I_{50} was higher for imazethapyr than for flumetsulam and metsulfuron-methyl herbicides (Yu et al., 2012; Pandolfo et al.,

2016). Different levels of resistance to herbicides may be related to the position of the mutation sites in relation to the herbicide-coupling site (Yu et al., 2012; Pandolfo et al., 2016).

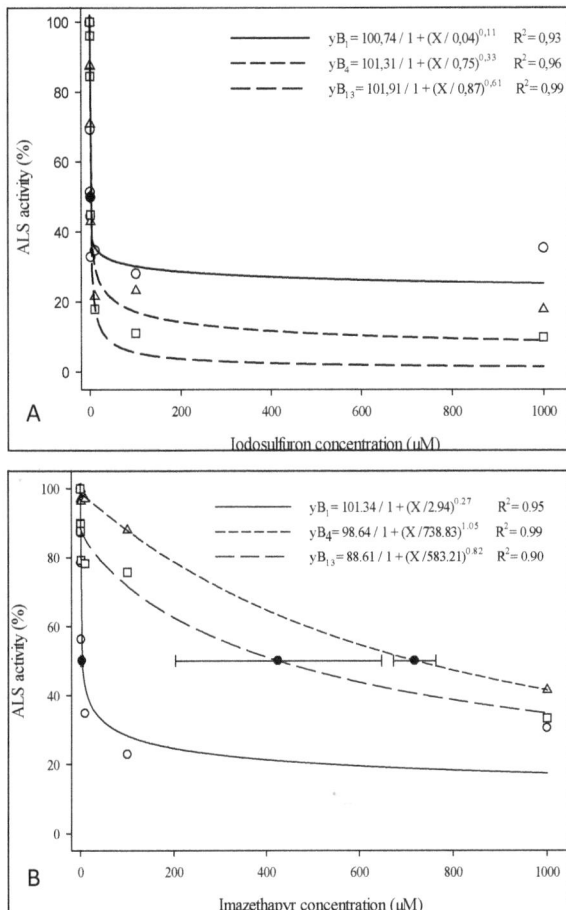

Figure 2. In vitro inhibition of ALS activity (%) of susceptible (B_1) and resistant (B_4 and B_{13}) radish biotypes subjected to different concentrations of iodosulfuron (A) and imazethapyr (B) herbicides (µM). Points represent the mean values of repetitions in each biotype.

Table 2. I_{50} values with confidance intervals (95% CI) and resistance factors (RF) of susceptible (B_1) and resistant (B_4 and B_{13}) radish biotypes subjected to different concentrations of iodosulfuron and imazethapyr herbicides.

Biotypes	I_{50}[1]		Resistance fator[2]
	µM	95 CI	(FR)
Iodosulfuron (B_1)	0.043	0.09 - (-)0.004	-
Iodosulfuron (B_4)	0.65	1.05 – 0.25	15
Iodosulfuron (B_{13})	0.82	1.00 – 0.64	19
Imazethapyr (B_1)	3.2	5.8 – 0.6	-
Imazethapyr (B_4)	718	763.04 – 672.96	224
Imazethapyr (B_{13})	425	646.94 – 203.06	133

[1]I_{50} = dose required to inibit 50% of ALS enzyme activity.
[2]RF obtained by division I_{50} of the resistant biotype by I_{50} of the susceptible biotype.

ALS gene sequencing

A 1758 bp fragment of the ALS gene with five conserved regions was sequenced from the cDNA of susceptible (B_1) and resistant (B_4 and B_{13}) radish

biotypes. This region includes all domains with mutation points that were previously identified in biotypes of *R. raphanistrum* L. resistant to ALS inhibitor herbicides (Tan & Medd, 2002; Yu et al., 2003; 2012; Han et al., 2012). The partial sequence presented a single nucleotide change of TGG to TTG, which led to a Trp-574-Leu substitution in resistant biotypes (Figure 3).

This mutation was identified in different weeds resistant to ALS inhibitor herbicides, including a recent discovery in biotypes of *R. sativus* L. (Pandolfo et al., 2016). Mutation of the ALS gene can compromise the herbicide coupling to the target site of the enzyme and affect the weed control with ALS inhibitor herbicides (McCourt et al., 2006). The Trp-574-Leu substitution is considered the most relevant mutation because it confers resistance to all five chemical groups of the ALS inhibitor herbicides (Pandolfo et al., 2016). However, this mutation has not been reported in resistant biotypes of *R. sativus* L. in Brazil; this can be considered the first identified case. In biotypes of *R. raphanistrum* L., the most frequent mutations involve a proline at amino acid position 197 (Pro_{197}), which confers resistance to the chemical groups of sulfonylureas and triazolopyrimidines (Tan & Medd, 2002; Yu et al., 2003). However, the levels of cross-resistance will depend on the position where the mutation occurred and the modified amino acid (Yu et al., 2012; Han et al., 2012).

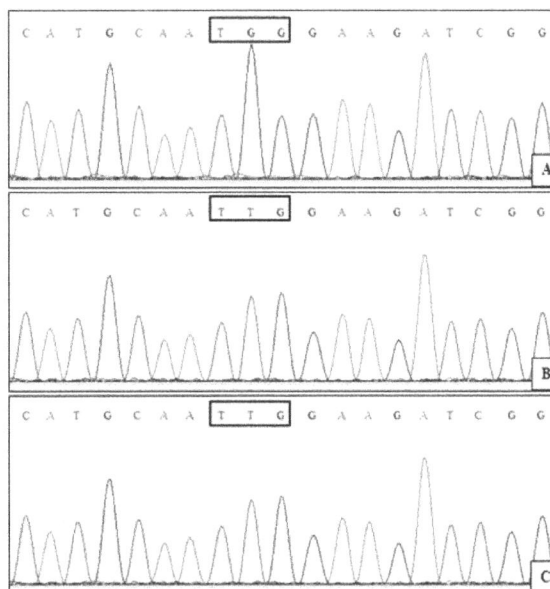

Figure 3. Comparison of partial ALS gene sequence of *Raphanus sativus* L. where the TGG codon for Trp-574 is the susceptible biotype (A) and the TTG codon for Leu-574 is present in biotypes resistant to iodosulfuron and imazethapyr herbicides (B and C).

Furthermore, all biotypes presented silent mutations at the 122 position that contains an alanine (Ala$_{122}$), where the codon GCC was substituted for GCT (data not shown) without resulting in a modification to the amino acid sequence. Amino acid substitutions at Ala-122 are more unlikely due to the necessity of two nucleotide alterations to modify the amino acid (Yu et al., 2012). However, in *R. raphanistrum* biotypes, the substitution of Alanine 122 to Tyrosine (Ala-122-Tyr) conferred resistance to three chemical groups of ALS inhibitors herbicides (Han et al., 2012).

The mutation diagnosed here (Trp-574-Leu) is one of the most important because it confers resistance to all chemical groups of ALS inhibitor herbicides, thus reducing the control options for resistant biotypes (Pandolfo et al., 2016; Cechin et al., 2016).

Metabolization

The results demonstrated that the application of cytochrome P$_{450}$ monooxygenase inhibitors followed by the application of iodosulfuron or imazethapyr herbicides did not efficiently control the resistant biotypes (Table 3). The isolated spraying of cytochrome P$_{450}$ monooxygenase inhibitors did not cause phytotoxicity in any of the radish biotypes (data not showed).

Table 3. Control (%) in susceptible (B$_1$) and resistant (B$_4$ and B$_{13}$) radish biotypes subjected to the application of iodosulfuron or imazethapyr herbicides alone or preceded by cytochrome P$_{450}$ monooxygenase inhibitors at 28 days after application (DAA).

Treatment	28 DAA B$_1$		28 DAA B$_4$		28 DAA B$_{13}$	
Iodosulfuron	99	aA	5.0	bA	0.0	bB
Malathion + iodosulfuron	100	aA	4.0	cA	21	bA
PBO + iodosulfuron	99	aA	2.0	bA	4.0	bB
Control	0.0	ns B	0.0	A	0.0	B
V.C. (%)	18.00					
Imazethapyr	97	aA	1.0	bA	1.0	bA
Malathion + imazethapyr	97	aA	0.0	bA	0.0	bA
PBO + imazethapyr	98	aA	2.0	bA	1.0	bA
Control	0.0	ns B	0.0	A	0.0	A
V.C. (%)	6.95					

*means followed by the same lowercase letter (line) and the same uppercase letter (column) do not differ by Duncan's test (p ≤ 0.05). ns = not significant (p > 0.05).

At 28 DAA, spraying of iodosulfuron and imazethapyr herbicides (isolated or combined with P$_{450}$ inhibitors) did not result in any significant differences in control for B$_1$, B$_4$, and B$_{13}$ biotypes (Table 3). However, for B$_{13}$ biotype, the application of iodosulfuron herbicide after the use of malathion as a P$_{450}$ inhibitor resulted in 21% control, in contrast to the other treatments (Table 3). The use of the cytochrome P$_{450}$ inhibitor in *Lolium rigidum* (Gaudin) biotypes before clorosulfuron herbicide showed a synergistic effect compared to the isolated

application of the herbicide (Yu & Powles, 2014). However, spraying imazethapyr herbicide either alone or in combination with PBO or malathion resulted in less than 2% control, suggesting that there is no differential metabolism by cytochrome P$_{450}$ monooxygenase in these biotypes.

The results for SDM did not show significant differences in dry matter accumulation for biotypes B$_1$, B$_4$, and B$_{13}$ when imazethapyr herbicide was sprayed (Table 4). Nevertheless, the use of malathion or PBO in the resitant biotype B$_{13}$ resulted in 66% and 48% reductions in SDM, respectively, compared with the isolated spraying of iodosulfuron (Table 4). Similar results were found in *Echinochloa phyllopogon* (Stapf) biotypes resistant to penoxsulan herbicide, where the use of malathion before herbicide application reduced 60% of the SDM content at a dose of 10 g a.i ha^{-1} of penoxsulam (Yasuor et al., 2009).

Table 4. Shoot dry matter (SDM) in susceptible (B$_1$) and resistant (B$_4$ and B$_{13}$) radish biotypes subjected to the application of iodosulfuron or imazethapyr herbicides alone or preceded by cytochrome P$_{450}$ monooxygenase inhibitors at 28 days after application (DAA).

Treatment	SDM B$_1$		SDM B$_4$		SDM B$_{13}$	
Iodosulfuron	0.43	bB	1.74	aAB	1.84	aA
Malathion + iodosulfuron	0.33	bB	1.58	aB	1.11	aB
PBO + iodosulfuron	0.25	bB	1.45	aB	1.24	aB
Control	1.88	ns A	2.01	A	1.87	A
V.C. (%)	20.18					
Imazethapyr	0.26	bB	3.55	bA	4.05	aA
Malathion + imazethapyr	0.15	bB	4.09	aA	4.33	aA
PBO + imazethapyr	0.21	cB	3.36	bA	4.31	aA
Control	3.52	ns A	3.51	A	3.89	A
V.C. (%)	11.36					

*means followed by the same lowercase letter (line) and the same uppercase letter (column) do not differ by Duncan's test (p ≤ 0.05). ns = not significant (p > 0.05).

However, in an experiment with ALS herbicide-resistant biotypes of *Poa annua* L., an increase in control by bispyribac-sodium herbicide was observed when malathion was used as an inhibitor of P$_{450}$ monooxygenase. However, these results were not observed for other ALS enzyme inhibitors, indicating that the mechanism of resistance involved cannot be attributed to differential metabolism by P$_{450}$ monooxygenase (Brosnan et al., 2016).

The differences found among the biotypes for control and SDM when malathion or PBO inhibitors where used indicate cytochrome P$_{450}$ may be involved in metabolic resistance, which may be specific to a given herbicide.

Therefore, the results of control and SDM observed in B$_4$ and B$_{13}$ biotypes did not indicate the involvement of cytochrome P$_{450}$ enzyme in resistance of radish to iodosulfuron and imazethapyr herbicides.

Conclusion

The kinetic parameters of ALS enzyme were not affected in resistant biotypes. A mutation was detected at position 574 of the ALS gene, which resulted in a substitution of tryptophan to leucine (Trp-574-Leu) in resistant radish biotypes. The results demonstrated that there was no differential metabolism of iodosulfuron and imazethapyr herbicides in resistant biotypes when cytochrome P_{450} was inhibited by malathion and piperonyl butoxide.

Acknowledgements

We thank the coordination of the Higher Education Personnel Training (CAPES) for the scholarship to the first author and also the Embrapa/Monsanto Partnership.

References

Ahmad-Hamdani, M. S., Yu, Q., Han, H., Cawthray, G. R., Wang, S. F., & Powles, S. B. (2013). Herbicide resistance endowed by enhanced rates of herbicide metabolism in wild oat (*Avena* spp.). *Weed Science*, 61(1), 55-62.

Ashigh, J., & Tardif, F. (2007). An Ala$_{205}$Val substitution in acetohydroxyacid synthase of Eastern black nightshade (*Solanum ptychanthum*) reduces sensitivity to herbicides and feedback inhibition. *Weed Science*, 55(6), 558-565.

Bradford, M. M. (1976). A rapid and sensitive method for the quantitation of microgram quantities of protein utilizing the principle of protein-dye binding. *Analytical Biochemistry*, 72(5), 248-254.

Brosnan, J. T., Vargas, J. J., Breeden, G. K., Grier, L., Aponte, R. A., Tresch, S., & Laforest, M. (2016). A new amino acid substitution (Ala-205-Phe) in acetolactate synthase (ALS) confers broad spectrum resistance to ALS-inhibiting herbicides. *Planta*, 243(1), 149-159.

Cechin, J., Vargas, L., Agostinetto, D., Zimmer, V., Pertile, M., & Garcia, J. R. (2016). Resistence of radish biotypes to iodosulfuron and alternative control. *Planta Daninha*, 34(1), 151-160.

Dal Magro, T., Rezende, S. T., Agostinetto, D., Vargas, L., Silva, A. A., & Falkoski, D. L. (2010). Propriedades enzimáticas da enzima ALS de *Cyperus difformis* e mecanismo de resistência da espécie ao herbicida pyrazosulfuron-ethyl. *Ciência Rural*, 40(12), 2439-2445.

Délye, C., Jasieniuk, M., & Le Corre, V. (2013). Deciphering the evolution of herbicide resistance in weeds. *Trends in Genetics*, 29(11), 1-10.

Duggleby, R. G., McCourt, J. A., & Guddat, L. W. (2008). Structure and mechanism of inhibition of plant acetohydroxyacid synthase. *Plant Physiology and Biochemistry*, 46(3), 309-324.

Endo, M., Shimizu, T., Fujimori, T., Yanagisawa, S., & Toki, S. (2013). Herbicide-resistant mutations in acetolactate synthase can reduce feedback inhibition and lead to accumulation of branched-chain amino acids. *Food and Nutrition Sciences*, 4(5), 522-528.

Frans, R., & Crowley, H. (1986). Experimental design and techniques for measuring and analyzing plant responses to weed control practices. In N. D. Camper (Ed.), *Research methods in weed science*. (3th ed.). Champaign, IL: Southern Weed Science Society.

Gerwick, B. C., Mireles, L. C., & Eilers, R. J. (1993). Rapid diagnosis of ALS/AHAS inhibitor herbicide resistant weeds. *Weed Technology*, 7(2), 519-524.

Hall, L. M., Stromme, K. M., & Horsman, G. P. (1998). Resistance to acetolactato sintase inhibithors and quinclorac in biotypes of false clover (*Gallium sourium*). *Weed Science*, 46(3), 390-396.

Han, H., Yu, Q., Purba, E., Walsh, M., Friesen, S., & Powles, S. B. (2012). A novel amino acid substitution Ala-122-Tyr in ALS confers high-level and broad resistance across ALS-inhibiting herbicides. *Pest Management Science*, 68(3), 1164-1170.

McCourt, J. A., & Duggleby, R. G. (2006). Acetohydroxyacid synthase and its role in the biosynthetic pathway for branched-chain amino acids. *Amino Acids*, 31(2), 173–210.

McCourt, J. A., Panf. S. S., King-Scott, J., Guddat, L. W., & Duggleby, R. G. (2006). Herbicide-binding sites revealed in the structure of plant acetohydroxyacid synthase. *Proceedings of the National Academy of Sciences*, 103(3), 569-573.

Nelson, D. L., & Cox, M. M. (2008). Enzymes. In *Lehninger principles of biochemistry*. (5th ed., p. 194-204). New York, US: Freeman.

Pandolfo, C. E., Presotto, A., Moreno, F., Dossou, I., Migasso, J. P., Sakima, E., & Cantamutto, M. (2016). Broad resistance to acetohydroxyacid-synthase-inhibiting herbicides in feral radish (*Raphanus sativus* L.) populations from Argentina. *Pest Management Science*, 72(2), 354-361.

Pandolfo, C. E., Presotto, A., Poverene, M., & Cantamutto, M. (2013). Limited occurrence of resistant radish (*Raphanus sativus*) to ahas-inhibiting herbicides in Argentina. *Planta Daninha*, 31(3), 657-666.

Powles, S. B., & Yu, Q. (2010). Evolution in action: Plants resistant to herbicides. *Annual Review of Plant Biology*, 61(4), 317-347.

Rigoli, R. P., Agostinetto, D., Schaedler, C. E., Dal Magro, T., & Tironi, S. P. (2008). Habilidade competitiva relativa do trigo (*Triticum aestivum*) em convivência com azevém (*Lolium multiflorum*) ou nabo (*Raphanus raphanistrum*). *Planta Daninha*, 26(1), 93-100.

Seefeldt, S. S., Jensen, J. E., & Fuerst, E. P. (1995). Log-logistic analysis of herbicide dose-response relationships. *Weed Technology*, 9(2), 218-227.

Singh, B. K., Stidham, M. A., & Shaner, D. L. (1988). Assay of acetohydroxyacid synthase. *Analytical Biochemistry*, 171(1), 173-179.

Tan, M. K., & Medd, R. W. (2002). Characterisation of the acetolactate synthase (ALS) gene of *Raphanus raphanistrum* L. and the molecular assay of mutations associated with herbicide resistance. *Plant Science*, *163*(2), 195-205.

Yasuor, H., Osuna, M. D., Ortiz, A., Saldaín, N. E., Eckert, J. W., & Fischer, A. J. (2009). Mechanism of resistance to penoxsulam in late watergrass (*Echinochloa phyllopogon* (Stapf) Koss). *Journal of Agricultural and Food Chemistry*, *57*(9), 3653-3660.

Yu, Q., & Powles, S. B. (2014). Metabolism-based herbicide resistance and cross-resistance in crop weeds: A threat to herbicide sustainability and global crop production. *Plant Physiology*, *166*(3), 1106-1118.

Yu, Q., Han, H., & Vila-Aiub, M. M. (2010). AHAS herbicide resistance endowing mutations: effect on AHAS functionality and plant growth. *Journal of*

Experimental Botany, *61*(14), 3925-3934.

Yu, Q., Han, H., Li, M., Purba, E., Walsh, M. J., & Powles, S. B. (2012). Resistance evaluation for herbicide resistance–endowing acetolactate synthase (ALS) gene mutations using *Raphanus raphanistrum* populations homozygous for specific ALS mutations. *Weed Research*, *52*(2), 178-186.

Yu, Q., Hashem, A., Walsh, M. J., & Powles, S. B. (2003). ALS gene proline (197) mutations confer ALS herbicide resistance in eight separated wild radish (*Raphanus raphanistrum*) populations. *Weed Science*, *51*(6), 831-838.

Yuan, J. S., Tranel, P. J., & Stewart, C. N. (2007). Non-target-site herbicide resistance: a family business. *Trends in Plant Science*, *12*(1), 6-13.

13

Cryopreservation of *Byrsonima intermedia* embryos followed by room temperature thawing

Luciano Coutinho Silva[1*], Renato Paiva[1], Rony Swennen[2,3], Edwige Andrè[3] and Bart Panis[3]

[1]*Departamento de Biologia, Setor de Fisiologia Vegetal, Universidade Federal de Lavras, Cx. Postal 3037, 372000-000, Lavras, Minas Gerais, Brazil.* [2]*Division of Crop Biotechnics, Laboratory for Tropical Crop Improvement, Katholieke University Leuven, Leuven, Belgium.* [3]*Bioversity International, Katholieke University Leuven, Leuven, Belgium. *Author for correspondence. E-mail: lucoutsilva@yahoo.com.br*

ABSTRACT. *Byrsonima intermedia* is a shrub from the Brazilian Cerrado with medicinal properties. The storage of biological material at ultra-low temperatures (-196°C) is termed cryopreservation and represents a promising technique for preserving plant diversity. Thawing is a crucial step that follows cryopreservation. The aim of this work was to cryopreserve *B. intermedia* zygotic embryos and subsequently thaw them at room temperature in a solution rich in sucrose. The embryos were decontaminated and desiccated in a laminar airflow hood for 0-4 hours prior to plunging into liquid nitrogen. The embryo moisture content (% MC) during dehydration was assessed. Cryopreserved embryos were thawed in a solution rich in sucrose at room temperature, inoculated in a germination medium and maintained in a growth chamber. After 30 days, the embryo germination was evaluated. No significant differences were observed between the different embryo dehydration times, where they were dehydrated for at least one hour. Embryos with a MC between 34.3 and 20.3% were germinated after cryopreservation. In the absence of dehydration, all embryos died following cryopreservation. We conclude that *B. intermedia* zygotic embryos can be successfully cryopreserved and thawed at room temperature after at least one hour of dehydration in a laminar airflow bench.

Keywords: long-term storage, native plant, zygotic embryos, rapid freezing, desiccation.

Introduction

The Brazilian Cerrado shelters numerous plant species with high fruitful and/or medicinal potentials. The Cerrado harbours the largest savannah plant biodiversity on the planet (KLINK; MACHADO, 2005), with more than 10,000 plant species (RATTER et al., 1997), which is persistently threatened by the intensive agricultural use of the land (PEREIRA; GAMA 2010). Studies on the propagation of native species are scarce, and numerous species that represent an ecological and/or pharmaceutical interest could become extinct before being properly studied.

Byrsonima intermedia A. Juss. is a medicinally valuable species that belongs to the Malpighiaceae family from the Cerrado, and, such as numerous species of this genus, it presents seeds with a coat dormancy that are difficult to propagate (NOGUEIRA et al., 2004). A wide variety of medicinal properties are attributed to this plant, including anti-inflammatory (MOREIRA et al., 2011; ORLANDI et al., 2011), anti-ulcer (SANNOMIYA et al., 2007; SANTOS et al., 2012), gastroprotective (SANTOS et al., 2009, 2012), healing (SANTOS et al., 2009), duodenal antimicrobial and antidiarrheal effects (SANTOS et al., 2012).

Plant biodiversity conservation can be performed by protecting the natural habitats (in situ) or by growing the plants in field collection (GONZÁLEZ-BENITO et al., 2003; ENGELMANN, 2011; WATANAWIKKIT et al., 2012). These two types of germplasm conservation tactics are costly and prevent the exchange of material due to the risk of disease and pathogen spreading (ENGELMANN, 1997) in addition to the risk of plague attacks and natural disasters (GONZÁLEZ-BENITO et al., 2003). Germplasm conservation using in vitro cultures is a technique that effectively conserves plant biodiversity; however, it is also extremely laborious and costly to maintain a large collection. Conversely, seed cryopreservation is an accessible, efficient and one of the most promising techniques for long-term and safe storage (ENGELMANN, 2004; JOHNSON et al., 2012; N'NAN et al., 2012).

Zygotic embryos or embryonic axes from a wide variety of plant species can be dessicated (ENGELMANN, 1992). Water removal is essential for preventing injury during freezing as well as for maintaining post-thaw viability (PANIS et al., 2001). During cooling or thawing, intracellular water may form lethal ice crystals that lead to cell death through a crystallisation or recrystallization process, respectively (MAZUR, 1984). For example, crystallisation is typically avoided by a rapid cooling by directly plunging the embryos into liquid nitrogen (LN) (GONZALEZ-ARNAO et al., 2008). However, during rewarming, living cells may suffer damage by recrystallization (FKI et al., 2012), a phenomenon that may occur if the thawing is not performed properly (GONZALEZ-ARNAO et al., 2008; HOPKINS et al., 2012). To avoid recrystallization, thawing must be rapid (HOPKINS et al., 2012; MAZUR, 1984).

The classical thawing method involves plunging an enclosed cryovial tube containing explants into a sterile water bath (~ 40°C) for a short period of time (JOHNSON et al., 2012; N'NAN et al., 2012; WEN; WANG, 2010). However, this procedure requires specific equipment and precise time control. Conversely, using a small volume of a sterile solution to thaw the embryos at room temperature may be advantageous during embryo cryopreservation. In this context, we aimed to establish a cryopreservation protocol for B. intermedia zygotic embryos by rapidly freezing desiccated embryos and subsequently thawing them at room temperature.

Material and methods

Ripe B. intermedia fruits were collected from a natural population in southern Minas Gerais State, Brazil, located at 918.0 m altitude, 21°14'S and 44.9°00'W GRW. After harvesting, the pulp was removed, and the seeds were soaked in 0.1 M sodium hydroxide (NaOH) for five minutes and washed in unsterile running water for 10 minutes. Endocarps were dried for three days at room temperature over filter paper and stored at 4°C prior to decontamination.

Embryos were decontaminated by two methods: (i) Method 1 - Endocarps were opened in non-aseptic conditions using forceps, and the embryos were extracted. In a laminar airflow hood (LAF), the embryos were plunged into 70% alcohol v/v for 30 seconds and subsequently into a NaOCl solution (1% of active chlorine), pH 8.5, plus one drop of Tween for five minutes; and (ii) Method 2 - In a LAF (aseptic conditions), the seeds were subjected to 95% alcohol for two minutes and to NaOCl (2% active chlorine) plus one drop of Tween for 20 minutes. They were washed three times in sterile water, and the endocarps were opened using forceps. The embryos were extracted and plunged into 70% alcohol v/v for 30 seconds and then into a NaOCl solution (1% active chlorine), pH 8.5, plus one drop of Tween for five minutes. For both treatments (Method 1 and 2), after the NaOCl immersion, the embryos were washed three times in sterile water, the integuments (endotegmen and exotesta) were removed and the embryos were inoculated on MS medium (MURASHIGE; SKOOG, 1962) with 0.09 M sucrose, 3 g L^{-1} Phytagel® and with pH-adjusted to 5.8. The cultures were maintained in a growth room at 25 ± 2°C at a 16h photoperiod. For each treatment, 50 seeds were inoculated with one seed per test tube, where each tube represented one replicate. After 30 days, the number of decontaminated embryos and germination events, which are characterised by the presence of normal seedlings (NS) that displayed both roots and shoots, were evaluated.

For embryo desiccation, germination and cryopreservation, the seeds were decontaminated and the embryos were extracted using Method 2.

For the control germination, the embryos were desiccated in a LAF over a filter paper disc (ø 60 mm) in an uncovered Petri dish (90 × 15 mm) for 0-4 hours at 25°C and inoculated in MS medium (as described above). Ten embryos were inoculated per each desiccation time. To obtain an accurate MC, three replicates of 10 embryos each were used. After desiccation, the embryos were completely dried in an oven at 70°C for 72 hours, and the MC was calculated using the following equation: % MC (FWb) = [(FW − DW)/FW]*100, where: % MC (FWb) represents the percentage of moisture on a fresh weight basis, FW represents the fresh weight (mg) and DW represents the dry weight (mg). For cryopreservation, dehydrated embryos (24 per desiccation time) were placed in a 2.0 cm^3 cryotube (6 embryos per cryotube) and plunged into and maintained in LN for 60 minutes. The cryotubes were only closed after the LN plunge. We used open cryotubes to allow the embryos to directly contact the LN, and thus an ultra-rapid cooling rate of approximately 130°C min.$^{-1}$ was obtained (GONZÁLEZ-BENITO et al., 2003).

Before thawing in a LAF, the cryotubes were opened, the LN was withdrawn and the embryos were thawed at room temperature for 15 minutes in a low volume (5 cm^3) of filter-sterilised unloading solution (US) (SAKAI et al., 1991) in a sterile Petri dish (60 × 15 mm). Next, the embryos were inoculated in MS medium with 0.3 M sucrose, maintained for 24h in the dark, transferred to MS basal medium and maintained in the dark for six days. After seven days, the control and cryopreserved embryos were placed in light conditions at a photoperiod of 16h at 25 ± 2°C. After 30 days, the number of embryos displaying roots, shoots or normal seedlings (NS) were evaluated. The statistic tool R (R DEVELOPMENT CORE TEAM, 2012) was used for analysis, and the data were subjected to ANOVA, Chi-squared and Scott-Knot tests (p ≤ 0.05).

Results and discussion

A significant difference between the two decontamination methods was observed (p < 0.0001). Decontamination Method 2 was more efficient, resulting in a higher percentage of contaminant-free explants (86%) vs. Method 1 (36%). The percentage of germinated embryos was also significantly higher (p < 0.0001). Approximately half (44%) of the contamination-free embryos processed using Method 2 germinated vs. 4% using Method 1 (Figure 1).

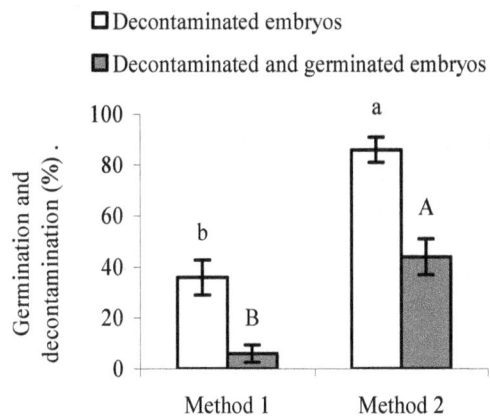

Figure 1. Germination and decontamination efficiencies of *Byrsonima intermedia* embryos. The bars represent the means ± SE of 50 seeds per treatment. The bars displaying the same letters for the same treatments are not significantly different according to the Chi-squared test (p ≤ 0.05).

Factors that may account for this low germination are: (i) the *Byrsonima* genus displays a high embryonic dormancy (LORENZI, 2002), (ii) the embryo extraction procedure is harmful due to an indehiscent and highly lignified endocarp layer (LORENZI, 2002) and (iii) the embryo cotyledons harbour phenolic compounds (SOUTO; OLIVEIRA, 2005). The inhibitory effect of phenolic compounds on seed germination has been well documented (TESIO et al., 2011; TOKUHISA et al., 2007; VICENTE; PLASENCIA 2011). However, we hypothesise that the low percentage of growing NS after decontamination is due to the procedure used to extract embryos, which involves the use of forceps. When the extremely hard endocarp is broken, the sensitive embryonic axis may be damaged. Moreover, integuments are extremely difficult to remove without harm.

Embryos that were not plunged into LN (control) did not statistically differ in the incidence of roots (p = 0.7839), shoots (p = 0.8991) and NS (p = 0.3283), which averaged 56, 42 and 26%, respectively. A desiccation of up to four hours did not decrease the germination percentages of the control embryos (Figure 2A). The reduction in the % MC (FWb) after four hours of desiccation was significant (p < 0.0001). We observed a high initial MC (62.7%) just after the embryos were decontaminated (Figure 2A and B).

We observed a significant difference in the number of embryos displaying roots (p = 0.0003), shoots (p = 0.0195), and NS (p = 0.0311) with averages of 39, 27 and 22%, respectively, between desiccated and non-desiccated embryos after cryopreservation. Non-desiccated embryos did not survive exposure to LN (Figure 2B). This result was expected when the MC was too high (62.7%), which allowed for lethal ice crystals to form. No differences were observed between the desiccated and cryopreserved embryos for all parameters evaluated. Compared to the controls (Figure 2A), desiccated and cryopreserved embryos (Figure 2B) displayed similar regeneration rates. Nogueira et al. (2011) previously reported a 10% germination incidence, only characterised by root protrusion, for *B. intermedia* zygotic embryos with 15 or 25% MC after being plunged into LN and thawed in a water bath at 37°C for 5 minutes. Compared to those results, we observed a four-fold higher root protrusion (Fig. 2B). NS after 30 days of cryopreservation are illustrated in Figure 3.

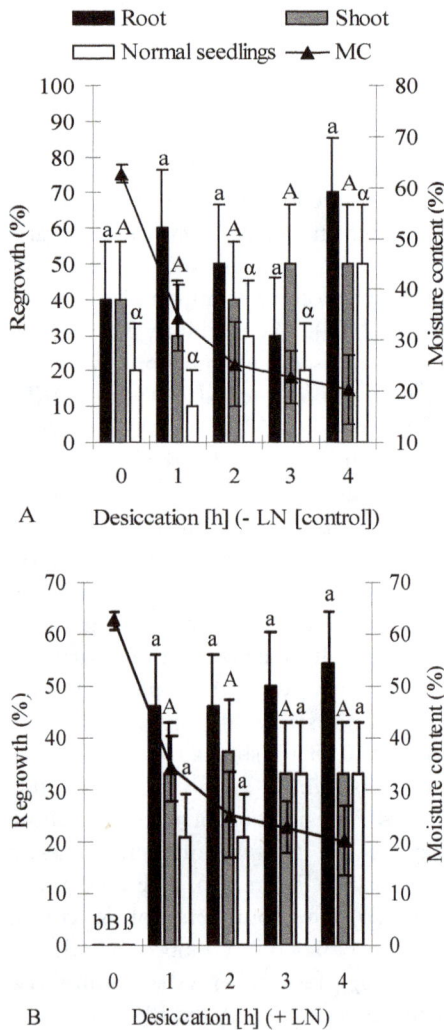

Figure 2. Moisture content (±SE, n = 30) and regrowth (%) of the roots, shoots or normal seedlings (+SE, n = 10) after embryo desiccation for different periods of time (hours), cryopreservation and thawing at room temperature. Embryos were not plunged (A) or plunged (B) into liquid nitrogen (-196°C). The bars followed by the same letter are not significantly different according to the Skott-Knott test (p ≤ 0.05). MC (moisture content); -LN (no LN plunge); + LN (LN plunge).

Figure 3. Normal seedlings after 30 days of cryopreservation were obtained from embryos desiccated for one to four hours, plunged into liquid nitrogen (-196°C), thawed at 25°C and inoculated in MS basal medium. The roots are indicated by an arrow (3-4h). Bars = 0.5 cm.

Zygotic embryos are large and complex structures with a heterogeneous cellular composition (ENGELMANN, 1992) and can display differential sensitivities to desiccation and cryopreservation. The embryonic root poles appear to be more resistant than the shoot poles (ENGELMANN, 2004), which may explain the high percentage of root protrusion compared to NS regeneration (Figure 2).

Given that excised and non-decontaminated embryos displayed a MC of approximately 6.6% (data not shown), the high initial MC percentage observed may be due to water imbibition. The seeds were in contact with water for varying periods: five minutes during decontamination in NaOCl, three fast washes in sterile water, and, for extracting embryo integuments (endotegmen and exotesta), they were placed over moistened sterile filter paper for several minutes. Integument extraction is laborious and time-consuming, and thus, from the first until the last embryo collected for dehydration, at least four hours were required, and imbibition was thus unavoidable during this period. Imbibition occurs during the phase I (PI) of germination, which is characterised by a fast water influx (BEWLEY, 1997; WEITBRECHT et al., 2011), where even dead seeds can soak (KRISHNAN et al., 2004). Therefore, the PI duration is variable (4-24h for most species). In soybeans, viable and non-viable seeds reached the end of PI after 12h of imbibition (KRISHNAN et al., 2004). Schopfer and Plachy (1984) demonstrated a PI of 6 – 9 for *Brassica napus* germination. For *Arabidopsis thaliana*, the PI terminates at ~ 4h (WEITBRECHT et al., 2011). The PI of tobacco seeds is ~12h (MANZ et al., 2005) and 24h for *Tabebuia impetiginosa* (SILVA et al., 2004). According to the behaviour of the species cited above, we believe this period was sufficient for the embryos to absorb a substantial amount of water. At the end of the desiccation period (4h), the embryos displayed a 20.3% MC (FWb) (Figure 2).

We did not follow the classical cryopreservation procedure for thawing zygotic embryos, where the embryos enclosed in a cryotube are typically rewarmed by a short exposure to a sterile water bath (~ 40°C), e.g., *Sabal spp* at 40°C / 1 min. (WEN; WANG, 2010), *Cocos nucifera* L. at 40°C / 2 min. (N'NAN et al., 2012) and *Xyris tennesseensis* Kral. at 37°C / 1-2 min. (JOHNSON et al., 2012). Our thawing approach was successfully performed at room temperature using a filter-sterilised US composed of 1.2 M sucrose dissolved in MS medium, pH 5.8. Performing the thawing procedure without a water bath may reduce the risk of contamination. Despite this, this solution is widely used for thawing, in room temperature, PVS2 (plant vitrification solution 2 [SAKAI et al., 1991]) treated and cryopreserved meristems (CONDELLO et al., 2011; PANIS et al., 2005; SANT et al., 2008). This study was the first to successfully apply this technique for thawing zygotic embryos.

Conclusion

Byrsonima intermedia zygotic embryos can be successfully cryopreserved using a rapid freezing method after at least one hour of desiccation in a laminar airflow hood.

These embryos can be successfully thawed using a filter-sterilized unloading solution at room temperature.

Acknowledgements

The authors would like to thank the FAPEMIG, CNPq, CAPES and the Laboratory for Tropical Crop Improvement (KU Leuven).

References

BEWLEY, J. D. Seed germination and dormancy. **The Plant Cell**, v. 9, n. 7, p. 1055-1066, 1997.

CONDELLO, E.; CABONI, E.; ANDRÈ, E.; PIETTE, B.; DRUART, P.; SWENNEN, R.; PANIS, B. Cryopreservation of apple *in vitro* axillary buds using droplet-vitrification. **CryoLetters**, v. 32, n. 2, p. 175-185, 2011.

ENGELMANN, F. Cryopreservation of embryos. In: DATTÉE, Y.; DUMAS, C.; GALLAIS, A. (Ed.). **Reproductive biology and plant breeding**. Berlin: Springer Verlag, 1992. p. 281-290.

ENGELMANN, F. *In vitro* conservation methods. In: FORD-LLOYD, B. V.; NEWBURRY, J. H.; CALLOW, J. A. (Ed.). **Biotechnology and plant genetic resources**: conservation and use. Wallingford: CABI, 1997. p. 119-162.

ENGELMANN, F. Plant cryopreservation: progress and prospects. *In Vitro* **Cellular and Development Biology-Plant**, v. 40, n. 5, p. 427-433, 2004.

ENGELMANN, F. Use of biotechnologies for the conservation of plant biodiversity. *In Vitro* **Cellular and Development Biology-Plant**, v. 47, n. 1, p. 5-16, 2011.

FKI, L.; BOUAZIZ, N.; CHKIR, O.; BENJEMAA-MASMOUDI, R.; RIVAL, A.; SWENNEN, R.; DRIRA, N.; PANIS B. Cold hardening and sucrose treatment improve cryopreservation of date palm meristems. **Biologia Plantarum**, v. 57, n. 2, p. 1-5, 2012.

GONZALEZ-ARNAO, M. T.; PANTA, A.; ROCA, W. M.; ESCOBAR, R. H.; ENGELMANN, F. Development and large scale application of cryopreservation techniques for shoot and somatic embryo cultures of tropical crops. **Plant Cell Tissue and Organ Culture**, v. 92, n. 1, p. 1-13, 2008.

GONZÁLEZ-BENITO, M. E.; SALINAS, P.; AMIGO, P. Effect of seed moisture content and cooling rate in liquid nitrogen on legume seed germination and seedling vigour. **Seed Science and Technology**, v. 31, n. 2, p. 423-434, 2003.

HOPKINS, J. B.; BADEAU, R.; WARKENTIN, M.; THORNE, R. E. Effect of common cryoprotectants on critical warming rates and ice formation in aqueous solutions. **Cryobiology**, v. 65, n. 3, p. 169-178, 2012.

JOHNSON, T.; CRUSE-SANDERS, J. M.; PULLMAN, G. S. Micropropagation and seed cryopreservation of the critically endangered species Tennessee yellow-eye grass, *Xyris tennesseensis* Kral. *In Vitro* **Cellular and Developmental Biology-Plant**, v. 48, n. 3, p. 369-376, 2012.

KLINK, C. A.; MACHADO, R. B. Conservation of the Brazilian Cerrado. **Conservation Biology**, v. 19, n. 3, p. 707-713, 2005.

KRISHNAN, P.; JOSHI, D. K.; NAGARAJAN, S.; MOHARIR, A. V. Characterization of germinating and non-viable soybean seeds by nuclear magnetic resonance (NMR) spectroscopy. **Seed Science Research**, v. 14, n. 4, p. 355-362, 2004.

LORENZI, H. **Árvores brasileiras**: manual de identificação e cultivo de plantas arbóreas nativas do Brasil. São Paulo: Nova Odessa, 2002. v. 2.

MANZ, B.; MÜLLER, K.; KUCERA, B.; VOLKE, F.; LEUBNER-METZGER, G. Water uptake and distribution in germinating tobacco seeds investigated *in vivo* by nuclear magnetic resonance imaging. **Plant Physiology**, v. 138, n. 3, p. 1538-1551, 2005.

MAZUR, P. Freezing of living cells: mechanisms and applications. **American Journal of Physiology - Cell Physiology**, v. 247, n. 3, p. C125-C142, 1984.

MOREIRA, L. Q.; VILELA, F. C.; ORLANDI, L.; DIAS, D. F.; SANTOS, A. L. A.; SILVA, M. A.; PAIVA, R.; ALVES-DA-SILVA, G.; GIUSTI-PAIVA, A. Anti-inflammatory effect of extract and fractions from the leaves of *Byrsonima intermedia* A. Juss. in rats. **Journal of Ethnopharmacology**, v. 138, n. 2, p. 610-615, 2011.

MURASHIGE, T.; SKOOG, F. A revised medium for rapid growth and bioassays with tobacco tissue cultures. **Physiologia Plantarum**, v. 15, n. 3, p. 473-497, 1962.

N'NAN, O.; BORGES, M.; KONAN KONAN, J.-L.; HOCHER, V.; VERDEIL, J.-L.; TREGEAR, J.; N'GUETTA, A. S. P.; ENGELMANN, F.; MALAURIE, B. A simple protocol for cryopreservation of zygotic embryos of ten accessions of coconut (*Cocos nucifera* L.). *In Vitro* **Cellular and Develompmental Biology-Plant**, v. 48, n. 2, p. 160-166, 2012.

NOGUEIRA, R. C.; PAIVA, R.; CASTRO, A. H.; VIEIRA, C. V.; ABBADE, L. C.; ALVARENGA, A. A. Germinação *in vitro* de murici-pequeno (*Byrsonima intermedia* A. Juss.). **Ciência e Agrotecnologia**, v. 28, n. 5, p. 1053-1059, 2004.

NOGUEIRA, G. F.; PAIVA, R.; CARVALHO, M. A. F.; SILVA, D. P. C.; SILVA, L. C.; NOGUEIRA, R. C. Cryopreservation of *Byrsonima intermedia* A. Juss. embryos using different moisture contents. **Acta Horticulturae**, v. 908, p. 199-202, 2011.

ORLANDI, L.; VILELA, F. C.; SANTA-CECÍLIA, F. V.; DIAS, D. F.; ALVES-DA-SILVA, G.; GIUSTI-PAIVA, A. Anti-inflammatory and antinociceptive effects of the stem bark of *Byrsonima intermedia* A. Juss. **Journal of Ethnopharmacology**, v. 137, n. 3, p. 1469-1476, 2011.

PANIS, B.; PIETTE, B.; SWENNEN, R. Droplet vitrification of apical meristems: a cryopreservation protocol applicable to all *Musaceae*. **Plant Science**, v. 168, n. 1, p. 45-55, 2005.

PANIS, B.; SWENNEN, R.; ENGELMANN, F. Cryopreservation of plant germplasm. **Acta Horticulturae**, v. 560, p. 79-86, 2001.

PEREIRA, A. C.; GAMA, V. F. Anthropization on the Cerrado biome in the Brazilian Uruçuí-Una Ecological Station estimated from orbital images. **Brazilian Journal of Biology**, v. 70, n. 4, p. 969-976, 2010.

R DEVELOPMENT CORE TEAM. **R**: A language and environment for statistical computing. Vienna: R Foundation for Statistical Computing, 2012.

RATTER, J. A.; RIBEIRO, J. F.; BRIDGEWATER, S. The Brazilian Cerrado vegetation and threats to its biodiversity. **Annals of Botany**, v. 80, n. 3, p. 223-230, 1997.

SAKAI, A.; KOBAYASHI, S.; OIYAMA, I. Cryopreservation of nucellar cells of navel orange (*Citrus sinensis* Osb. var.

brasiliensis Tanaka) cooled to −196°C. **Journal of Plant Physiology**, v. 137, n. 1, p. 465-470, 1991.

SANNOMIYA, M.; CARDOSO, C. R. P.; FIGUEIREDO, M. E.; RODRIGUES, C. M.; SANTOS, L. D.; SANTOS, F. V.; SERPELONI, J. M.; CÓLUS, I. M. S.; VILEGAS, W.; VARANDA, E. A. Mutagenic evaluation and chemical investigation of *Byrsonima intermedia* A. Juss. leaf extracts. **Journal of Ethnopharmacology**, v. 112, n. 2, p. 319-326, 2007.

SANT, R.; PANIS, B.; TAYLOR, M.; TYAGI, A. Cryopreservation of shoot-tips by droplet vitrification applicable to all taro (*Colocasia esculenta* var. *esculenta*) accessions. **Plant Cell, Tissue and Organ Culture**, v. 92, n. 1, p. 107-111, 2008.

SANTOS, R. C.; SANNOMIYA, M.; PELLIZZON, C. H.; VILEGAS, W.; HIRUMA-LIMA, C. A. Aqueous portion of *Byrsonima intermedia* A. Juss (Malpighiaceae): indication of gastroprotective and healing action of a medicinal plant from Brazilian Cerrado. **Planta Medica**, v. 75, n. 9, p. 933-933, 2009.

SANTOS, R. C.; KUSHIMA, H.; RODRIGUES, C. M.; SANNOMIYA, M.; ROCHA, L. R.; BAUAB, T. M.; TAMASHIRO, J.; VILEGAS, W.; HIRUMA-LIMA, C. A. *Byrsonima intermedia* A. Juss.: gastric and duodenal anti-ulcer, antimicrobial and antidiarrheal effects in experimental rodent models. **Journal of Ethnopharmacology**, v. 140, n. 2, p. 203-212, 2012.

SCHOPFER, P.; PLACHY, C. Control of seed germination by abscisic acid. II. Effect on embryo water uptake in *Brassica napus* L. **Plant Physiology**, v. 76, n. 1, p. 155-160, 1984.

SILVA, E. A. A.; DAVIDE, A. C.; FARIA, J. M. R.; MELO, D. L. B.; ABREU, G. B. Germination studies on *Tabebuia impetiginosa* Mart. seeds. **Cerne**, v. 10, n. 1, p. 1-9, 2004.

SOUTO, L. S.; OLIVEIRA, D. T. Morfoanatomia e ontogênese do fruto e semente de *Byrsonima intermedia* A. Juss. (Malpighiaceae). **Revista Brasileira de Botânica**, v. 28, n. 4, p. 697-712, 2005.

TESIO, F.; WESTON, L. A.; FERRERO, A. Allelochemicals identified from Jerusalem artichoke (*Helianthus tuberosus* L.) residues and their potential inhibitory activity in the field and laboratory. **Scientia Horticulturae**, v. 129, n. 3, p. 361-368, 2011.

TOKUHISA, D.; SANTOS DIAS, D. C. F.; ALVARENGA, E. M.; HILST, P. C.; DEMUNER, A. J. Phenolic compound inhibitors in papaya seeds (*Carica papaya* L.). **Revista Brasileira de Sementes**, v. 29, n. 3, p. 180-188, 2007.

VICENTE, M. R.-S.; PLASENCIA, J. Salicylic acid beyond defence: its role in plant growth and development. **Journal of Experimental Botany**, v. 62, n. 10, p. 3321-3338, 2011.

WATANAWIKKIT, P.; TANTIWIWAT, S.; BUNN, E.; DIXON, K. W.; CHAYANARIT, K. Cryopreservation of *in vitro*-propagated protocorms of *Caladenia* for terrestrial orchid conservation in Western Australia. **Botanical Journal of the Linnean Society**, v. 170, n. 2, p. 277-282, 2012.

WEITBRECHT, K.; MÜLLER, K.; LEUBNER-METZGER, G. First off the mark: early seed germination. **Journal of Experimental Botany**, v. 62, n. 10, p. 3289-3309, 2011.

WEN, B.; WANG, B. Pretreatment incubation for culture and cryopreservation of *Sabal* embryos. **Plant Cell, Tissue and Organ Culture**, v. 102, n. 2, p. 237-243, 2010.

14

Effect of *Anacardium humile* St. Hill (Anacardiaceae) Aqueous Extract on *Mahanarva fimbriolata* (Stal, 1854) (Hemiptera: Cercopidae)

Melissa Gindri Bragato Pistori[1,2]**, Antonia Railda Roel**[1*]**, José Raul Valério**[2]**, Marlene Conceição Monteiro Oliveira**[2,3]**, Eliane Grisoto** [4] **and Rosemary Matias**[5]

[1]*Programa de Pós-graduação em Biotecnologia, Universidade Católica Dom Bosco, Av. Tamandaré, 6000, 78117 900, Campo Grande, Mato Grosso do Sul, Brazil.* [2]*Laboratório de Entomologia de Plantas Forrageiras Tropicais, Empresa Brasileira de Pesquisa Agropecuária, Campo Grande, Mato Grosso do Sul, Brazil.* [3]*Agência de Desenvolvimento Agrário e Extensão Rural, Campo Grande, Mato Grosso do Sul, Brazil.* [4]*Escola Superior de Agricultura "Luiz de Queiroz", Piracicaba, São Paulo, Brazil.* [5]*Programa de Pós-graduação em Meio Ambiente e Desenvolvimento Regional, Universidade Anhanguera, Campo Grande, Mato Grosso do Sul, Brazil. *Author for correspondence. E-mail: arroel@ucdb.br*

ABSTRACT. The study of plant-derived substances for the control of insect pests is desirable in the attempt to discover less toxic insecticides that are safe for the environment. Indeed, extracts from the cashew of the savannah, *Anacardium humile*, have shown insecticidal activities against certain insects. The sugarcane root spittlebug *Mahanarva fimbriolata* is considered an important pest of sugarcane, causing severe damage and significant yield reductions. The aim of this study was to evaluate the effect of the aqueous extract of *A. humile* (0.05, 0.4 and 1.0%) on *M. fimbriolata*. The application of the aqueous extract of *Anacardium humile* resulted in 53.1% nymphal mortality at a concentration of 1.0%, which was significantly higher than that observed in the negative control. The nymphal period and longevity of *M. fimbriolata*, however, were not affected by the aqueous *A. humile* extract.

Keywords: insecta, insecticidal plants, plant active ingredients, sugarcane pests.

Introduction

Brazil is the world's largest sugar cane producer, and the sugar and ethanol produced supply both the domestic and international needs (ÚNICA, 2012). In the last few years the State of Mato Grosso do Sul, Brazil, has increased its sugar cane plantation areas. However, this crop is attacked by numerous insect species, including the sugarcane root spittlebug *Mahanarva fimbriolata* (Stål, 1854), an insect that causes considerable damage to both sugar cane plants and pastures (GALLO et al., 2002). The management and cultural practices in sugar cane fields have been modified, aiming for higher productivity with a reduced impact on the environment (ALMEIDA et al., 2003). The main modification in the sugarcane management system is with regard to the harvesting process: instead of the traditional burning prior to manual harvest, growers have been forced to avoid it, performing only the mechanical harvest.

Although desirable, this new harvesting process, termed "greencane", results in a significant pest problem, as it creates a habitat that is conducive to spittlebug development. Without burning, the greencane harvesting process favors the buildup of a straw thatch layer on the soil surface. Such litter creates a dark and humid microclimate, stimulating the emission of surface roots and providing an ideal habitat for the development of spittlebug nymphs (DINARDO-MIRANDA et al., 2004, GARCIA et al., 2007a and b).

The proximity of large areas of sugarcane to the fragile Pantanal biome in Mato Grosso do Sul State is worrisome because there is a great risk of contaminating both the water and soil due to the common use of fertilizer and biocides in sugarcane production. However, the use of substances of plant origin that present insecticidal activity provides a sustainable alternative to control insect pests in agriculture (RATAN, 2010).

Anacardium humile St. Hill, a common species of plant in the Pantanal and Cerrado, is widely used by the local populations for medicinal purposes (ALMEIDA et al., 2003). This species of plant has also been the subject of studies conducted by research groups in search of compounds with insecticidal properties (ANDRADE FILHO et al., 2010; PORTO et al., 2008).

The purpose of this work was to evaluate the effect of an aqueous extract of *A. humile* leaves on the survival and biology of the sugarcane root spittlebug.

Material and methods

The bioassays were conducted in the Entomology Laboratory of Tropical Forage Plants, at Embrapa Beef Cattle Research Center, from December 2008 to June 2009 using the susceptible sugarcane cultivar RB86-7515. The mean temperature and relative humidity in the laboratory were $24.9 \pm 1.4°C$ and $54.8 \pm 11.3\%$.

The nymphs

To obtain the nymphs for conducting the trial, eggs of the spittlebug *M. fimbriolata* were obtained according to methodology described by Valério (1993). After being removed from the oviposition substrate (agar-water) and superficially sterilized with Clorox solution, the eggs were kept in Petri dishes in an incubator at 27°C until the nymphs hatched.

The treatments

The aqueous extract of *A. humile* leaves was prepared following the methodology cited by Andrade Filho et al. (2010). The extract was diluted in water to three concentrations (0.05, 0.4 and 1.0%). The insecticide Thiamethoxam (Actara® 250 WG) was used for the positive control, and distilled water only was used for the negative control. Each experimental unit consisted of a plastic pot with a lid that had a central circular opening for fitting and hanging a smaller plastic pot containing a sugarcane seedling. The bottom of this seedling pot was cut to expose the three-month-old sugarcane plant roots. After the application of the plant extracts to these roots, six newly hatched spittlebug nymphs were

introduced into the unit. The outer pot was covered with opaque plastic sheeting to ensure darkness, a condition that is optimal for both the roots and nymphs. There were ten replications for each treatment and control.

Nymphal development

The duration of nymphal period was estimated through daily observations, as based on the emergence of *M. fimbriolata* adults in each treatment. The adults showing malformations were recorded and photographed.

Longevity of females and males

The newly emerged adults were maintained in treatment-identified cages containing a sugarcane plant. Through daily observations, the female and male longevities were obtained by recording and removing the dead insects from the cages.

Fecundity

To determine the total number of eggs laid per female in each treatment, the eggs laid throughout the life span of each female were collected and counted at various time intervals.

Egg viability

For each treatment, ten samples of twenty eggs were maintained in Petri dishes in a climatic chamber at 27°C. All of the hatched nymphs were recorded through daily observations. For the statistical analysis, the data were transformed to a percentage of the viable and nonviable eggs.

Statistical analysis

The experimental design was completely randomized. The factors in the aqueous extract treatment consisted of three concentration levels (0.05, 0.4 and 1.0%) using spray applications. Ten replicates were used per treatment. Each experimental unit was infested with six newly hatched nymphs. The data analysis was performed using the Sanest program, applying an F test by ANOVA. The means were compared using the Tukey test ($p < 0.05$), and the results were expressed as the mean $\pm$ standard error of the mean. The results were transformed into the square root ($x + 0.5$) for the statistical analysis of the variables fertility and egg viability.

Results and discussion

Nymphal mortality

The mortality caused by the aqueous extracts of *A. humile* ranged from 23.2 at the 0.05% concentration after 5 days of treatment to 53.1 at the

1.0% concentration 15 and 20 days after the application (Table 1). In contrast, the nymphal mortality rate ranged from 14.7 to 21.3% for the negative control (distilled water). These percentages were higher than the natural mortality of 6% recorded by Garcia et al. (2006a and b).

Table 1. Mortality (%) ± standard error of *Mahanarva fimbriolata* nymphs in sugarcane, at the 5, 10, 15 and 20th day after the application of aqueous extract of *Anacardium humile* in different concentrations. Temperature: 24.9 ± 1.4°C, RH 54.8 ± 11.3%.

Treatment	5th Day	10th Day	15th Day	20th Day
0.05%	23.2 ± 7.54 bc[1]	24.9 ± 7.56 bc	29.8 ± 7.78 bc	29.8 ± 7.78 bc
0.4%	31.3 ± 7.64 bc	32.9 ± 8.24 bc	32.9 ± 8.24 bc	32.9 ± 8.24 bc
1.0%	48.1 ± 9.44 b	51.5 ± 10.08 b	53.1 ± 10.18 b	53.1 ± 10.18 b
Negative control[2]	14.7 ± 4.61 c	17.9 ± 3.88 c	21.3 ± 3.55 c	21.3 ± 3.55c
Positive control[3]	100 ± 0 a	100 ± 0 a	100 ± 0 a	100 ± 0 a
C.V. (%)	48,8	48,3	46,6	46,6

[1]Means followed by same letter in the column are not different by tukey test (p < 0.05).
[2]Distilled water. [3]Thiametoxam.

Nymphal mortality rate increased with the rise in the aqueous extract concentration, although only at the highest concentration (1%) a significant difference in relation to negative control was observed. Andrade Filho et al. (2010) reported that the same extract, in concentrations ranging from 0.05 to 2.0% presented insecticidal activity to the whitefly nymph of *Bemisia tuberculata* (Bondar, 1923). Porto et al. (2008) however, found no toxic effect of aqueous extract of *A. humile* in concentrations from 0.0125 to 1% on *Aedes aegypti* (Linnaeus, 1762).

Nymphal development and the longevity of males and females

Data regarding the duration of nymphal period in the different concentrations are presented in Table 2. No statistical difference was observed among treatments and the negative control, whose average nymphal period was of 35.2 days. Such data agree with the findings of Garcia et al. (2006b). This author registered a nymphal period of 37 days when nymphs of *M. fimbriolata* was reared on sugar cane seedlings, at 25°C.

Table 2. Nymphal period (days), longevity (days) of males and females of the root spittlebug, *Mahanarva fimbriolata* in sugarcane, treated with aqueous extract of *Anacardium humile* in different concentrations. Temperature: 24.9 ± 1.4°C, RH 54.8 ± 11.3%.

Treatment	Nymphal Period[1]	Longevity	
		Males	Females
0.05%	34.0 ± 0.43a	13.8 ± 0.74 a	17.3 ± 2.18 a
0.4%	34.6 ± 0.38 a	13.6 ± 1.82 a	12.1 ± 1.71 a
1.0%	34.8 ± 0.57 a	15.2 ± 2.56 a	17,6 ± 2.59 a
Negative control[2]	35.2 ± 0.43 a	13.1 ± 1.71 a	15.2 ± 1.33 a
C.V. (%)	7.5	35.8	32,2

[1]Means followed by same letter in the column are not different by tukey test (p < 0.05).
[2]Distilled water.

There were no significant differences for the longevity of the males and females among all of the tested concentrations and the negative control: the average lifespan for the males was 13 days and 15 days for the females (Table 2). Similar results were reported by Garcia et al. (2006b), with an average longevity of 17.2 days for *M. fimbriolata* reared under conditions similar to those described in the present work. These researchers also noted that there was a reduction in the spittlebug longevity, regardless of the sex, in relation to the control. It was reported that the males exposed to neem *Azadirachta indica* products and aqueous extracts showed longevity reductions of approximately 50%, whereas the reductions were a slightly higher for females (55-60%).

Fecundity, viability and number of fertile females

The average number of eggs per female in the negative control was 165.4 (Table 3). Studying the same insect species at a similar temperature, Garcia et al. (2006b) reported an average of 342.1 eggs per female, though different sugarcane varieties were used. Similar data were found by Gallo et al. (2002), who stated that each female can produce, on average, 340 eggs during her lifetime. In the present study, the lowest tested aqueous extract concentration (0.05%) did not alter the number of eggs laid by each female (145.7) compared to the negative control (165.4) in which only a water spray was used. However, the females from the nymphs treated with the aqueous *A. humile* extract at concentrations of 0.4 and 1% produced 70.7 and 78.1% fewer eggs, respectively, compared to both, the negative control and the 0.05% concentration. A similar result was observed by Schmutterer (1992) when measuring the effect of a neem extract, verifying that the fecundity of Hemiptera was strongly influenced by the neem extract.

Although all ten of the females used in the negative control laid eggs, this was not true for the tested aqueous extract concentrations. When the concentrations of 0.05, 0.4 and 1.0% were used, eight, six and five females laid eggs, respectively. This finding suggests that the *A. humile* aqueous extract impaired the *M. fimbriolata* female fecundity.

Regarding the viability of the obtained eggs, only 40.5% of the eggs were viable in the negative control. Conversely, Garcia et al. (2006b) found that 81% of the eggs were viable. Such a difference could be explained by considering that the authors used different sugarcane varieties than that used in the present study. However, another more probable explanation is that the present work was conducted during the final phase of the infestation period, a time when females tend to oviposit increasing proportions of diapausing eggs.

Table 3. Fecundity (average number) and egg viability (%) ± standard error of the root spittlebug, *Mahanarva fimbriolata*, in sugarcane after application of aqueous extract of *Anacardium humile* in different concentrations. Temperature: 24.9 ± 1.4°C, RH 54.8 ± 11.3%.

Treatment	Fecundity	Viability (%)	Fertile females[1]
0.05%	145.7 ± 38.5 ab[2]	22.0 ± 10.4 ab	8
0.4%	70.7 ± 44.0 b	54.5 ± 9.0 a	6
1.0%	78.1 ± 45.6 b	4.5 ± 1.3 b	5
Negative control[3]	165.4 ± 14.8 a	40.5 ± 12.0 a	10
C.V. (%)	61.4	80.0	

[1]Initial number of couples. [2]Means followed by same letter in the column are not different by tukey test (p < 0.05). [3]Distilled water.

A very low egg viability (4.5%), however, was observed at the 1.0% concentration, confirming an important effect of the *A. humile* aqueous extract on the biology of *M. fimbriolata*. Possible effects on both the female fecundity and viability of the eggs suggest a possible cumulative effect on the *M. fimbriolata* females originating from nymphs reared on *A. humile* aqueous extract-treated plants. Evaluating the effect of an aqueous extract of *Azadirachta indica* A. Juss (neem), Garcia et al. (2006a) found an 80% reduction in egg fertility plus morphological and physiological changes in approximately 10% of the eggs of this spittlebug. This species (*A. indica*) of the Meliaceae family contains active triterpenoids ingredients; the Anacardiaceae *A. humile*, also contains theses ingredients and, may cause similar effects. Specifically, the aqueous extract of *A. humile* leaves contains tannins, reducing sugars and saponins. Saponins are complex molecules of high molecular weight that consist of triterpenes or steroids (ANDRADE FILHO et al., 2010). Moreover, Correia et al. (2006) indicated that flavonoids, phenolic lipids and triterpenes are the main substances in several species of the Anacardiaceae family. Although in very low numbers, the emergence of deformed teneral *M. fimbriolata* adults was observed in all three of the *A. humile* aqueous extract concentrations, but not in the negative control. Of the 60 nymphs in each treatment, the numbers of malformed individuals were two, one and one, respectively, for the 1, 0.4 and 0.05% concentrations.

Conclusion

The aqueous extract of *A. humile* sprayed onto the roots of sugarcane caused reductions in the *M. fimbriolata* nymphal survival rate and additional detrimental effects such as decreased fecundity and egg viability. Further investigations of the chemical composition of the extract could determine the active compound acting as an insecticide and could

help direct the synthesis of molecules for the control of insect pests.

Acknowledgements

To Pantanal Research Center (Portuguese acronym CPP), National Institute for Science and Technology in Wetlands (INAU), National Council for Scientific and Technological Development (CNPq), Ministry of Science and Technology (MCT), Foundation of Teaching, Science and Technology Development of Mato Grosso do Sul State (Fundect/MS), Embrapa Beef Cattle Research Center (Cnpgc).

References

ALMEIDA, J. E. M.; BATISTA FILHO, A.; SANTOS, A. S. Efficiency of Isolates of *Metarhizium anisopliae* for the Control of Sugarcane Root Spittlebug *Mahanarva fimbriolata* (Hom.: Cercopidae). **Arquivos do Instituto Biológico**, v. 70, n. 1, p. 101-103, 2003.

ANDRADE FILHO, N. N.; ROEL, A. R.; PORTO, K. R. A.; SOUZA, R. O.; COELHO, R. M.; PORTELA, A. Toxicity of aqueous extract of leaves of *Anacardium humile* St. Hill (Anacardiaceae) on *Bemisia tuberculata* (Bondar, 1923) (Hemiptera: Aleyrodidae). **Ciencia Rural**, v. 40, n. 8, p. 1689-1694, 2010.

CORREIA, S. J.; DAVID, J. P.; DAVID, M. J. Secondary metabolites from species of Anacardiaceae. **Química Nova**, v. 29, n. 6, p. 1287-1300, 2006.

DINARDO-MIRANDA, L. L.; COELHO, A. L.; FERREIRA, J. M. G. Influence of Time and Rate of Application of Insecticides on *Mahanarva fimbriolata* (Stål) (Hemiptera: Cercopidae) Control and on Quality and Yield of Sugarcane. **Neotropical Entomology**, v. 33, n. 1, p. 91-98, 2004.

GALLO, D.; NAKANO, O.; SILVEIRA NETO, S.; CARVALHO, R. P. L.; BATISTA, G. C.; BERTI FILHO, E.; PARRA, J. R. P.; ZUCCHI, R. A.; ALVES, S. B.; VENDRAMIM, J. D. **Entomologia Agrícola**, Piracicaba: Fealq, 2002.

GARCIA, J. F.; GRISOTO, E.; VENDRAMIM, J. D.; BOTELHO, P. S. Bioactivity of neem, *Azadirachta indica*, against spittlebug *Mahanarva fimbriolata* (Hemiptera: Cercopidae) on sugarcane. **Journal of Economic Entomology**, v. 99, n. 6, p. 2010-2014, 2006a.

GARCIA, J. F.; BOTELHO, P. S. M.; PARRA, J. R. P. Biology and fertility life table of *Mahanarva fimbriolata* (Hemiptera: Cercopidae) in sugarcane. **Scientia Agricola**, v. 63, n. 4, p. 317-320, 2006b.

GARCIA, J. F.; BOTELHO, P. S. M.; PARRA, J. R. P. Feeding site of the spittlebug *Mahanarva fimbriolata* (Hemiptera: Cercopidae) on sugarcane. **Scientia Agricola**, v. 64, n. 5, p. 555-557, 2007a.

GARCIA, J. F.; BOTELHO, P. S. M.; PARRA, J. R. P. Laboratory rearing technique of *Mahanarva fimbriolata*

(Hemiptera: Cercopidae). **Scientia Agricola**, v. 64, n. 1, p. 73-76, 2007b.

PORTO, K. R. A.; ROEL, A. R.; SILVA, M. M.; COELHO, R. M.; SCHELEDER, E. J. D.; JELLER, A. H. The effect of *Anacardium humile* Saint. Hill (Anacardiaceae) on *Aedes aegypti* (Linnaeus, 1762) (Diptera, Culicidae) larvae. **Revista da Sociedade Brasileira de Medicina Tropical**, v. 41, n. 6, p. 586-589, 2008.

RATAN, R. S. Mechanism of action of insecticidal secondary metabolites of plant origin. **Crop Protection**, v. 29, n. 9, p. 913-920, 2010.

SCHMUTTERER, H. Potencial of azadirachtin-containing pesticides for integrated pestcontrol in developing and industrialized countries. **Journal of Insect Physiology**, v. 34, n. 7, p. 713-719, 1992.

ÚNICA-União da Indústria de Cana-de-Açúcar. **Dados e cotações**. Available from: <http://www.unica.com.br>. Access on: June 15, 2012.

VALÉRIO, J. R. Obtenção de ovos de cigarrinhas (Homoptera: Cercopidae) em ágar-água. **Anais da Sociedade Entomológica do Brasil**, v. 22, n. 3, p. 583-590, 1993.

15

Control of coffee berry borer, *Hypothenemus hampei* (Ferrari) (Coleoptera: Curculionidae: Scolytinae) with botanical insecticides and mineral oils

Flávio Neves Celestino[1*], Dirceu Pratissoli[2], Lorena Contarini Machado[2], Hugo José Gonçalves dos Santos Junior[2], Vagner Tebaldi de Queiroz[3] and Leonardo Mardgan[2]

[1]*Instituto Federal de Educação, Ciência e Tecnologia do Espírito Santo, Rod. ES-130, km 01, 29980-000, Montanha, Espírito Santo, Brazil.* [2]*Departamento de Produção Vegetal, Universidade Federal do Espírito Santo, Vitória, Espírito Santo, Brazil.* [3]*Departamento de Química e Física, Universidade Federal do Espírito Santo, Vitória, Espírito Santo, Brazil. *Author for correspondence. E-mail: fncelestino@yahoo.com.br*

ABSTRACT. The objective of this study was to evaluate botanical oils, mineral oils and an insecticide that contained azadirachtin (ICA) for the control of *Hypothenemus hampei*, in addition to the effects of residual castor oil. We evaluated the effectiveness of the vegetable oils of canola, sunflower, corn, soybean and castor, two mineral oils (assist® and naturol®), and the ICA for the control of *H. hampei*. The compounds were tested at a concentration of 3.0% (v v^{-1}). The median lethal concentration (LC$_{50}$) was estimated with Probit analysis. The oil of castor bean and extract of castor bean cake were also evaluated at concentrations of 3.0% (v v^{-1}) and 3.0% (m v^{-1}), respectively. The mortality rates for *H. hampei* caused by the ICA and the castor oil were 40.8 and 53.7%, with LC$_{50}$ values of 6.71 and 3.49% (v v^{-1}), respectively. In the castor oil, the methyl esters of the fatty acids were palmitic (1.10%), linoleic (4.50%), oleic (4.02%), stearic (0.50%) and ricinoleic acids (88.04%). The extract of the castor bean cake was not toxic to *H. hampei*. The persistence of the castor oil in the environment was low, and the cause of mortality for *H. hampei* was most likely the blockage of the spiracles, which prevented the insects from breathing.

Keywords: *Ricinus communis*, *Azadirachta indica*, median lethal concentration, fatty acids, toxicity.

Introduction

The coffee berry borer, *Hypothenemus hampei* (Ferrari) (Coleoptera: Curculionidae: Scolytinae), is one of the most important phytosanitary problems for the coffee crop (Vega Infante, Castillo, & Jaramillo, 2009; Vega et al., 2014). The damage caused by this beetle pest to the coffee fruits decreases the weight of the beans and changes the type, classification and flavour of the coffee beverage (Vega et al., 2009).

The phytosanitary management of the pest indicates the use of control tactics that are based on a cost-benefit analysis and a low operational effect on the agroecosystem. Thus, because of the choice of society to reduce the environmental effects from the management of pests, the research on the use of botanical insecticides has increased in recent years (Isman & Grieneisen, 2014). Among the substances extracted from plants, the azadirachtin obtained from *Azadirachta indica* A. Juss (Meliaceae) is an important alternative in pest management (Isman, 2006; Janini et al., 2011; Pinto, Barros, Torres, & Neves, 2013). The chemical constituent of this pesticide interferes with the synthesis and the release of hormones (ecdysteroids) from the prothoracic gland, which leads to incomplete ecdysis in immature insects (Isman, 2006). Furthermore, the azadirachtin has an antifeedant effect for a wide range of insects and may cause sterility in female insects (Isman, 2006). The emulsifiable oil, the extracts of leaves and the seeds of neem were repellent to and increased the mortality of the coffee berry borer (Depieri & Martinez, 2010).

The derivatives of the castor bean (*Ricinus communis* L., Euphorbiaceae) also have potential in pest management (Ramos-López, Pérez, Rodríguez-Hernández, Guevara-Fefer, & Zavala-Sánchez, 2010; Tounou et al., 2011), with products that showed insecticidal activity against *Spodoptera frugiperda* (J.E. Smith) (Lepidoptera: Noctuidae) through effects on the duration and viability of the larval and pupal stages and on the weight reduction of the pupae (Ramos-López et al., 2010). The insecticidal action of the castor bean derivatives was also demonstrated for *Helicoverpa zea* (Boddie) (Lepidoptera: Noctuidae), *Plutella xylostella* L. (Lepidoptera: Plutellidae) and *H. hampei* (Bestete, Pratissoli, Queiroz, Celestino, & Machado, 2011; Rondelli et al., 2011; Tounou et al., 2011; Pérez, Zayas, Villa, Puentes, & García, 2012).

In addition to the derivatives of vegetables, mineral oils are also used to control insect pests (Nicetic, Cho, & Rae, 2011). The primary cause of death of arthropods with mineral oil is likely the blockage of the trachea or spiracles, and the lack of oxygen causes death to the insects (Stadler, Zerba, & Buteler, 2002; Stadler & Buteler, 2009). However, the mineral oils also affect the integument, with the ruptures of cell membranes and browning. The behaviour of insects is affected by these oils with typically repellent effects on oviposition and feeding, in addition to effects on the nervous system (Najar-Rodríguez, Walter, & Mensah, 2007; Najar-Rodríguez, Lavidis, Mensah, Choy, & Walter, 2008; Stadler & Buteler, 2009). Thus, the objectives of this study were to evaluate the botanical insecticides, the mineral oils, an insecticide containing azadirachtin (ICA) and the effects of residual castor oil for the control of the coffee berry borer.

Material and methods

The experiment was conducted at the Entomology sector of the Nucleus of Scientific & Technological Development in Phytosanitary Management (Nudemafi) of the Agrarian Sciences Centre of the Federal University of Espírito Santo (CCA-UFES). The environmental conditions of the experiment were a temperature of $25\pm1°C$, a relative humidity (RH) of $60\pm10\%$ and a photoperiod of 12:12 h light:dark. The rearing and the maintenance of *H. hampei* were according to Dalvi and Pratissoli (2012).

Identification of the chemical constituents of castor oil

The transesterification reaction was performed using 150 mg of castor oil and methyl-benzene alcohol at the ratio of 4:1, according to the methodology proposed by Folch, Lees and Sloane-Stanley (1957) and modified by Lepage and Roy (1986). The mixture was heated at 100°C for 60 min. Then, an aliquot of 10 μL of the transesterified sample was dissolved in 1000 μL of dichloromethane, and the sample was injected, in duplicate, into a gas chromatograph coupled with a mass spectrometer (GC/MS). The analysis was performed using the electron impact ionization (70 eV energy), with helium as the carrier gas (flow rate = 1.47 mL min.$^{-1}$) and a fused silica capillary column with the stationary phase RTX-5MS; the column was 30 m long with an internal diameter of 0.25 mm. We used the following temperature program: an initial oven temperature of 100°C (5 min.), which was followed by an increase of 10°C min.$^{-1}$ up to 280°C for 10 min. The temperature of the injector and the detector were maintained at 250°C. The 1.0 μL sample was injected in split mode, with a split ratio of 1:5, and in a continuous scan mode (interval m z^{-1} 33-808 a.m.u.). The identification of the compounds was performed with comparisons of the mass spectra in the NIST library (National Institute of Standards and Technology [NIST], 2015).

The quantification of the chemical constituents of the castor oil was performed on a gas chromatograph (SHIMADZU GC-2010 Plus, SINC do Brazil Scientific Instruments CO., São Paulo, São Paulo State, Brazil) equipped with a flame ionization detector (GC-FID); the column type was identical to the one that was previously described. The carrier gas was nitrogen (0.80 mL min.$^{-1}$; 85.2 kPa), and temperatures of the injector and the detector were 250 and 280°C, respectively. The temperature programme in the oven was identical to that used in the analysis with the GC-MS. The 1.0 μL sample was injected in split mode, with a split ratio of 1:30.

Tests of insecticidal activity of botanical insecticides, mineral oils and the insecticide containing azadirachtin (ICA)

For the experiments, we used the vegetables oils of soybean (Bunge Alimentos S.A.), sunflower (Bunge Alimentos S.A.), canola (Olvebra Indústria S.A.), corn (Bunge Alimentos S.A.) and castor; the insecticide that contained the azadirachtin (ICA) (Sempre Verde Killer Neem®; Active ingredient: 0.3% azadirachtin); and the mineral oils assist® (BASF – The Chemical Company) and naturol® (Farmax – Distribuidora Amaral LTDA). The compounds were acquired in the stores of the segment, with the exception of the castor oil. The castor seeds (R. communis), IAC 80, were acquired by the Agronomic Institute of Campinas (IAC) and were cold-pressed to extract the oil; this product was stored in a container wrapped with foil and hermetically sealed.

The experimental unit was a gerbox® (6 cm diameter x 2 cm) that was lined with filter paper that contained 15 newly emerged female coffee berry borers. Each treatment had 5 replications. In this bioassay, the oils were tested at a concentration of 3.0% (v v^{-1}), and in the preparation of the solutions, an adhesive surfactant 0.01% (w v^{-1}) (PS Tween® 80; Dynamic Contemporary Chemistry LTDA) and acetone 2% (v v^{-1}) were added. The control was prepared with deionized water and the adhesive surfactant 0.01% (w v^{-1}) and the acetone 2% (v v^{-1}). The female coffee berry borers were sprayed with a Potter Spray Tower® at a pressure of 15 pounds per square inch (PSI), which applied a volume of 5.5 mL per replicate. After the spraying, 0.15 g of ground coffee/gerbox® was offered as food. The mortality analysis was performed seven days after the release of the females, with a correction according to Abbott's formula (Abbott, 1925). The experimental design was completely randomized, and the data were subjected to analysis of variance. The means were grouped using the Scott-Knott grouping method, at 5% probability.

The LC$_{50}$ values were estimated for the treatments that caused corrected mortality that exceeded 40.0%. To estimate the lethal concentrations, a series of eight concentrations on the logarithmic scale were used. The lower limit (the concentration that caused the deaths of approximately 15% of the insects) and the upper limit (the concentration that caused the deaths of more than 50% of the insects; mortality above 95% did not occur, and therefore, the LC$_{90}$ was not calculated) were determined in preliminary tests, along with the respective controls. The lower and upper limits of the LC$_{50}$ for the ICA were 0.5 and 10.0% (v v^{-1}), respectively; whereas for the castor oil, the lower and upper limits were 0.5 and 5.0% (v v^{-1}), respectively. The LC$_{50}$ value was estimated with the Probit analysis using the POLO-PC program, with a confidence interval of 95% (Leora Software, 1987).

The derivatives of castor oil and cake were compared for insecticidal activity. The extract of the castor bean cake was obtained from 20 g of ground cake in a crucible. The vegetable material was transferred to a beaker that contained a KH$_2$PO$_4$ buffer solution, NaCl, and Na$_2$HPO$_4$, with masses of 3.40, 0.88, and 3.55 g L^{-1} of deionized water (pH 6.0), respectively. The mixture was placed on a magnetic stirrer for 30 minutes and then was filtered through a tissue 'voile'. The extract was diluted to the concentration of 3% (w v^{-1}) of the mass of the cake using the previously described buffered solution. In the control, only the buffered solution was used. The experimental unit was as described previously, and each treatment had 5 replications. The mortality analysis was performed seven days after the release of the females, with a correction according to Abbott's formula (Abbott, 1925). The experiment consisted of two treatments in a completely randomized experimental design, and the data were subjected to analysis of variance. The means were compared with F-tests at 5% probability.

Effect of residual castor oil

In this experiment, the experimental unit was identical to that described above, and each treatment had 5 replications. The treatment concentrations of castor oil were 0.5, 1.0, 1.5, 2.0, 2.5 and 3.0% (v v^{-1}), with the control without the oil (0.0% v v^{-1}). In the preparation of the solutions, the adhesive surfactant, Tween® 80 PS, 0.05% (v v^{-1}), was added. The ground coffee (0.15 g of ground coffee/gerbox®) was sprayed with a Potter Spray Tower® at a pressure of 15 pounds per square inch (PSI), which applied a volume of 5.5 mL per replicate. We examined the infestations of the coffee berry borer females at 0 (immediately after application), 1, 2, 3, 4 and 5 days after application. The mortality analysis was performed seven days after the release of the females, with a correction according to Abbott's formula (Abbott, 1925). The experimental design was completely randomized in a split-plot design in time with 6 concentrations of castor oil and 6 sample times after application (concentration of castor oil x time after application). The data were subjected to analysis of variance. The data were also subjected to regression analysis at 5% probability to determine the effects of the castor oil concentrations and the times after the application on the mortality of the coffee berry borers.

Results and discussion

Several features of the castor oil separated this compound from the others, such as the lowest iodine index and the highest viscosity, which are properties that are directly related to the chemical composition (Table 1). The saponification index indicates the mean molecular weight of the fatty acids that are esterified to glycerol in the triacylglycerol molecule, i.e., a high saponification index indicates the presence of fatty acids with low molecular weights, and vice versa. According to Nabil and Yasser (2012), the insecticidal activity of *Jatropha curcas* L. (Euphorbiaceae) seed oil on *Sitophilus granarius* L. (Coleoptera: Curculionidae) was related to the presence of high molecular weight fatty acids, which might also be an important factor in the insecticidal activity of the castor oil.

Among the tested vegetable oils, only the castor oil contained the ricinoleic fatty acid ester, which also had the lowest levels of the esters of palmitic, oleic, linoleic and linolenic fatty acids (Table 1). The profile of the fatty acids in the castor oil was obtained after the chromatographic analysis of the compounds in the mixture of the transesterification reactions (Figure 1). The chromatographic profile showed consistency, as expected for the castor oil, because of the fatty acid methyl esters that were identified, i.e., palmitic ($C_{16:0}$; 1.10%), linoleic ($C_{18:2}$; 4.50%), oleic ($C_{18:1}$; 4.02%), stearic ($C_{18:0}$; 0.50%) and ricinoleic (12-OH 9-$C_{18:1}$; 88.04%). The ricinoleic acid content, the primary component of castor oil, was similar to the contents that were reported for castor oil in other studies, 84.2 and 90.2% (Conceição et al., 2007; Salimon, Mohd Noor, Nazrizawati, Mohd Firdaus, & Noraishah, 2010).

The mortality of the coffee berry borers caused by the vegetable oils, the mineral oils and the ICA was significantly different ($F_{7:39}$ = 6.49, P = 0.0001; Table 2). The lowest mortality of *H. hampei* was caused with the two mineral oils (naturol® and assist®) and the vegetable oils of canola, corn, soybean and sunflower, and the mortality ranged from 13.6 to 30.3% (Table 2). Although the difference between them was not significant, the ICA (40.8%) and the castor oil (53.7%), with both at a concentration of 3.0% (v v⁻¹), caused the highest mortalities of the coffee berry borer (Table 2). The extract of the castor bean cake had no insecticidal effect on the coffee berry borers.

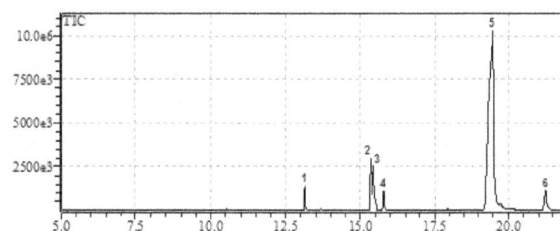

Figure 1. Chromatogram of the castor oil, variety IAC 80, after the transesterification reaction. 1 – $C_{16:0}$; 2 – $C_{18:2}$; 3 – $C_{18:1}$; 4 – $C_{18:0}$; 5 – 12-OH $C_{18:1}$; and 6 – UI (unidentified).

The ICA and the castor oil had LC_{50} values of 6.71 and 3.49% (v v⁻¹), respectively, for the coffee berry borer (Table 2). Although the difference was not significant between the LC_{50} values of the ICA and the castor oil, the ratio of the castor oil toxicity to the ICA toxicity was 1.92-fold higher (Table 2).

The effects of the mineral oils (naturol® and assist®) and the vegetables oils of canola, corn, soybean and sunflower likely occurred because the oils covered the openings of the spiracles of the insects, which led to asphyxiation and death. In some cases, researchers reported that the asphyxiation was caused by a blockage of spiracles and/or trachea as the primary mode of action of the mineral and vegetable oils (Law-Ogbomo & Egharevba, 2006; Stadler & Buteler, 2009; Egwurube, Magaji, & Lawal, 2010).

Table 1. Physico-chemical characteristics of the vegetable oils.

Chemical composition (%)[1]	Canola[2]	Sunflower[2]	Corn[2]	Soybean[2]	Castor bean[3]
Palmitic acid ($C_{16:0}$)	3.75	5.70	10.47	9.90	1.10
Stearic acid ($C_{18:0}$)	1.87	4.79	2.02	3.94	0.50
Oleic acid ($C_{18:1}$)	62.41	15.26	24.23	21.35	4.02
Ricinoleic acid ($C_{18:1}$)	-	-	-	-	88.04
Linoleic acid ($C_{18:2}$)	20.12	71.17	60.38	56.02	4.50
Chemical constants					
Saponification (mg KOH g⁻¹)	168-181	186-198	187-195	189-195	180.30
Iodine (g I₂ 100 g⁻¹)	94-120	136-148	103-135	124-139	62.76
Physical constants					
Factor of refraction (40°C)	1.465-1.469	1.467-1.470	1.465-1.468	1.466-1.470	1.479
Viscosity (cP) (30°C)	50.50	41.30	47.40	41.20	332.00

[1]Number of carbon atoms: number of double bonds; [2]References: Conceição et al. (2007); Zambiazi, Przybylski, Zambiazi and Mendonça (2007); Brock et al. (2008); Salimon et al. (2010); and Codex Alimentarum (2014); [3]Oil composition used in the experiment.

Table 2. Corrected mortality (%) of the female coffee berry borers caused by 3.0% (v v^{-1}) vegetable oils, mineral oils and the insecticide containing azadirachtin. The median lethal concentration (LC$_{50}$) values were determined at 25±1°C, a RH of 60±10% and a photoperiod of 12:12 hours light:dark.

Oils	Mortality[1]	N[2]	Slope±SE[3]	LC$_{50}$[4] (% v v^{-1})	RT$_{50}$[5]	χ^2 (DF)[6]
Naturol®	13.6±2.99 b	-	-	-	-	-
Assist®	16.0±5.71 b	-	-	-	-	-
Canola	25.4±5.54 b	-	-	-	-	-
Corn	26.5±4.10 b	-	-	-	-	-
Soybean	29.0±2.89 b	-	-	-	-	-
Sunflower	30.3±7.46 b	-	-	-	-	-
ICA[7]	40.8±5.68 a	574	1.71±0.25	6.71 (5.35 – 8.87) a	-	4.05 (5)
Castor bean	53.7±4.68 a	521	1.65±0.33	3.49 (2.65 – 5.52) a	1.92	2.08 (4)

[1]Means (± SE) followed by the same lowercase letters in a column do not differ at 5% probability by the Scott-Knott grouping method; [2]Number of insects used in the LC$_{50}$ bioassay; [3]Standard error (means followed by same letter in a column do not differ by the standard error); [4]Lethal concentration (LC$_{50}$) and confidence interval of the LC$_{50}$ at 95% probability (95% CI); [5]Reason toxicity = higher LC$_{50}$/lower LC$_{50}$; [6]Chi-square and degrees of freedom; and [7]Insecticide containing azadirachtin (ICA), with the trade name of Sempre Verde Killer Neem® (SVKN).

The castor oil was physicochemically differentiated from the other vegetable oils primarily because of the high ricinoleic acid content and the high viscosity (Table 1). The castor oil consisted of 88.04% ricinoleic acid; and the effects of the ricinoleic acid on the morphophysiology of the ovaries and the salivary glands of *Rhipicephalus sanguineus* (Latreille) (Acari: Ixodidae) led to the avoidance of two important processes: reproduction and feeding (Arnosti et al., 2011a, 2011b; Sampieri et al., 2013).

The insecticidal activity of castor oil was demonstrated on pests such as *S. frugiperda*, *H. zea* and *P. xylostella* (Ramos-López et al., 2010; Bestete et al., 2011; Rondelli et al., 2011; Tounou et al., 2011). However, for the coffee berry borer, only the extract of the green leaves of castor oil plants (2.2 g L^{-1} of boiled water) was evaluated, which reduced the number of individuals per sample, when analysed 72 hours after application (Pérez et al., 2012). For the neem oil (Dalneem®, 0.1% (v v^{-1}) of azadirachtin; Dalneem Brazil), the mortality of and the repellency to *H. hampei* caused by neem oil resulted in a reduced number of grains damaged by this pest (Depieri & Martinez, 2010). The authors found mortality of *H. hampei* up to 88.3% when the females and the beans were sprayed with a 1% (v v^{-1}) aqueous solution of emulsifiable neem oil. They also found that the number of brocades of coffee beans was lower when the neem oil (13.8%) was applied compared with the control (64.6%).

An interaction was found between the concentration of castor oil and the days after application on the mortality of the coffee berry borers (F$_{25:179}$ = 2.22; P = 0.0018; Figures 2 and 3). The mortality of the coffee berry borer females

released 0, 1, 3, 4 and 5 days after the application of castor oil was adjusted to the linear model, i.e., an increase in mortality occurred with the increase in the concentration of the castor oil (Figure 2). However, on the second day after application of the product, the coffee berry borer mortality remained constant as a function of concentration, with a mean of 10.83% (Figure 2). This result was related to the low mortality of the coffee berry borers, which occurred even at the higher concentrations of castor oil.

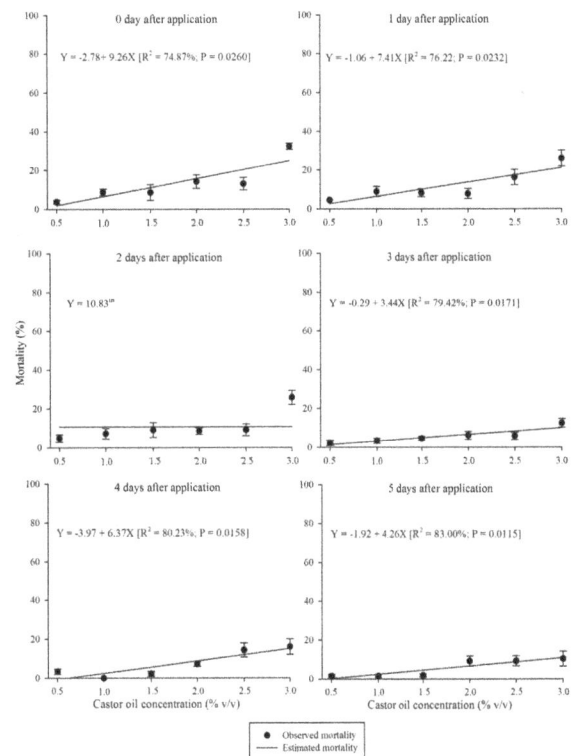

Figure 2. Effects of residual castor oil on the mortality of the coffee berry borer with days after application as a function of the concentration. The experiment was conducted at 25±1°C, a RH of 60±10% and a photoperiod of 12:12 hours light: dark.

The coffee berry borer mortality caused by the residual castor oil at the concentrations of 0.5, 2.0 and 2.5% (v v^{-1}), which was not adjusted to any model, remained constant over time (Figure 3). The constant mortality with time was associated with the low mortality caused by these castor oil concentrations and also with the similar values until the 5th day after application. However, for mortality of the coffee berry borers at the concentrations of 1.0, 1.5 and 3.0% (v v^{-1}), which was adjusted to the linear model, there was a reduction in the mortality as a function of time (Figure 3). In this case, it was notable that the initial mortality for the concentrations of 1.0 and 1.5% (v v^{-1}) was low, unlike for the concentration of 3.0% (v v^{-1}) of castor oil, which had a higher initial value of mortality for the coffee berry borers.

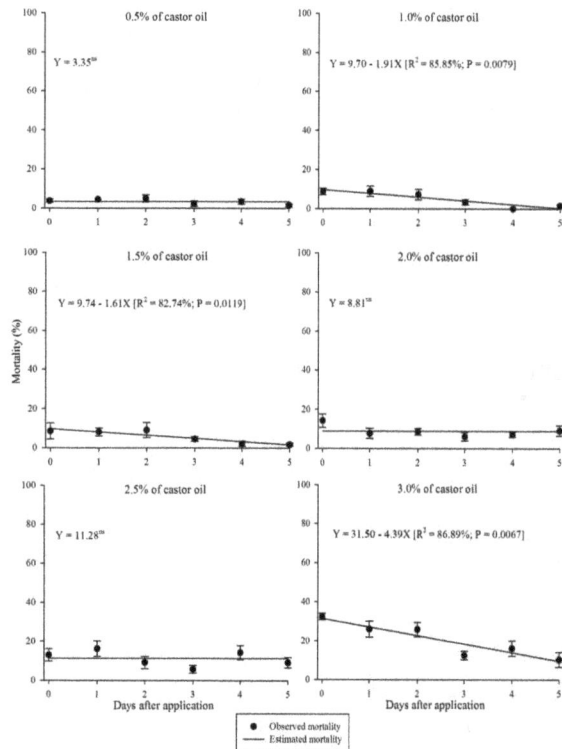

Figure 3. Effect of residual castor oil on the mortality of the coffee berry borer with concentration as a function of days after application. The experiment was conducted at 25±1°C, a RH of 60±10% and a photoperiod of 12:12 hours light:dark.

As in the present work, the residual effects of castor oil resulted in a significant decrease in the larval mortality of *P. xylostella* during the time between the application and the release of insects (Tounou et al., 2011). This result might be related to the low persistence of castor oil, which requires further research to improve the persistence. When a botanical insecticide is exposed to the elements, such as light, temperature and air, the insecticide is quickly degraded and becomes unstable in the environment (Gangwar, 2012; Radwan & El-Shiekh, 2012; Turek & Stintzing, 2013) because of the removal of the chemical compound protection compartment in the plant. Additionally, the insecticides are affected by destructive extraction methods, which cause oxidative damage and chemical and/or polymerization reactions (Miresmailli & Isman, 2014). Thus, aged plant extracts tend to lose the quality of some of the

attributes, including odour, flavour, colour and consistency (Turek & Stintzing, 2013). The diversity in the composition of the botanical extracts and the instability of the constituents might make some of these botanicals unsuitable for application, particularly when residual effects are desired for long periods of time (Gahukar, 2014; Miresmailli & Isman, 2014).

Understanding the persistence of botanical insecticides in the environment is extremely important, because the coffee berry borers remain within the coffee fruit for nearly the entire life cycle, in addition to a cryptic habit; thus, this pest is difficult to control (Vega et al., 2009). Therefore, the insecticides must reach this pest from September to December, because during this period, the coffee berry borer females leave the fruit to search for food (the female transit period). Therefore, studies on the effects of botanical insecticide residuals are an important factor for success in the management of this pest with plant-derived products.

Conclusion

The primary component of castor oil is ricinoleic acid. The insecticide that contained azadirachtin (ICA) and the castor oil caused mortality of 40.8 and 53.7% of the coffee berry borers, respectively. The castor seed cake extract was not toxic to *H. hampei* at a concentration of 3.0% (w v^{-1}). Moreover, the castor oil had low persistence in the environment, which complicated the control of the coffee berry borers.

Acknowledgements

The authors are grateful to the Conselho Nacional de Desenvolvimento Científico e Tecnológico (CNPq), the Fundação de Amparo à Pesquisa do Espírito Santo (Fapes), the Coordenação de Aperfeiçoamento de Pessoal de Nível Superior (Capes) and the Núcleo de Desenvolvimento Científico e Tecnológico em Manejo Fitossanitário (Nudemafi) for their financial support of this study.

References

Abbott, W. S. (1925). A method of computing the effectiveness of an insecticide. *Journal of Economic Entomology, 18*(1), 265-267.

Arnosti, A., Brienza, P. D., Furquim, K. C. S., Chierice, G. O., Bechara, G. H., Calligaris, I. B., & Camargo-Mathias, M. I. (2011b). Effects of ricinoleic acid esters from castor oil of *Ricinus communis* on the vitellogenesis of *Rhipicephalus sanguineus* (Latreille,

1806) (Acari: Ixodidae) ticks. *Experimental Parasitology, 127*(2), 575-580.

Arnosti, A., Brienza, P. D., Furquim, K. C. S., Gilberto, O. C. B., Claro Neto, S., Bechara, G. H., ... Camargo-Mathias, M. I. (2011a). Effects of *Ricinus communis* oil esters on salivary glands of *Rhipicephalus sanguineus* (Latreille, 1806) (Acari: Ixodidae). *Experimental Parasitology, 127*(2), 569-574.

Bestete, L. R., Pratissoli, D., Queiroz, V. T., Celestino, F. N., & Machado, L. C. (2011). Toxicidade de óleo de mamona a *Helicoverpa zea* e a *Trichogramma pretiosum*. *Pesquisa Agropecuária Brasileira, 46*(8), 791-797.

Brock, J., Nogueira, M. R., Zakrzevski, C., Corazza, F. C., Corazza, M. L., & Oliveira, J. V. (2008). Determinação experimental da viscosidade e condutividade térmica de óleos vegetais. *Ciência e Tecnologia de Alimentos, 28*(3), 564-570.

Codex Alimentarum. (2014). *Codex standard for named vegetable oils - Codex Stan 210 (Amended 2003, 2005)*. Recuperado de http://www.justice.gov.md/file/Centrul%20de%20armonizare%20a%20legislatiei/Baza%20de%20date/Materiale%202008/Legislatie/Codex%20STAN%20210. pdf

Conceição, M. M., Candeia, R. A., Silva, F. C., Bezerra, A. F., Fernandes Jr., V. J., & Souza, A. G. (2007). Thermoanalytical characterization of castor oil biodiesel. *Renewable and Sustainable Energy Reviews, 11*(5), 964-975.

Dalvi, L. P., & Pratissoli, D. (2012). Técnica de criação de *Hypothenemus hampei* (Ferrari, 1867) (Coleoptera: Scolytidae). In: D. Pratissoli (Ed.), *Técnicas de criação de pragas de importância agrícola, em dietas naturais* (p. 297- 305). Vitória, ES: Edufes.

Depieri, R. A., & Martinez, S. S. (2010). Redução da sobrevivência da broca-do-café, *Hypothenemus hampei* (Ferrari) (Coleoptera: Scolytidae), e do seu ataque aos frutos de café pela pulverização com nim em laboratório. *Neotropical Entomology, 39*(4), 632-637.

Egwurube, E., Magaji, B. T., & Lawal, Z. (2010). Laboratory evaluation of neem (*Azadirachta indica*) seed and leaf powders for the control of khapra beetle, *Trogoderma granarium* (Coleoptera: Dermestidae) infesting groundnut. *International Journal of Agriculture & Biology, 12*(4), 638-640.

Folch, J., Lees, M., & Sloane-Stanley, G. H. (1957). A simple method for the isolation and purification of total lipids from animal tissues. *Journal of Biological Chemistry, 226*(1), 497 -509.

Gahukar, R. T. (2014). Factors affecting content and bioefficacy of neem (*Azadirachta indica* A. Juss.) phytochemicals used in agricultural pest control: A review. *Crop Protection, 62*, 93-99.

Gangwar, S. K. (2012). Experimental study to find the effect of different neem (*Azadirachta indica*) based products against moringa hairy caterpillar (*Eupterote mollifera* Walker. *Bulletin of Environment, Pharmacology and Life Sciences, 1*(8), 35-38.

Isman, M. B. (2006). Botanical insecticides, deterrents, and repellents in modern agriculture and an increasingly regulated world. *Annual Review of Entomology, 51*, 45-66.

Isman, M. B., & Grieneisen, M. L. (2014). Botanical insecticide research: many publications, limited useful data. *Trends in Plant Science - Cell Press, 19*(3), 140-145.

Janini, J. C., Boiça Júnior, A. L., Jesus, F. G., Silva, A. G., Carbonell, S. A., & Chiorato, A. F. (2011). Effect of bean genotypes, insecticides, and natural products on the control of *Bemisia tabaci* (Gennadius) biotype B (Hemiptera: Aleyrodidae) and *Caliothrips phaseoli* (Hood) (Thysanoptera: Thripidae). *Acta Scientiarum. Agronomy, 33*(3), 445-450.

Law-Ogbomo, K. E., & Egharevba, R. K. A. (2006). The Use of vegetable oils in the control of *Callosobruchus maculatus* (F.) (Coleoptera: Bruchidae) in three cowpea varieties. *Asian Journal of Plant Sciences, 5*(3), 547-552.

Leora Software. (1987). *POLO-PC: a User's Guide to Probit or Logit Analyses* [Software]. Berkeley, CA: Leora Software.

Lepage, G., & Roy, C. C. (1986). Direct transesterification of all classes of lipids in a one-step reaction. *Journal of Lipid Research, 27*(1), 114-120.

Miresmailli, S., & Isman, M. B. (2014). Botanical insecticides inspired by plant–herbivore chemical interactions. *Trends in Plant Science - Cell Press, 19*(1), 29-35.

Nabil, A. E. A., & Yasser, A. M. K. (2012). *Jatropha curcas* oil as insecticide and germination promoter. *Journal of Applied Sciences Research, 8*(2), 668-675.

Najar-Rodríguez, A. J., Lavidis, N. A., Mensah, R. K., Choy, P. T., & Walter, G. H. (2008). The toxicological effects of petroleum spray oils on insects – Evidence for an alternative mode of action and possible new control options. *Food and Chemistry Toxicology, 46*(9), 3003-3014.

Najar-Rodríguez, A. J., Walter, G. H., & Mensah, R. K. (2007). The efficacy of a petroleum spray oil against *Aphis gossypii* Glover on cotton. Part 1: Mortality rates and sources of variation. *Pest Management Science, 63*(6), 586-595.

National Institute of Standards and Technology. (2015). *NIST Virtual Library*. Gaithersburg, MD: NIST. Recuperado de http://www.nist.gov/nvl/

Nicetic, O., Cho, Y. R., & Rae, D. J. (2011). Impact of physical characteristics of some mineral and plant oils on efficacy against selected pests. *Journal of Applied Entomology, 135*(3), 204-213.

Pérez, Y. O., Zayas, D. V., Villa, O. V., Puentes, R. A., & García, S. T. (2012). Aplicación de extractos de hojas de *Ricinus communis* L. en el control de la Broca del cafeto. *Centro Agrícola, 39*(1), 85-90.

Pinto, E. S., Barros, E. M., Torres, J. B., & Neves, R. C. S. (2013). The control and protection of cotton plants using natural insecticides against the colonization by

Aphis gossypii Glover (Hemiptera: Aphididae). *Acta Scientiarum. Agronomy, 35*(2), 169-174.

Radwan, O. A., & El-Shiekh, Y. W. A. (2012). Degradation of neem oil 90% EC (azadirachtin) under storage conditions and its insecticidal activity against cotton leafworm, *S. littoralis. Researcher, 4*(3), 77-83.

Ramos-López, M. A., Pérez, G. S., Rodríguez-Hernández, C., Guevara-Fefer, P., & Zavala-Sánchez, M. A. (2010). Activity of *Ricinus communis* (Euphorbiaceae) against *Spodoptera frugiperda* (Lepidoptera: Noctuidae). *African Journal of Biotechnology, 9*(9), 1359-1365.

Rondelli, V. M., Pratissoli, D., Polanczyk, R. A., Marques, E. J., Sturm, G. M., & Tiburcio, M. O. (2011). Associação do óleo de mamona com *Beauveria bassiana* no controle da traça-das-crucíferas. *Pesquisa Agropecuária Brasileira, 46*(2), 212-214.

Salimon, J., Mohd Noor, D. A., Nazrizawati, A. T., Mohd Firdaus, M. Y., & Noraishah, A. (2010). Fatty acid composition and physicochemical properties of malaysian castor bean *Ricinus communis* L. seed oil. *Sains Malaysiana, 39*(5), 761-764.

Sampieri, B. R., Arnosti, A., Furquim, K. C. S., Chierice, G. O., Bechara, G. H., Carvalho, P. L. P. F. ... Camargo-Mathias, M. I. (2013). Effect of ricinoleic acid esters from castor oil (*Ricinus communis*) on the oocyte yolk components of the tick *Rhipicephalus sanguineus* (Latreille, 1806) (Acari: Ixodidae). *Veterinary Parasitology, 191*(3-4), 315-322.

Stadler, T., & Buteler, M. (2009). Modes of entry of petroleum distilled spray-oils into insects: a review. *Bulletin of Insectology, 62*(2), 169-177.

Stadler, T., Zerba, M. I., & Buteler, M. (2002). Toxicity and cuticle softening effect by mineral and vegetable oils to the cotton boll weevil *Anthonomus grandis* Boheman (Coleoptera: Curculionidae). In: G. A. C. Beattie, D. M. Watson, M. L. Stevens, D. J. Rae, & R. N. Spooer-Hart, (Eds.), *Spray Oils Beyond 2000: Sustainable Pest and Disease Management* (p. 152-155). Sydney, NSW: University Western Sydney.

Tounou, A. K., Gbénonchi, M., Sadate, A., Komi, A., Dieudonné, G. Y. M., & Komla, S. (2011). Bio-insecticidal effects of plant extracts and oil emulsions of *Ricinus communis* L. (Malpighiales: Euphorbiaceae) on the diamondback, *Plutella xylostella* L. (Lepidoptera: Plutellidae) under laboratory and semi-field conditions. *Journal of Applied Biosciences, 43*(3), 2899-2914.

Turek, C., & Stintzing, F. C. (2013). Stability of essential oils: a review. *Comprehensive Reviews in Food Science and Food Safety, 12*(1), 40-53.

Vega, F. E., Infante, F., Castillo, A., & Jaramillo, J. (2009). The coffee berry borer, *Hypothenemus hampei* (Ferrari) (Coleoptera: Curculionidae): a short review, with recent findings and future research directions. *Terrestrial Arthropod Reviews, 2*(2), 129-147.

Vega, F. E., Simpkins, A., Bauchan, G., Infante, F., Kramer, M., & Land, M. F. (2014). On the eyes of male coffee berry borers as rudimentary organs. *PLoS ONE, 9*(1), e85860.

Zambiazi, R. C., Przybylski, R., Zambiazi, M. W., & Mendonça, C. B. (2007). Fatty acid composition of vegetable oils and fats. *B.CEPPA, 25*(1), 111-120.

16

The bagging of *Annona crassiflora* fruits to control fruit borers

Germano Leão Demolin Leite[1*], Manoel Ferreira Souza[1], Patrícia Nery Silva Souza[1], Márcia Michelle Fonseca[1] and José Cola Zanuncio[2]

[1]*Insetário G.W.G. de Moraes, Instituto de Ciências Agrárias, Universidade Federal de Minas Gerais, Av. Universitária, 1000, Cx. Postal 135, 39404-006, Montes Claros, Minas Gerais, Brazil.* [2]*Departamento de Entomologia, Universidade Federal de Viçosa, Viçosa, Minas Gerais, Brazil.* *Author for correspondence. E-mail: gldleite@ig.com.br*

ABSTRACT. The objective of this work was to evaluate the use of plastic bags to protect the fruits of *Annona crassiflora* (Annonaceae) against *Cerconota* sp. (Lepidoptera: Oecophoridae). As protection against this fruit-boring insect, 100 fruits were enclosed in plastic bags. Another 100 fruits were not bagged. The fruits were selected from the following five ranges of diameters: $1 = 0.5 – 1.99$; $2 = 2.00 – 3.99$; $3 = 4.00 – 7.90$; $4 = 8.00 – 11.90$; and $5 = 12.00 – 16.00$ cm. The bagged fruits of various diameters were attacked less frequently by the pest. The bagged fruits with a diameter of less than two cm were not attacked. The percentage of fruits attacked and the number of larvae/fruit increased as the diameter of fruits increased in both treatments. The bagged fruits initially less than two cm in diameter showed the greatest final diameter and height.

Keywords: araticum, *Cerconota* sp., plastic bag, cultural control.

Introduction

The culture of "araticum", *Annona crassiflora* (Mart.) (Annonaceae), is very important in the Brazilian savanna, especially in the northern region of Minas Gerais, where the collection of the fruits of this plant and of *Caryocar brasiliense* (Camb.) (Caryocaraceae) represents an important source of income for the local communities (ALMEIDA et al., 1998; FERNANDES et al., 2004; LEITE et al., 2006, 2007, 2009, 2011; MELO et al., 2002). The fruit pulp of the *Annona crassiflora* contains 1.28% protein. Approximately 80% of the fatty acids are monounsaturated. Saturated fatty acids constitute 16% of the total, and polyunsaturated fatty acids constitute 4% of the total (ALMEIDA et al., 1998). The primary monounsaturated, saturated, and polyunsaturated fatty-acid constituents are oleic acid, palmitic acid,

and linolenic acid, respectively. The fruits have a high total sugar content (56.4%) and a low tannin content (0.38%) The fruits also contain vitamin C (ALMEIDA et al., 1998). However, fruit collectors in northern Minas Gerais report that attacks on fruits by insect pests have often harmed the region's level of production.

Among the fruit-boring insects known or suspected to be associated with *A. crassiflora*, *Bephratelloides pomorum* (Fabricius, 1908) (Hymenoptera: Eurytomidae) and *Cerconota anonella* (Sepp, 1830) (Lepidoptera: Oecophoridae) are considered the main pests of the Annonaceae (BROGLIO-MICHELETTI; BERTI FILHO, 2000; BROGLIO-MICHELETTI et al., 2001; SILVA et al., 2006).

The use of waxed-paper or translucent plastic bags to protect the fruits when they are still small from attacks by fruit-boring insects is one of the oldest and most effective control practices (FAORO,

2003; ROSA, 2002; SÃO JOSÉ et al., 1997). A number of studies have demonstrated the effectiveness of these control tactics in preventing the attacks of fruit borers on other Annonaceae, such as *A. muricata* L. (BROGLIO-MICHELETTI; BERTI FILHO, 2000; BUSTILLO; PEÑA, 1992; CARNEIRO; BEZERRIL, 1993; MANICA, 1994; McCOMIE, 1987).

Given the scarcity of scientific reports on insects associated with *A. crassiflora*, the objectives of this study were to identify the borer(s) of fruits in the northern region of Minas Gerais and to evaluate the effect of bagging on fruits of different sizes to identify an alternative control method for producers and gatherers of this plant.

Material and methods

The study was conducted at the "Olhos D'Água" rural community (Latitude: 16° 53 '45.2 "S Longitude: 43° 53' 21.6" W, altitude: 990 m), which is located 30 km from the Institute of Agrarian Sciences, Federal University of Minas Gerais (ICA/UFMG), Montes Claros, Minas Gerais State, Brazil, from 2006 through 2008. The study was requested by the producers, who had been observing production losses caused by the pests of *A. crassiflora*. The producers asked for a practical and inexpensive control method that would be feasible under their modest financial circumstances.

The fruits were enclosed in transparent plastic bags with a capacity of 5 L (35 cm wide and 50 cm long) with two small holes at the base to allow aeration of the fruit. Alternatively, Kraft paper bags with a 5 L capacity were used. Both types of bags were tied with cord cotton.

For this experiment, 100 trees with a high production of fruit were selected. The experimental unit was represented by 300 fruit. In all, 100 fruits were bagged with transparent plastic bags, 100 fruits with Kraft paper bags, and 100 were not bagged (control) in each of the following five diameter categories (20 fruits in each category): 1 = 0.5 - 1.99 cm; 2 = 2.00 - 3.99 cm; 3 = 4.00 - 7.90 cm; 4 = 8.00 - 11.90 cm; and 5 = 12.00 - 16.00 cm on the selected trees.

The diameter and height of each fruit were measured weekly with a ruler. Also on a weekly basis, we evaluated the effectiveness of treatments by counting the number of infected fruits and the number of insects inside the fruits of each category (diameter).

The damaged fruits were collected and taken to the Insectarium G.W.G. Moraes of the ICA/UFMG. The fruits were packed in plastic pots covered with white cloth and placed in an incubator (temperature 25°C) to allow the emergence of insects. The insects were identified by Dr. Camargo Amabílio (Embrapa Cerrados). The data were analyzed with an ANOVA and a Tukey test with a 5% significance level.

Results and discussion

The fruit-boring insects found by the study were identified as *Cerconota* sp. (Lepidoptera: Oecophoridae). The genus *Cerconota* includes important pests of other Annonaceae (JUNQUEIRA et al., 1996; BROGLIO-MICHELETTI; BERTI FILHO, 2000).

The bagging of fruits using Kraft paper bags was not effective because the bags were not rain-resistant. During the fruiting period of *A. crassiflora*, the paper bags on all experimental plots were ripped or torn open (data not shown). However, the fruits bagged with transparent plastic bags showed fewer attacks by the borers than the control (Figures 1 and 2). This difference was observed for fruits of all the diameters investigated. These findings agree with the results of a study by Broglio-Micheletti et al. (2001) that verified the efficiency of fruit bagging with plastic bags and perforated plastic bags for controlling insect borers in the fruit of *A. muricata*.

The bagged fruit less than two cm in diameter were not attacked by *Cerconota* sp., whereas 5% of the non-bagged fruit were attacked by this pest (Figure 1). The entire crop of *A. crassiflora* fruit was lost if the fruits were not bagged. When the fruits reached the harvesting stage, 100% of them had already been attacked by the borers. On average, 11 caterpillars of *Cerconota* sp. were observed per fruit (Figure 1).

Oliveira et al. (2001) investigated the control of the seed borer *B. pomorum* in the fruits of *A. muricata* and found that the fruits should be bagged soon after the flower petals fall. For *A. crassiflora*, bagging is not necessary when the petals drop, but the fruits should be bagged when they are less than two cm in diameter. Broglio-Micheletti and Berti Filho (2000) observed that fruits of *A. muricata* bagged with microperforated plastic bags were attacked 25% less frequently by *C. anonella* than fruits treated only with the insecticide Trifumuron.

We observed an increase in the percentage of damaged fruits and of larvae in the fruits as the fruit diameter increased in the different treatments (Figure 1). This observation suggests that oviposition had occurred on the fruits that were previously attacked. To avoid this problem, a recommended practice is the destruction of fruits damaged by this pest because adults continue to emerge from the fruits, even after the fruits have fallen to the ground (BROGLIO-MICHELETTI et al., 2001).

The bagged fruit with an initial diameter of below two cm showed larger final diameters and heights compared with the control (Figure 2a). The final size (diameter and height) of the bagged fruit that were more than two cm in initial diameter did not differ significantly from that of the control (Figure 1).

Oliveira (1998) found that the presence or absence of bagging did not produce a statistically significant difference in the weight of bunches of bananas. These authors observed that the effectiveness of bagging for increasing production may depend on the location where the crop is grown. The only general consensus is that bagging significantly improves the external appearance of the banana fruit.

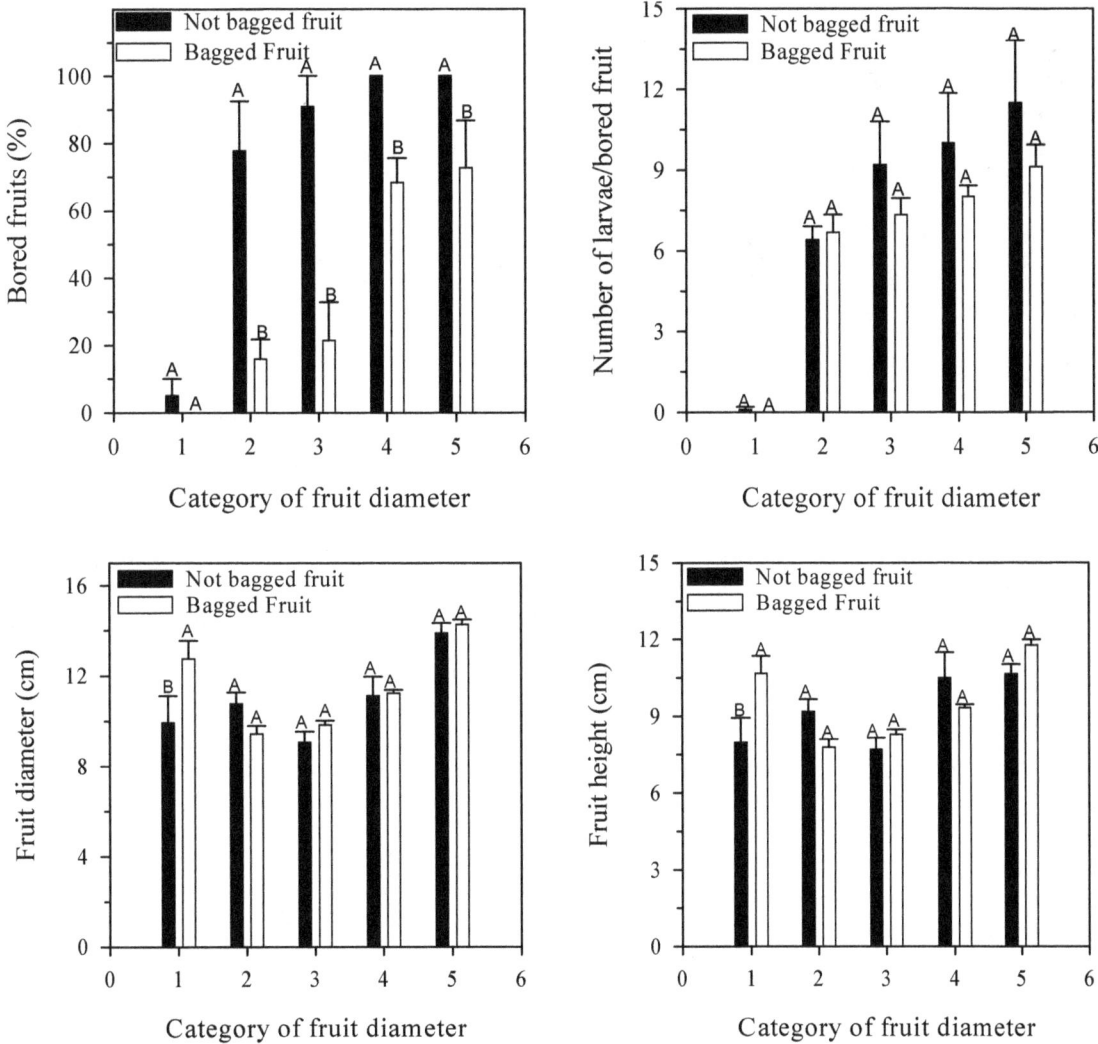

Figure 1. Effects of bagging fruits of *Annona crassifllora* to control the fruit borer *Cerconota* sp. Final diameter (cm) and height of fruit (cm) for different initial categories of fruit diameter (1 = 0.5 – 1.99; 2 = 2.0 – 3.99; 3 = 4.0 – 7.9; 4 = 8.0 – 11.9; and 5 = 12.0 – 16.0 cm). Montes Claros, Minas Gerais State. The averages, which are identified by the same letter in the histogram pairs, do not differ significantly (Tukey test, p < 0.05).

Figure 2. Fruit less than two cm in diameter (A); bagged fruit (B); undamaged fruit (bagged fruit) (C), and damaged fruit (not bagged fruit) (D).

Conclusion

The fruit-boring insect was identified as *Cerconota* sp. (Lepidoptera: Oecophoridae). The fruits of *A. crassiflora* should be bagged when they are less than two cm in diameter with a transparent plastic bag.

References

ALMEIDA, S. P.; PROENÇA, C. E. B.; SANO, S. M.; RIBEIRO, J. F. **Cerrado**: espécies vegetais úteis. Planaltina: Embrapa/CPAC, 1998.

BROGLIO-MICHELETTI, S. M. F.; BERTI FILHO, E. Controle de *Cerconota anonella* em pomar de gravioleira. **Scientia Agricola**, v. 57, n. 3, p. 557-559, 2000.

BROGLIO-MICHELETTI, S. M. F.; AGRA, A. G. S. M.; BARBOSA, G. V. S.; GOMES, F. L. Controle de *Cerconota anonella* (Sepp.) (Lep.: Oecophoridae) e de *Bephratelloides Pomorum* (Fab.) (Hym.: Eurytomidae) em Frutos de Graviola (*Annona muricata* L.) **Revista Brasileira de Fruticultura**, v. 23, n. 3, p. 722-725, 2001.

BUSTILLO, A. E.; PEÑA, J. E. Biology and control of the *Annona* fruit borer *Cerconota anonella* (Lepidoptera: Oecophoridae). **Fruits**, v. 47, n. 1, p. 81-84, 1992.

CARNEIRO, J. S.; BEZERRIL, E. F. Controle das brocas dos frutos (*Cerconota anonella*) e das sementes (*Bephrateloides maculicolis*) da graviola no Planalto da Ibiapaba, CE. **Anais da Sociedade Entomológica do Brasil**, v. 22, n. 1, p. 155-160, 1993.

FAORO, I. D. Técnica e custo para o ensacamento de frutos de pêra japonesa. **Revista Brasileira de Fruticultura**, v. 25, n. 2, p. 339-340, 2003.

FERNANDES, L. C.; FAGUNDES, M.; SANTOS, G. A.; SILVA, G. M. Abundância de insetos herbívoros associados ao pequizeiro (*Caryocar brasiliense* Cambess.). **Revista Árvore**, v. 28, n. 6, p. 919-924, 2004.

JUNQUEIRA, N. T. V.; CUNHA, M. M.; OLIVEIRA, M. A. S.; PINTO, A. C. Q. **Graviola para exportação**: aspectos fitossanitários. Brasília: Embrapa – SPI, 1996.

LEITE, G. L. D.; VELOSO, R. V. S.; CASTRO, A. C. R.; LOPES, P. S. N.; FERNANDES, G. W. Efeito do AIB sobre a qualidade e fitossanidade dos alporques de *Caryocar brasiliense* Camb (Caryocaraceae). **Revista Árvore**, v. 31, n. 2, p. 315-320, 2007.

LEITE, G. L. D.; VELOSO, R. V. S.; SILVA, F. W. S.; GUANABENS, R. E. M.; FERNANDES, G. W. Within tree distribution of a gall-inducing *Eurytoma* (Hymenoptera, Eurytomidae) on *Caryocar brasiliense* (Caryocaraceae). **Revista Brasileira de Entomologia**, v. 53, n. 4, p. 643-648, 2009.

LEITE, G. L. D.; VELOSO, R. V. S.; ZANUNCIO, J. C.; FERNANDES, L. A.; ALMEIDA, C. I. M. Phenology of *Caryocar brasiliense* in the Brazilian Cerrado Region. **Forest Ecology and Management**, v. 236, n. 2-3, p. 286-294, 2006.

LEITE, G. L. D.; VELOSO, R. V. S.; ZANUNCIO, J. C.; ALVES, S. M.; AMORIM, C. A. D.; SOUZA, O. F. F. Factors affecting *Constrictotermes cyphergaster* (Isoptera: Termitidae) nesting on *Caryocar brasiliense* trees in the Brazilian savanna. **Sociobiology**, v. 57, n. 1, p. 165-180, 2011.

MANICA, I. **Fruticultura**: cultivo das anonáceas - ata, cherimólia e graviola. Porto Alegre: Evangraf, 1994.

McCOMIE, L. D. The soursop (*Annona muricata* L.) in Trinidad, its importance, pests and problems associated with pest control. **Journal of the Agricultural Society of Trinidad and Tobago**, n. 87, p. 42-55, 1987.

MELO, J. D.; SALVIANO, A.; SILVA, J. A. **Produção de mudas e plantio de araticum**. Planaltina: Embrapa – Cerrados, 2002.

OLIVEIRA, P. E. Fenologia e biologia reprodutiva das espécies de cerrado. In: SANO, S. M.; ALMEIDA, S. P. (Ed.). **Cerrado**: ambiente e flora. Planaltina: Embrapa – CPAC, 1998. p. 288-556.

OLIVEIRA, M. A. S.; JUNQUEIRA, N. T. V.; ALVES, R. T.; ICUMA, I. M.; OLIVEIRA, J. N. S.; ANDRADE, G. A. **Broca-da-semente da graviola no Distrito Federal**. Planaltina: Embrapa – SAC, 2001.

ROSA, J. I. **Ensacamento de frutos**. Porto Alegre: Emater, 2002.

SÃO JOSÉ, A. R.; SOUZA, I. V. B.; MORAIS, O. M.; REBOUÇAS, T. N. H. **Anonáceas**: produção e mercado - pinha, graviola, atemóia e cherimólia. Vitória da Conquista: UESB, 1997.

SILVA, E. L.; CARVALHO, C. M.; NASCIMENTO, R. R.; MENDONÇA, A. L.; SILVA, C. E.; GONCALVES, G. B.; FREITAS, M. R. T.; SANT'ANA, A. E. G. Reproductive behaviour of the *Annona* fruit borer, *Cerconota anonella*. **Ethology**, v. 112, n. 10, p. 971-976, 2006.

17

Effect of foliar fungicide and plant spacing on the expression of lipoxygenase enzyme and grain rot in maize hybrids

Elizandro Ricardo Kluge[1*], Marcelo Cruz Mendes[2], Marcos Ventura Faria[3], Leandro Alvarenga Santos[4], Heloisa Oliveira dos Santos[5] and Kathia Szeuczuk[1]

[1]Departamento de Agronomia, Setor de Ciências Agrárias e Ambientais, Universidade Estadual do Centro-Oeste, Rua Simeão Camargo Varela de Sá, Vila Carli, 03, Guarapuava, Paraná, Brazil. [2]Departamento de Agronomia, Setor de Fitotecnia e Grandes Culturas, Universidade Estadual do Centro-Oeste, Guarapuava, Paraná, Brazil. [3]Departamento de Agronomia, Setor de Genética e Melhoramento, Universidade Estadual do Centro-Oeste, Guarapuava, Paraná, Brazil. [4]Departamento de Agronomia, Setor de Fitopatologia., Universidade Estadual do Centro-Oeste, Guarapuava, Paraná, Brazil. [5]Departamento de Agricultura, Setor de Tecnologia de Sementes, Universidade Federal de Lavras, Lavras, Minas Gerais, Brazil. *Author for correspondence. E-mail: elizandrokluge@gmail.com

ABSTRACT. This study was carried out with the objective of evaluating the effect of fungicide application on grain rot in commercial maize hybrids and the relation between grain rot and the expression of lipoxygenase enzyme in grain in conventional row spacing of 0.70 m and reduced row spacing of 0.45 m. Treatments were made in a 3 x 8 factorial scheme, using three forms of management with fungicide (Trifloxystrobin + Prothioconazole) and eight maize hybrids divided into two groups (tolerant and susceptible) with three repetitions, totaling 72 plots in each environment (conventional and reduced spacing) in the 2013/2014 crop. The following characteristics were evaluated: grain rot percentage and lipoxygenase enzyme expression (LOX) in the grain. The hybrid and the fungicide utilized influenced the grain rot percentage. Grain rot percentage was reduced by the use of the fungicide, and the highest reduction was in susceptible hybrids with two applications, V8 and V8+VT. There was higher expression of LOX enzyme in maize hybrids that belong to the group tolerant of fungi that cause grain rot .The use of the fungicide in two applications, V8 (eight leaves) and VT (tasseling), increased the intensity of the LOX enzyme, which was more evident for the reduced spacing.

Keywords: LOX, spindle rot, chemical control, reduced spacing, Zea mays.

Introduction

Maize (*Zea mays* L.) is one of the most important and ancient crops of the world (Werle, Nicolay, Santos, Borsoi, & Secco, 2011). This species is cultivated in many different environments and climates at different latitudes, from Russia to Argentina. The culture has a wide geographic range of cultivation and can be developed in many different soil and climate conditions; consequently, it is possible to use various commercial hybrids that have sundry tolerance levels to leaf and grain pathogens (Pozar, Butruille, Diniz, & Viglioni, 2009). The occurrence of these pathogens causes reductions in grain yield and grain health quality because infection by these fungi results in paralysis of the normal process of grain filling and reduces maize cob weight.

Northern leaf blight (caused by *Exserohilum turcicum*), common rust (caused by *Puccinia sorghi*) and gray leaf spot (caused by *Cercospora zeae-maydis*) are important foliar diseases of maize (Juliati & Souza, 2005). Mendes, Von Pinho, Machado, Albuquerque, and Falquete (2011) highlights the rot of the cob caused mainly by fungi present in the field such as *Fusarium verticilioides*, *Stenocarpella maydis*, and *Stenocarpella macrospora*, which cause grain rot in maize. The indiscriminate use of susceptible hybrids, intensive planting systems and the improper use of high technology, associated with the occurrence of favorable climate for epidemic development, have contributed to an increase of the diseases of importance in maize and the use of fungicides (Brito, Von Pinho, Pozza, Pereira, & Faria Filho, 2007). Thus, it becomes increasingly important to select the correct genetic material to be used (Mendes, Pereira, Von Pinho, & Balestre, 2012), and in recent years, the discussion of management strategies to reduce the disease in a sustainable manner has increased, using crop rotation, genotype and especially the adoption of chemical control.

Sequential degradation of lipids, which are primary products of the lipoxygenase reaction, occurs when plant tissues are damaged by pathogens or mechanically. These enzymes are activated and oxidize fatty acids, producing a certain concentration of aldehydes and volatile compounds that inhibit the formation and development of fungus in grain (Mendes et al., 2012).

Currently, the adoption of reduced spacing associated with the use of modern and not modern maize hybrids changes the spatial arrangement of maize plants in the field. This can change the microclimate in a positive way, increasing the photosynthetically active radiation interception by the canopy; it can also be observed that there is an increase in the absorption efficiency of nutrients and water, exerting influence on maize grain yield, possibly negatively influencing the hybrids' tolerance to grain diseases and the efficiency of control by the active component of fungicides.

The efficiency of chemical control in the management of grain rot in maize is still a matter of doubt in relation to the efficiency of fungicides and number of applications, and research that can clarify the effects of fungicides on pathogens associated with cob diseases and the relation to the susceptibility of the hybrid used is lacking.

Thus, due to the new maize hybrids that have been launched in the market every year with high yield potential, it is evident the importance of this research to elucidate the results obtained from the use of fungicide in maize, its association with the activity of the specific enzymes, the presence of pathogen agents that cause cob rot and different plant spacing. Therefore, the objective of this work was to evaluate the effect of fungicide application on grain rot in commercial maize hybrids, and the relation between grain rot and lipoxygenase enzyme expression in the grain, in conventional and reduced spacing.

Material and methods

Two experiments were conducted in the crop season of 2013/2014, in Guarapuava, State Paraná, Brazil. The first experiment (environment 1) was conducted at Cedeteg with conventional row spacing of 0.70 m and at a latitude of 25° 23' 04.83" south, a longitude of 51° 29' 44.32" west, and an altitude of 1,028 m. The second experiment (environment 2) was installed at Três Capões Farm with reduced row spacing of 0.45 m and at a latitude of 25° 26' 57.79" south, a longitude of 51° 38' 29.18" west, and an altitude of 948 m. In both experiments, a population of 75,000 plants per hectare was used. This experiment was conducted on the no-tillage system, in areas where there was white oat (*Avena sativa*) as ground cover. The topography is considered plane. The region has a humid mesothermal subtropical climate. The climate classification proposed by Köppen is Cfb type without a defined dry season, cool summers and winters with severe and frequent frosts. The annual average temperature is 16.8°C, ranging from 6.8°C (minimum average) to 36°C (maximum average), and the annual average total rainfall is 1,500 mm with an annual average relative humidity of 77.9%. The soil is classified as Dystrophic Haplohumox Oxisol, clayey texture (*Empresa Brasileira de Pesquisa Agropecuária* [Embrapa], 2006).

The experimental design was in random blocks, with three replications, in a 3 x 8 factorial scheme, totaling 24 treatments and 72 plots in each environment (conventional and reduced spacing) in the crop season of 2013/2014. The first factor consisted of three levels of foliar fungicide application (Trifloxystrobin 150.0 g L^{-1} (15.0% m/v) + Prothioconazole 175.0 g L^{-1} (17.5% m/v): at stage V8 (eight expanded leaves) with 0.4 L ha^{-1}, at stage VT (tasseling) with 0.5 L ha^{-1} and control (without fungicide). The second factor had eight maize hybrids divided into two groups according to their reaction to the causative fungus of the grain rot complex: tolerant (AG 9045PRO, AG 8041PRO, DKB 245PRO2, and 2B707PW) and susceptible (P 32R48H, DKB 390PRO, P 30F53H, and P 30R50H).

The experiments were installed in the first fortnight of October 2013, and the harvest occurred in the second fortnight of March 2014, after physiological maturity. The first experiment had a row spacing of 0.70 m and a length of 5 m, four rows and a total area of 14 m² per plot. The second experiment had a row spacing of 0.45 m and a length of 5 m, four rows and a total area of 9 m². In both experiments, the two central rows of each plot were used.

The fungicide applications were carried out with the aid of a CO_2 pressurized costal sprayer, equipped with four nozzles, hollow cone spray nozzles (0.3) spaced at 0.5 m, consumption of 200 L ha⁻¹ and 3.6 km h⁻¹ of speed displacement. Climatic variables at the time of application (start and end) were monitored by a digital anemometer. Meteorological data were collected every 10 days at the IAPAR/UNICENTRO/ CEDETEG experimental station located approximately 500 meters from the area of the first experiment.

The percentage of grain rot and the evaluation of the lipoxygenase enzyme were performed. The percentage of grain rot was determined according to the procedure proposed in ordinance No. 11 of April 12, 1996 (Brasil, 1996). The electrophoretic enzyme analyses, in both environments, were performed in the central laboratory of seeds belonging to the Department of Agriculture of the Universidade Federal de Lavras, Lavras, Minas Gerais State, Brazil. To perform the analysis, composite grain samples derived from three replicates of each treatment were used, and the gel analysis was made as in Alfenas (2006). To carry out the electrophoretic run, 50 mL of each supernatant was applied in the gel groove, and the running was at 4°C at 120 V for approximately 8 hours.

After that, the gels were revealed in the presence of a substrate specific for the particular enzyme (Alfenas, 1998).

Percentage of grain rot was submitted to homogeneity of variances testing by Harley's test (Ramalho, Ferreira, & Oliveira, 2000). After that, individual and joint variance analyses involving the cultivation environments were performed, and the averages were grouped by the Scott-Knott test at the 5% probability level using SISVAR® software as the statistical program (Ferreira, 2011).

Results and discussion

There was an accumulated rainfall level of 1,008 mm in environment 1 (CEDETEG) throughout the culture cycle. During the initial phase of the experiment implementation, shortly after the sowing in October and November, 230 mm of accumulated rainfall volume was verified, ensuring good initial development for the culture. Cumulatively, 778 mm of rainfall was observed during the fungicide application and in the subsequent months until the harvest. In environment 2 (Três Capões Farm) with reduced spacing, there was an accumulated rainfall of 896 mm during the culture cycle. During the initial phase of implementation, shortly after the sowing in October and November, there was good rainfall volume; 215 mm of accumulated rainfall volume was observed, ensuring good initial development for the culture. During the fungicide application and in the subsequent months until the harvest, rainfall was observed for the study area with accumulation of 681 mm (Figure 1).

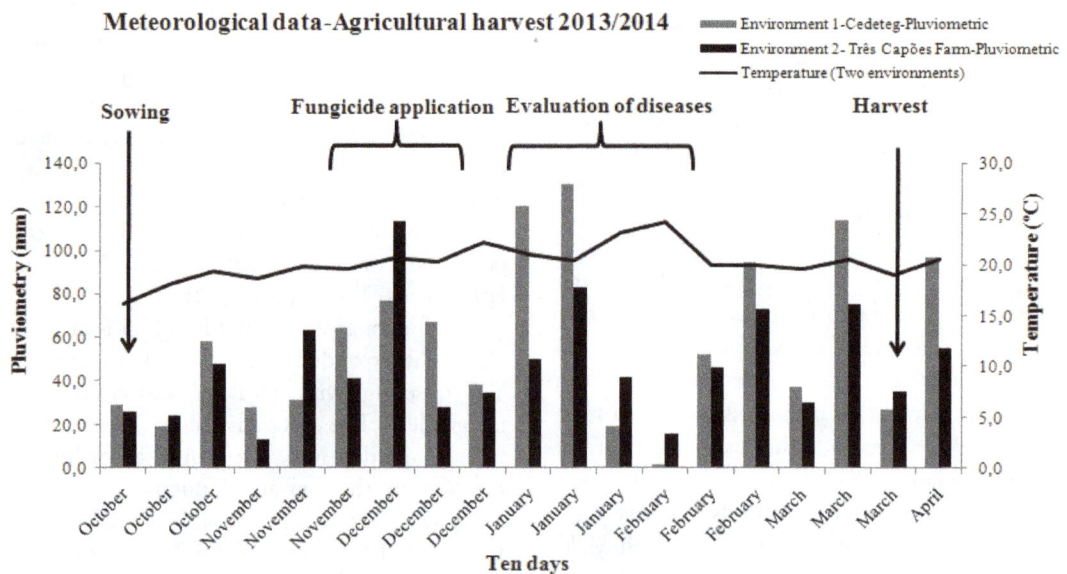

Figure 1. Pluviometry and temperature for ten days in Guarapuava, Paraná State, Brazil, CEDETEG (environment 1) and Três Capões Farm (environment 2) from October to April in the 2013/2014 crop season.

Regarding the results obtained in relation to the minimum and maximum temperatures, it is important to note that the two experiments are in the same municipality and that the season is close in the two experiments, so it was decided to use the average temperature. According to the results of the joint variance analyses, for the percentage of grain rot, a significant effect was observed (p < 0.01) in the interaction hybrid x environment (Table 1).

Table 1. Percentage of grain rot obtained for the different hybrids and treatments with fungicide (C-Control, V8-eight expanded leaves and VT-tasseling) in two locations in Guarapuava, Paraná State, Brazil, crop season of 2013/2014.

Hybrids	Environment 1				Environment 2			
	C	V8	V8+VT	Mean	C	V8	V8+VT	Mean
AG 9045 PRO	12.1 b	14.6 a	6.6 b	11.1 bA	13.1 a	12.4 a	6.7 b	10.8 aA
DKB 245 PRO2	7.5 b	10.5 b	4.5 b	7.4 cA	10.5 a	6.0 b	4.2 b	6.9 bA
AG 8041 PRO	10.0 b	3.4 c	8.7 b	7.4 cA	8.8 a	6.3 b	5.5 b	6.8 bA
2B 707 PW	3.4 b	3.4 c	4.0 b	3.6 cA	6.5 a	2.9 b	3.6 b	4.3 bA
Group 1★	8.3 bA	7.8 bA	6.0 bA		9.7 aA	6.9 bB	5.0 bB	
P 32R48H	23.0 a	17.1 a	28.0 a	22.7 aA	15.4 a	16.2 a	13.2 a	14.9 aB
DKB 390 PRO	26.6 a	22.6 a	21.7 a	23.6 aA	9.6 a	11.8 a	15.3 a	12.3 bB
P 30F53H	17.0 a	12.0 b	8.1 b	12.4 bA	11.7 a	11.9 a	5.7 b	9.8 bA
P 30R50H	8.8 b	8.9 b	3.7 b	7.2 cA	4.7 a	11.4 a	4.7 b	7.0 bA
Group 2★	18.8 aA	15.2 Aa	15.4 aA		10.4 aA	12.8 aA	9.7 aA	

Means followed by the same lower case letter in columns for each fungicide treatment and capital letter in line for treatments with fungicide and the average of each environment do not differ statistically from each other by Scott-Knott Test at the 5% probability level. Environment 1 Cedeteg (conventional spacing - 0.70 m) and Environment 2 Três Capões Farm (reduced spacing 0.45 m), both in Guarapuava, Paraná State, Brazil. ★ Group 1: Hybrids considered tolerant to grain rot; and Group 2: Hybrids considered susceptible to grain rot.

In environment 1, a significant difference was verified between the hybrids within each treatment with fungicide; in the check treatment, for grain rot, there was a higher percentage in DKB 390PRO, P 32R48H and P 30F53H (susceptible hybrids), all belonging to group 2 (Table 1). In Brito, Pereira, Von Pinho, and Balestre (2012), the use of fungicides (Azoxystrobin + Cyproconazole) in maize possibly caused a reduction in the grain rot percentage.

In treatment V8, there was a higher percentage for hybrid AG 9045PRO (group 1) but it did not differ statistically from hybrids DKB 390PRO and P 30F53H (belonging to group 2). It is possible to note that hybrids AG 8041PRO and 2B707PW (belonging to group 1) had the lowest percentage of grain rot. In relation to V8 + VT, taking into consideration the grain rot characteristic, there was a higher percentage for hybrids P 32R48H and DKB 390PRO (belonging to group 2) (Table 1). According to Casa, Reis, and Zambolim (2006), the fungus *Stenocarpella* spp. is mainly associated with grain rot, and it may be the principal agent of grain rot in maize, which may have occurred in this research. The use of fungicides is recommended for maize in hybrids that present susceptibility to the disease (Costa Cota, Silva, Lanza, & Figueiredo, 2012).

Evaluating the utilized treatments in the hybrid groups, in groups 1 and 2, there was no significant difference between treatments (Table 1). However, when the hybrid groups are compared for each treatment, there was a significant difference in all treatments with evaluated fungicide (check, V8 and V8 + VT); group 2 (susceptible) presented higher percentage of grain rot in environment 1 (conventional spacing), emphasizing the importance of the genotype evaluated (Table 1). Juliatti, Zuza, Souza, and Polizel (2007), which used the fungicides Pyraclostrobin + Epoxiconazole in two applications, reported the efficiency of the products at reducing the percentage of grain rot. The performance of two foliar applications (V8 + pre-tasseling) or one application at pre-tasseling of the fungicide azoxystrobin + cyproconazole resulted in a lower percentage of the fungus *Fusarium* sp. in the harvested grain in the two sowing periods (Stefanello, Bachi, Gavassoni, Hirata, & Pontim, 2012).

In environment 2, it was verified that in the check treatment, there was no significant difference when analyzing the hybrids (Table 1). One of the factors that should be further studied is the maize culture response to the arrangement of plants in the area. It is possible to distribute the plants in many ways in an area, variation of the spacing between rows and between plants being responsible for the different plant arrangements. This practice has been intensified in producer regions, yet the different hybrids respond differently to variations in the arrangement and density of plants (Demetrio, Filho, Cazetta, & Cazetta, 2008) and also in relation to the disease severity.

In the V8 treatment, there was a higher percentage for hybrid AG 9045PRO (group 1), but it did not differ statistically from hybrids P 32R48H, DKB 390PRO, P 30F53H, and P 30R50H hybrids (belonging to group 2). It can be noted that in hybrid AG 9045PRO of group 1 (tolerant), even the foliar fungicide application did not show effectiveness at reducing the grain rot percentage (Table 1). Studies by Duarte, Juliatti, Lucas, and Freitas (2009) corroborate the results obtained in this study, and the foliar application of fungicides in some maize genotypes did not result in fungus percentage control.

In the V8 + VT treatment, there was higher percentage for the P 32R48H and DKB 390PRO hybrids (belonging to the group 2) (Table 1). These data corroborate Brito et al. (2012), who noted that the use of foliar fungicide application makes a reduction of the grain rot percentage possible.

Comparing the groups of hybrids with fungicide treatments, in group 1 (tolerant), there was a significant difference between treatments for grain rot, where the higher percentage occurred in the check treatment compared to the V8 and V8 + VT treatments; therefore, the fungicide application reduced the grain rot percentage (Table 1). It is possible to favor a microclimate, using reduced spacing, with the increase or decrease in plant population, and this favors the percentage of disease. There was no significant difference between treatments from group 2 (susceptible) evaluated for grain rot (Table 1). Silva, Cunha Junior, Assis, and Imolesi (2008) found that hybrid P30K75, when it is cultivated with reduced spacing (50 cm of row spacing), showed higher severity index of foliar diseases. For group 2 (susceptible), there was no significant difference between the evaluated treatments for grain rot (Table 1).

However, when the groups of hybrids in each treatment were compared, there was a significant difference for grain rot, and in the V8 and V8 + VT treatments, group 1 showed a lower grain rot percentage compared to group 2; in the check treatment, there was no significant difference between the groups of evaluated hybrids (Table 1). Fungicide application is effective in the control of foliar diseases, provides higher grain yield and reduces the percentage of grain rot. An important factor in foliar disease development is the effect of climate, and the environment is an important component of host-pathogen-environment interactions. Cunha and Pereira (2009) used the fungicide Pyraclostrobin + Epoxiconazole, which provided disease control, as reflected in the yield, which was on average 16.3% higher than the evaluated check.

When the P 30R50H hybrid was evaluated for grain rot (environment 2), a similar behavior to the previous was observed a lower percentage of grain rot only in treatment V8 + VT; there was a significant difference when compared with the check, showing that the response to the fungicide application (Trifloxystrobin + Prothioconazole) may be higher susceptibility due to the reduced spacing, presenting a lower response to fungal treatments in a preventive way. In environment 1, this hybrid showed a lower percentage of grain rot in all treatments, even in the check treatment, possibly with interference from the space, which is higher in environment 1, with less percentage of grain rot (Table 1).

The high percentage of grain rot from group 2 (susceptible) is closely related to the susceptibility of these hybrids to the diseases of the grain rot complex. Duarte et al. (2009) tested fungicides on maize and reported that there was a reduction in the percentage of grain rot due to the foliar application and fungicide association (Azoxystrobin + Cyproconazole).

Another factor that may have contributed to this result of higher percentage of grain rot in the hybrids

of group 2 is the high volume of rainfall at the flowering stage of the culture, corresponding to the period of December and January, which are months considered critical for the occurrence of grain rot. There is evidence that the use of fungicide in maize culture enables the reduction in the percentage of grain rot; according to Juliatti et al. (2007), the foliar application of triazole and strobilurinfungicides (Pyraclostrobin + Epoxiconazole, Azoxystrobin + Cyproconazole and Azoxystrobin) resulted in a lower percentage of grain rot. In recent years, research has shown the effectiveness of fungicide application in the management of foliar diseases and in reducing the damage caused by them to productivity (Cunha, Silva, Boller, & Rodrigues, 2010).

It was also verified that for grain rot, there was a higher percentage in almost all groups of maize hybrids from group 2, in both environments evaluated, with the exception of the P 30F53H hybrid; regarding environment 1, in the treatments V8 and V8 + VT, there was a significant difference, and a lower percentage of grain rot occurred compared with the check treatment, showing a response to the Trifloxystrobin + Prothioconazole fungicide application. In many cases, the damage caused by foliar diseases in maize is considered indirect, leaving the plant weakened by reducing the leaf area and thus vulnerable to the entry of pathogens that cause rot of the stalk, grains and roots (Jardine and Laca-Buendía, 2009).

There was no significant difference for grain rot in environment 2; the hybrid P 30F53H showed a lower percentage only for the V8 + VT treatment. There may be higher susceptibility due to the reduced spacing, having a lower response to the fungicide treatments in a preventive way (Table 1). The use of Azoxystrobin + Cyproconazole in foliar application at pre-tasseling reduced the percentage of grain rot 5.12%, beyond an increase in productivity by 12.4% of different hybrids cultivated under high disease severity with and without fungicide application (Brito, Von Pinho, Souza Filho, & Altoé, 2008). Juliatti, Nascimento, and Rezende (2010) observed that all treatments with fungicide (strobilurin + triazole) provided an increase in thousand grain weight compared to the check treatment, showing a direct relation of the control of diseases with grain filling.

In relation to the averages of grain rot in the treatments used in environments 1 and 2, it was observed that there was a significant difference between the environments; analyzing the hybrids, a higher percentage of grain rot occurred in the P 32R48H and DKB 390PRO hybrids (both belonging to group 2) in environment 1 (Table 1). Additionally, as shown in Table 1, the percentage was high and similar in environments 1 and 2. Compared with the data obtained for precipitation (Figure 1), it can be observed that there was no interference of precipitation and temperature.

However, it was found that there was a higher percentage of grain rot in maize hybrids belonging to group 2 in both evaluated environments, but in environment 2, a positive response was noted only with two fungicide applications of Trifloxystrobin + Prothioconazole (V8 and V8 + VT); therefore, there was environmental interference in the percentage of grain rot, which possibly the reduced the microclimate conditions caused by spacing, causing an increase of causative agent attack. When the plant population is larger, there is higher demand for nutrients and water, resulting in plants more sensitive to pathogen infection. Therefore, hybrid yield potential must be evaluated using the recommended population. Increases in population, thousand grain weight and plant weight favor higher grain yields, which are decreased by the increased occurrence of foliar diseases (Silva, Teixeira, Martins, Simon, & Francischini, 2014).

Comparing the data obtained in environments 1 and 2 and the grain rot shown in Table 1, it can be verified that in the check, V8 and V8 + VT treatments, the hybrids belonging to group 1 (tolerant to rot grains) showed a lower percentage of grain rot, except th e check treatment in environment 2, where there was no significant difference between the groups of hybrids.

LOX expression was determined for all hybrids studied and all the treatments evaluated in environment 1 (Figure 2). In the check treatment, it was possible to note higher expression of LOX enzyme for hybrids belonging to the group considered tolerant of grain rot.

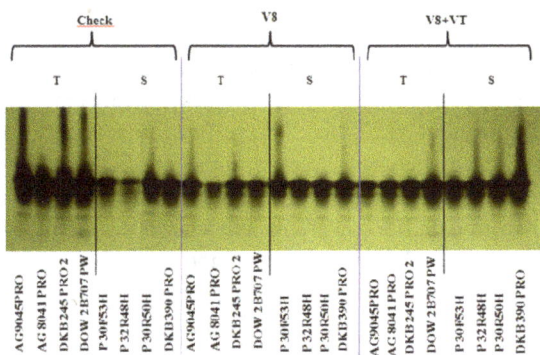

Figure 2. Lipoxygenase electrophoretic pattern in maize hybrids tolerant(T) and susceptible (S) to fungi that cause grain rot; the use of fungicide in stages V8, V8 + VT and the control produced in environment 1 in Guarapuava, Paraná State, Brazil, crop season of 2013/14

The hybrids belonging to group 1 have a lower percentage of grain rot compared with group 2

(Table 1), proving the results obtained by Zeringue, Brown, Neucere, and Cleveland (1996), which included higher expression of LOX and a consequently lower percentage of grain rot. The same authors suggest that the LOX enzyme, by lipoxygenation of linoleic acid, would be able to produce volatile aldehydes with chains of 6 to 12 carbons, inhibiting or preventing fungus development and even mycotoxin formation. According to Mendes et al. (2012), genotypes of maize with a high activity of lipoxygenase enzyme have greater ability to resist a fungal attack, resulting in lower rates of grain rot in the harvest. In addition, the fungus infection affects the quality of maize grain by the production of mycotoxins that cause damage to health (human and animal) due to the toxic activity that can be exerted on the organism (Kumar, Basu, & Rajendran, 2008).

Studies have shown that the resistance of maize genotypes to grain rot is also correlated with the content of polyunsaturated fatty acids, specifically with linoleic acid, and the resistance mechanism is through the production of LOX. Lipoxygenase is related to the oxylipin formation through the oxygenation of fatty acids. The oxylipins are a group of signaling molecules such as jasmonate, mitel-jasmonate (mitel-jasmonato), etc., compounds that act in plant defense (Kazan & Manners, 2008). Other studies have reported that other derivatives of lipoxygenase act in plant defense through the regulation of gene expression, cell death or antimicrobial products (Kishimoto, Matsui, Ozawa, & Takabayashi, 2008).

In the V8 treatment, as shown in Figure 2, the same enzyme expression was noted. For the two groups of hybrids, the fungicide application of Trifloxystrobin + Prothioconazole did not change the expression of LOX.

However, for the V8 + VT treatment, the same did not occur; therefore, it was possible to notice a higher expression of LOX enzyme for hybrids belonging to the group considered susceptible to grain rot, hybrids P 30F53H, P 32R48H, P 30R50H, and DKB 390PRO (Figure 2).

Based on the data presented in Table 1, the P32R48H and DKB 390PRO hybrids are in the group with the highest percentage of grain rot, but for the P30F53H and P30R50H hybrids, the percentage of grain rot was comparable to the hybrids considered tolerant. In this sense, it can be inferred that one of the possible contributions of the fungicide used, based on Trifloxystrobin + Prothioconazole when applied in V8 + VT, is to the route of the expression of the LOX enzyme. Studies related to the influence of chemical compounds and the activity of specific enzymes in maize grain on the resistance of genotypes associated with the pathogens that cause grain rot are being carried out in Brazil.

Several controlling loci of quantitative traits (QTLs) with the additive action of genes are associated with the resistance, which shows that this resistance is a horizontal or polygenic type. These QTLs may be determined by the use of microsatellite markers (Juliatti et al., 2009) and SNPs (Pozar et al., 2009).

In research conducted by Pozar et al. (2009) in Brazil, QTLs related to gray leaf spot resistance have been mapped and characterized through SNP and SSR markers.

The US Department of Agriculture conducted several experiments to prove the resistance of some varieties of maize due to different levels of fatty acids, mainly linoleic acid associated with lipoxygenase enzyme presence (Zeringue et. al., 1996).

The expression of LOX was possible to observe for all hybrids studied and all treatments evaluated in environment 2. In the check treatment, it was possible to notice higher expression of the LOX enzyme for the hybrids belonging to the group considered tolerant of grain rot, represented in figure 3 by the letter T, when compared with the group considered susceptible to the fungus, represented by S, in this same treatment. These results corroborate Mendes et al. (2012), which evaluated groups of hybrids considered resistant and hybrids considered susceptible to the grain rot complex for the LOX enzyme; in this study, the electrophoretic profiles for lipoxygenase revealed a higher intensity of bands for the hybrids resistant to fungus that causes grain rot in maize.

In the V8 treatment (Figure 3), it was possible to note higher expression of LOX for the hybrids belonging to the group considered susceptible to grain rot, showing that the fungicide application at V8 interfered in the expression of LOX.

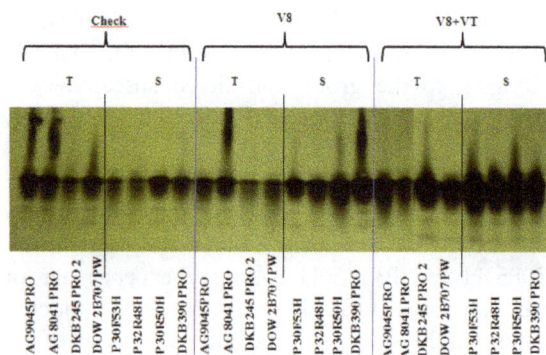

Figure 3. Electrophoretic pattern of lipoxygenase in maize hybrids considered tolerant and susceptible to fungi that cause grain rot; the use of fungicide in stages V8, V8 + VT and the control produced in environment 2 in Guarapuava, Paraná State, Brazil, crop season of 2013/14

In the V8 + VT treatment, it was possible to note higher expression of the LOX enzyme in the hybrids belonging to the group considered susceptible to grain rot, hybrids P 30F53H, P 32R48H, P 30R50H, and DKB 390PRO, represented in Figure 3 by the letter S, when compared to the group considered tolerant to fungus, represented by the letter T; however, even in the group considered susceptible to grain rot, strong expression of LOX was noted.

When the results in Table 1 are compared, hybrids P 30F53H and P 30R50H reduced the percentage of grain rot; therefore, it is economically viable, it affects the quality and price obtained for the grain, and this occurred in the reduced spacing and in the conventional spacing. Therefore, with two fungicide applications, it is possible to infer that there is a relation of the active principle's action and LOX expression. The effects of LOX action can contribute to defense reactions, inhibiting pathogen development and inducing the phytoalexins (Latunde-Dada & Lucas, 2001).

Thus, for the maize hybrids with high activity of the enzyme lipoxygenase, these genotypes have a greater ability to resist fungal attack at the end of the cycle; with that, lower production of grain rot occurs in the harvest. The lipoxygenase product can act either as an intermediate or as an end product of a metabolic route. In plants, lipoxygenases are involved in the biosynthesis of jasmonic acid and aldehyde that are responsible for defense mechanism signaling (Quaglia, Fabrizi, Zazzerini, & Zadra, 2012; Senthilraja, Anand, Kennedy, Raguchander, & Samiyappan, 2013).

Analyzing the data in environments 1 and 2 associated with grain rot (Table 1), it is possible to note that in the check, V8 and V8 + VT treatments, the hybrids belonging to group 1 (considered tolerant of grain rot) present a lower percentage of grain rot, except the check treatment in environment 2, where there was no significant difference between the groups of hybrids; the reduced spacing had a possible influence in this environment. However, in the check treatment, an increase of LOX enzyme expression occurred (Figures 2 and 3) for the hybrids belonging to the group considered tolerant of grain rot, as shown in Table 1. The hybrids belonging to group 1 have a lower percentage of grain rot when compared to group 2, and there was a direct relation between the lower percentage of grain rot and the higher expression of the lipoxygenase enzyme, proving the results obtained by Zeringue et al. (1996), which reported that higher expression of the LOX enzyme indicates a lower percentage of grain rot. The same authors suggest that the LOX enzyme, via lipoxygenation of linoleic acid, would be able to produce volatile aldehyde chains with 6 to 12 carbons, inhibiting or preventing development and even mycotoxin formation. Thus, the increase in enzyme expression may be related to the application of the fungicide Trifloxystrobin + Prothioconazole at V8 + VT reducing the cob rot.

Therefore, a positive response to the fungicide application (Trifloxystrobin + Prothioconazole) was observed, mainly in the percentage of grain rot in these hybrids; the reduction of grain rot was 2.3% and 3.4% in V8 + VT in the group of hybrids (tolerant and susceptible) in the 1st experiment, and in the 2nd experiment, the reduction was 3.4% and 0.7%, respectively, and in relation to lipoxygenase enzyme activity, evidencing the importance to continue this line of research and providing better understanding of active ingredient application and the action of the enzymes in the grain.

Conclusion

The fungicide (Trifloxystrobin + Prothioconazole) reduced the incidence of grain rot in two applications, V8 and V8 + VT, and these results were more evident in the hybrids considered susceptible.

There was higher expression of the enzyme LOX in maize hybrids belonging to the group considered tolerant of fungi that cause grain rot.

The use of the fungicide (Trifloxystrobin + Prothioconazole) in two applications, V8 (eight leaves) and VT (tasseling), increased the intensity of the LOX enzyme, which was more evident in the reduced spacing.

Acknowledgements

CAPES is acknowledged for providing a master's scholarship to the first author. CNPq, FINEP and Fundação Araucária are thanked for financial support.

References

Alfenas, A. C. (1998). *Eletroforese de isoenzimas e proteínas afins: fundamentos e aplicações em plantas e microrganismos.* Viçosa, MG: UFV.

Alfenas, A. C. (2006). *Eletroforese e marcadores bioquímicos em plantas e microrganismos.* Viçosa, MG: UFV.

Brasil (1996). Portaria n° 11, de 12 de abril de 1996 estabelece critérios complementares para classificação do milho. *Diário Oficial da União*, Brasília, n. 72.

Brito, A. H., Von Pinho, R. G., Pozza, E. A., Pereira, J. L. A. R., & Faria Filho, E. M. (2007). Efeito da Cercosporiose no rendimento de híbridos comerciais de milho. *Fitopatologia Brasileira, 32*(6), 472-479.

Brito, A. H., Von Pinho, R. G., Souza Filho, A. X., & Altoé, T. F. (2008). Avaliação da severidade da cercosporiose e rendimento de grãos em híbridos comerciais de milho. *Revista Brasileira de Milho e Sorgo, 7*(1), 19-31.

Brito, A. H., Pereira, J. L. A. R., Von Pinho, R. G., & Balestre, M. (2012). Controle químico de doenças foliares e grãos ardidos em milho (*Zea mays* L.). *Revista Brasileira de Milho e Sorgo, 11*(1), 49-59.

Casa, R. T., Reis, E. M., & Zambolim, L. (2006). Doenças do milho causadas por fungos do gênero *Stenocarpella*. *Fitopatologia Brasileira, 31*(5), 427-439.

Costa, R. V., Cota, L. V., Silva, D. D., Lanza, F. E. & Figueiredo, J. E. F. (2012). Eficiência de fungicidas para o controle de mancha branca no milho. *Revista Brasileira de Milho e Sorgo, 11*(3), 291-301.

Cunha, J. P. A. R., Silva, L. L., Boller, W., & Rodrigues, J. F. (2010). Aplicação aérea e terrestre de fungicida para o controle de doenças do milho. *Revista Ciência Agronômica, 41*(3), 366-372.

Cunha, J. P. A. R., & Pereira, R. G. (2009). Efeito de pontas e volumes de pulverização no controle químico de doenças do milho. *Revista Ciência Agronômica, 40*(4), 533-538.

Demetrio, C. S., Filho, D. F., Cazetta, J. O., & Cazetta, D. A. (2008). Desempenho de híbridos de milho submetidos a diferentes espaçamentos e densidades populacionais. *Pesquisa Agropecuária Brasileira, 43*(12), 1691-1697.

Duarte, R. P., Juliatti, F. C., Lucas, B. V., & Freitas, P. T. (2009). Comportamento de diferentes genótipos de milho com aplicação foliar de fungicida quanto à incidência de fungos causadores de grãos ardidos. *Bioscience Journal, 25*(4), 112-122.

Empresa Brasileira de Pesquisa Agropecuária [Embrapa]. (2006). *Sistema Brasileiro de Classificação de Solos* (2a ed.). Rio de Janeiro, RJ: Embrapa Solos.

Ferreira, D. F. (2011). Sisvar: a computer statistical analysis system. *Ciência e Agrotecnologia, 35*(6), 1039-1042.

Jardine, D. F., & Laca-Buendía, J. P. (2009). Eficiência de fungicidas no controle de doenças foliares na cultura do milho. *Fazu em Revista, 26*(6), 11-52.

Juliatti, F. C., & Souza, R. M. (2005). Efeitos de Épocas de Plantio na severidade de doenças foliares e produtividade de híbridos de milho. *Bioscience Journal. 21*(1), 103-112.

Juliatti, F. C., Zuza, J. L. M. F., Souza, P. P., & Polizel, A. C. (2007). Efeito do genótipo de milho e da aplicação foliar de fungicidas na incidência de grãos ardidos. *Bioscience Journal, 23*(2), 34-41.

Juliatti, F. C., Pedrosa, M. G., Silva, H. D., & Silva J. V. C. (2009). Genetic mapping for resistance to Gray leaf spot in maize. *Euphytica, 169*(2), 227-238.

Juliatti, F. C., Nascimento, C., & Rezende, A. A. (2010). Avaliação de diferentes pontas e volumes de pulverização na aplicação de fungicida na cultura do milho. *Summa Phytopathological, 36*(3), 216-221.

Kazan, K., & Manners, J. M. (2008). Jasmonate Signaling: toward na integrated view. *Plant Physiology, 146*(4), 1459-1468.

Kishimoto, K., Matsui, K., Ozawa, R., & Takabayashi, J. (2008). Direct fungicidal activities of C6-aldehydes are important constituents for defense responses in *Arabidopsis* against *Botrytis cinerea*. *Phytochemistry*, *69*(11), 2127-2132.

Kumar, V., Basu, M. S., & Rajendran, T. P. (2008). Mycotoxin research and mycoflora in some commercially important agricultural commodities. *Crop Protection*, *27*(6), 891-905.

Latunde-Dada, A. O., & Lucas, J. A. (2001). The plant defense activador acibenzolar-S-Methyl primes cowpea [(*Vigna unguiculata* (L.) Walp.)] seedlings for rapid induction of resistance. *Physiological and Molecular Plant Pathology, 58*(5), 199-208.

Mendes, M. C., Von Pinho, R. G., Machado, J. C., Albuquerque, C. J. B., & Falquete J. C. F. (2011). Qualidade sanitária de grãos de milho com e sem inoculação a campo dos fungos causadores de podridões de espiga. *Ciência e Agrotecnologia*, *35*(5), 931-939.

Mendes, M. C., Von Pinho, R. G., Von Pinho, E. V. R., & Faria, M. V. (2012). Comportamento de híbridos de milho inoculados com os fungos causadores do complexo grãos ardidos e associação com parâmetros químicos e bioquímicos. *Ambiência*, *8*(2), 277-279.

Pozar, G., Butruille, D., Diniz, H. S., & Viglioni, J. P. (2009). Mapping and validation of quantitative trait loci for resistance to *Cercospora* infection in tropical maize (*Zea mays* L.). *Theoretical and Applied Genetics*, *118*(3), 553-564.

Quaglia, M., Fabrizi, M., Zazzerini, A., & Zadra, C. (2012). Role of pathogen-induced volatiles in the Nicotiana tabacum and Golovinomyces cichoracearum interaction. *Plant Physiology and Biochemistry*, *52*(3), 9-20.

Ramalho, M. A. P., Ferreira, D. F., & Oliveira, A. C. (2000). *Experimentação em genética e melhoramento de plantas*. Lavras, MG: UFLA.

Senthilraja, G., Anand, T., Kennedy, J. S., Raguchander, T., & Samiyappan, R. (2013). Plant growth promoting rhizobacteria (PGPR) and entomopathogenic fungus bioformulation enhance the expression of defense enzymes and pathogenesis-related proteins in groundnut plants against leafminer insect and collar rot pathogen. *Physiological and Molecular Plant Pathology*, *82*(4), 10-19.

Silva, A. G., Cunha Junior, C. R., Assis, R. L., & Imolesi, A. S. (2008). Influência da população de plantas e do espaçamento entre linhas nos caracteres agronômicos do híbrido de milho P30K75 em Rio Verde, Goiás. *Bioscience Journal*, *24*(2) 89-96.

Silva, A. G., Teixeira, I. R., Martins, P. D. S., Simon, G. A., & Francischini, R. (2014). Desempenho agronômico e econômico de híbridos de milho na safrinha. *Revista Agroambiente On-line*, *8*(2), 261-271.

Stefanello, J., Bachi, L. M. A., Gavassoni, W. L., Hirata, L. M., & Pontim, B. C. A. (2012). Incidência de fungos em grãos de milho em função de diferentes épocas de aplicação foliar de fungicida. *Pesquisa Agropecuária Tropical*, *42*(4), 476-481.

Werle, A. J. K., Nicolay, R. J., Santos, R. F., Borsoi, A., & Secco, D. (2011). Avaliação de híbridos de milho convencional e transgênico (Bt), com diferentes aplicações de inseticida em cultivo safrinha. *Revista Brasileira de Tecnologia Aplicada nas Ciências Agrárias*, *4*(1), 150-159.

Zeringue, H. J., Brown, R. L., Neucere, J. N., & Cleveland, T. E. (1996). Relationships between C6-C12 alkanal and alkenal volatile contents and resistance of maize genotypes to *Aspergillus flavus* and aflatoxin prodution. *Journal of Agricultural and Food Chemistry*, *44*(2),403-407.

18

Control of *Brachiaria decumbens* and *Panicum maximum* by S-metolachlor as influenced by the occurrence of rain and amount of sugarcane straw on the soil

Núbia Maria Correia[1]*, Leonardo Petean Gomes[2] and Fabio José Perussi[2]

[1]*Departamento de Fitossanidade, Universidade Estadual Paulista, "Julio de Mesquista Filho", Via de Acesso Prof. Paulo Donato Castellane, s/n., 14884-900, Jaboticabal, São Paulo, Brazil.* [2]*Curso de Graduação em Agronomia, Universidade Estadual Paulista, "Julio de Mesquista Filho", Jaboticabal, São Paulo, Brazil. *Author for correspondence. E-mail: correianm@fcav.unesp.br*

ABSTRACT. With the objective to study the control of *Brachiaria decumbens* and *Panicum maximum* by herbicide S-metolachlor as influenced by the time interval between the herbicide application the occurrence of rain and the amount of sugarcane straw on the soil, two experiments were conducted in pots under greenhouse conditions. In the first, the factors were the amount of sugarcane straw left on the soil surface (0, 3, 6, 10, or 15 ton. ha^{-1}) and the S-metolachlor applied at doses of 0, 0.96, 1.44, 1.92, or 2.40 kg ha^{-1}. In the second, the factors were the amount of sugarcane straw left on the soil surface (0 or 10 ton. ha^{-1}), the interval of time elapsed between the application of S-metolachlor and simulated rain, which took place 1 day before and 0, 4, 8, 12, 16, or 20 days after the application. The herbicide doses were not affected by the amounts of straw left on the soil surface. When straw was not left on the soil surface, the control of the weeds by the herbicide was not influenced by the time interval up to 20 days after herbicide application. With 10 ton. ha^{-1} of straw, the control exerted by S-metolachlor was equally efficient whether it rained up to 12 days after the herbicide application. *P. maximum* was controlled even when the rain fell one day before the herbicide application.

Keywords: signal grass, buffalo grass, crop residues, herbicide leaching.

Introduction

When sugarcane plants are mechanically harvested without previously having burned them, the straw left on the soil surface may reduce the ability of herbicides to reach the soil surface. This capacity is dependent upon the physical and chemical characteristics of the herbicide such as solubility, vapor pressure, and polarity (RODRIGUES, 1993). After the herbicide is applied, the amount and the timing of rain or irrigation as well as the decomposition of plant residues are important factors in determining the retention of the herbicide by the straw (CORREIA et al., 2007). When retained by straw, herbicide losses likely occur due to photodegradation, volatilization, and adsorption by the plant residues. The adsorption by plant residues is dependent on their degree of decomposition and age (MERSIE et al., 2006).

In one study, the amount of the herbicide amicarbazone that was removed from the straw (5, 10, 15, and 20 ton. ha^{-1}) to the soil decreased with the time interval (7 and 14 days) between the application of the herbicide and the occurrence of simulated rain (CAVENAGHI et al., 2007). In another study, Carbonari et al. (2010) reported that the mixture of clomazone and hexazinone efficiently controlled *Brachiaria decumbens* if applied over, under or, in the absence of straw, independently of the length of time without rain (0, 3, 7, 15, 30, or 60 days). The authors observed, though, that if the period without rain exceeded 60 days, the weed control efficiency tended to be reduced, particularly when the herbicide had been applied to the soil or the straw surface. These results indicate that the herbicide applied to either the soil or straw surface undergoes degradation when exposed to weather conditions for an extended period of time if no rain occurs to leach the herbicide into the soil profile.

Studies of metolachlor report its possible retention by plant residues, which would reduce its efficacy (OLIVEIRA et al., 2001; TEASDALE et al., 2003; FONTES et al., 2004). The time interval between herbicide application and the first rain is another important factor; depending on how long that period is, the chances of herbicide loss by physical (such as volatilization) or chemical (such as photodegradation and adsorption) processes will be higher or lower. More detailed information about S-metolachlor is lacking. This herbicide is a metolachor stereoisomer (KURT-KARAKUS et al., 2010) that exhibits a high biological activity (MUNOZ et al., 2011). In Brazil, S-metolachlor is recommended for the pre-emergence control of monocotyledonous and some dicotyledonous species in soybean, corn, bean, cotton, and sugarcane crops. The herbicide is soluble in water (solubility = 480 mg L^{-1} at 25°C), poorly volatile (vapor pressure = 1.73×10^{-3} Pa at 20°C), non-ionic (pk = zero), and hydrophilic (K_{ow} = 3.05) (RODRIGUES; ALMEIDA, 2011). These values indicate that the molecule is stable with minimal losses to the environment and a moderate affinity with water.

B. decumbens and *P. maximum* are susceptible to control by S-metolachlor. These important weed species infest sugarcane fields. *B. decumbens* originated in Africa and was introduced to Brazil in 1950 as a forage crop. It is a perennial plant, decumbent, stoloniferous with roots arising from low nodes touching the soil surface. *P. maximum* originated in Africa and India and was introduced to Brazil during the slavery year s. It is a perennial species, robust, stoloniferous and with glaucous culms (KISSMANN, 1997).

This study tested the hypotheses that (i) sugarcane straw inhibits *B. decumbens* and *P. maximum* seedlings from emerging but not to the extent of dispensing with the need for herbicides such as S-metolachlor, (ii) the action of S-metolachlor is not affected by the presence of the mulch covering the soil surface, and (iii) S-metolachlor can withstand up to 20 rainless days after its application without losing its capacity to control these weeds. The objective of this work was to study the pre-emergence control of *B. decumbens* and *P. maximum* by the herbicide S-metolachlor as influenced by both the lag between its application and the occurrence of rain and the amount of sugarcane straw left covering the soil surface.

Material and methods

Two in-pot experiments were carried out under greenhouse conditions from January 12 to February 17, 2011 (the first experiment) and from April 2 to June 3, 2011 (the second experiment) at the Department of Phytosanitation of the Jaboticabal campus at Paulista State University (UNESP) in Jaboticabal, São Paulo State, Brazil.

Both experiments used a completely randomized design with four replications. In the first experiment (5 x 5 factorial), the first factor was the amount of sugarcane straw residue (0, 3, 6, 10, or 15 ton. ha^{-1}) left covering the soil surface, and the second factor was the applied dose (0, 0.96, 1.44, 1.92, or 2.40 kg ha^{-1}) of the herbicide S-metolachlor. In the second experiment (2 x 8 factorial), the first factor was the amount of sugarcane straw (0 or 10 ton. ha^{-1}) left on the soil surface, and the second factor was the time interval (1 day before, 0, 4, 8, 12, 16, or 20 days after herbicide application) between the herbicide application (at a rate of 1.92 kg ha^{-1}) and the occurrence of rain, and a control treatment without herbicide application.

Each experimental unit was formed by one 8-L plastic pot filled with a substratum formed by soil, sand, and an organic compound in a 3:1:1 ratio. After mechanically harvesting fourth-cut SP 903723 sugarcane plants for the first experiment and first-cut RB 835054 plants for the second one, the straw remaining on the soil surface was collected and taken to a greenhouse to dry.

The seeds of *B. decumbens* (1.3 g per pot) and of *P. maximum* (0.34 g per pot) were homogeneously distributed over the substratum in the pot and then covered with a 1-cm-thick layer of soil. In those treatments in which sugarcane straw was to be left on the substratum surface, the straw was cut into pieces short enough to fit into the pot and then placed over the substratum in a homogeneous layer in the appropriate amounts.

The bottoms of the pots were lined with a sheet of newspaper to prevent soil loss. Each pot was placed in a plastic tray with a diameter larger than that of the pot and without holes to ensure a consistent water supply. The soil moisture was monitored on a daily basis. Water was added to the containers as needed and was distributed through the soil by capillary action.

In both experiments the herbicide was sprayed on the weeds at the indicated doses during pre-emergence. A backpack sprayer equipped with two flat-fan nozzles (XR 110015) spaced 0.5 m apart and calibrated to deliver an equivalent of 200 L ha^{-1} was used at a constant pressure of 2.0 kgf cm^{-2} (maintained by CO_2). At the time of application, the soil was dry, and the following conditions were recorded: 95% relative humidity, 23.7°C air temperature, and 24.2°C soil temperature (at a depth of 5 cm). The dates, times, and the environmental conditions during the second experiment are presented in Table 1.

In the first experiment, one hour after the application of the herbicide, a 25-mm rainfall was simulated. In the second experiment, 30 mm of rainfall was simulated one day before herbicide application, soon after the application of the herbicide, or 4, 8, 12, 16, and 20 days later, depending on the assigned rainfall treatment.

The rain simulator was a circular device with a diameter of 0.2 m formed by flexible polyethylene tube containing seven nozzles (FL 10) that were 0.09 m from one another. This simulator was placed 2.5 m above the soil surface and uniformly sprayed a 1.5-m^2 area. Before the simulation, pluviometers were distributed over the entire area, which allowed calculations of how long the device needed to operate to attain the desired amount of rain (25 or 30 mm).

At 14 and 35 days after the application of the herbicide (DAA) in the first experiment and at 21 and 42 days after rain simulation (DARS) in the second, the number of emerged weed plants was counted. At 35 DAA and 42 DARS for the first and second experiments, respectively, the plants were trimmed close to the surface, placed inside paper bags and dried in a forced ventilation oven at 50°C to constant biomass to determine the dry matter of the plants' aerial parts.

An F test of the analysis of variance was used to test the effects of the amounts of straw and the herbicide dose (first experiment) and their interactions following a polynomial adjustment of the data. The effects of the rain intervals and those of soil covering (second experiment) and their interactions were compared using Tukey's test whit a 5% level of probability.

Results and discussion

The control of *Brachiaria decumbens* and *Panicum maximum* by the herbicide S-metolachlor in association with sugarcane plant straw covering the soil surface (first experiment)

There was a significant effect of the herbicide dose and an interaction between the straw and herbicide for all of the evaluated characteristics (Table 2). The amount of straw covering the soil surface had a significant effect on the number of *P. maximum* plants 35 DAA and also on the number and the dry matter of plants of *B. decumbens* at 14 and 35 DAA.

Table 1. Description of the treatments used in the experiment in addition to the dates, times and meteorological conditions at the moment of the S-metolachlor[1] application in the second experiment.

Rain simulation[2] - days after herbicide application	Straw (ton. ha^{-1})	Application of S-metolachlor					
		Date	Time	Temperature (°C)		Air relative humidity (%)	Wind speed (km h^{-1})	Nebulosity (%)
				Air	Soil			
One day before	0 / 10	04/22	10:00	27.7	31.1	61	0.8	0
Soon after	0 / 10	04/22	10:00	27.7	31.1	61	0.8	0
Four days later	0 / 10	04/18	7:10	21.1	25.1	79	0.0	0
Eight days later	0 / 10	04/14	9:25	23.3	28.1	90	0.0	0
Twelve days later	0 / 10	04/10	10:00	22.7	26.6	80	2.2	0
Sixteen days later	0 / 10	04/06	9:00	24.3	25.0	78	1.3	95
Twenty days later	0 / 10	04/02	13:00	29.4	29.1	66	3.5	90
Control treatment without herbicide application[3]	0 / 10	-	-	-	-	-	-	-

[1]Dose of 1.92 kg ha^{-1}. [2]30 mm of rain. [3]Rain simulated at the same day the herbicide was applied in the other treatments.

Table 2. Results of an F test of the analysis of variance for number of plants of *Brachiaria decumbens* and *Panicum maximum* at 14 and 35 days after the application (DAA) of S-metolachlor and the dry matter of the aerial parts of plants 35 DAA.

Sources of variation	B. decumbens			P. maximum		
	Number of plants		Dry matter	Number of plants		Dry matter
	Days after application					
	14	35	35	14	35	35
Straw	11.12**	17.24**	17.77**	2.07	4.39**	0.22
Herbicide	160.99**	250.94**	220.04**	229.39**	162.13**	153.73**
Straw x Herbicide	11.54**	10.55**	10.70**	5.21**	6.71**	2.22*
CV (%)	33.55	23.07	25.98	25.09	28.94	22.57

**, *Significant at 1 and 5% probability, respectively, by the F test of the analysis of variance.

Examining the interaction between straw and dose shows that in the control treatment without herbicide, the number of plants of *B. decumbens* was linearly reduced with increasing amounts of straw (Figure 1). The same behavior was observed when the aerial part of the plant dry matter was analyzed. The dry matter decreased in a polynomial form with increasing amounts of straw covering the soil surface. The number and the dry matter of *P. maximum* plants decreased linearly with increasing amounts of straw on the soil surface. Plant counts conducted 35 DAA showed that 15 ton. ha^{-1} of straw on the soil surface reduced *B. decumbens* and *P. maximum* plants emergence by 74 and 65%, respectively, compared to plots not covered with straw. Despite this reduction with the straw treatment (15 ton. ha^{-1} of straw), an average of 33 plants of *B. decumbens* and 50 of *P. maximum* per pot were found, and under field conditions, this situation would demand complementary control of those weeds.

For each of the five levels of straw, the number and the dry weight of *B. decumbens* and of *P. maximum* plants underwent polynomial reductions as the dose of S-metolachlor increased (Figure 2). S-metolachlor was efficient in controlling the weeds and, in general, the dose of 0.96 kg ha^{-1} was adequate for the efficient control of *B. decumbens* and *P. maximum* for all the amounts of straw (up to 15 ton. ha^{-1}). The results indicate that the 25 mm of simulated rain was sufficient to remove S-metolachlor from the straw to the soil because the biological control of *B. decumbens* and *P. maximum* was not affected.

The herbicide remaining in the straw (that is, the part that did not leach into the soil) may have been absorbed by the surviving seedlings when they were growing up through the straw layer. For S-metolachlor, this is possible because it is absorbed by the mesocotyl and coleoptile of seedlings before they emerge above soil surface. In the second experiment, *P. maximum* seedlings (but not those of *B. decumbens*) were capable of absorbing S-metolachlor from the straw. These data were sequentially presented.

Contrary to the observations made in this study, when beans in another study were directly sown under corn straw, metolachlor was not detected 15 days after its application in any of the soil layers (0-5, 5-10, and 10-15 cm) analyzed (FONTES et al., 2004). Teasdale et al. (2003) found that metolachlor had lower initial concentrations in the soil when the herbicide was sprayed over *Vicia villosa* straw, and the low concentrations resulted in poor control of *Panicum dichotomiflorum*. In both studies, the authors ascribed the results to the retention of the herbicide by the plant residues covering the soil surface and also to increased losses of the herbicides to the surrounding environment. S-metolachlor is a stereoisomer of metolachlor (KURT-KARAKUS et al., 2010) that has a high biological activity (MUNOZ et al., 2011). The amount of rain after the herbicide was applied may have had an important effect on the retention of the herbicide. Oliveira et al. (2001) speculated that the low precipitation during the first days after a mixture of atrazine and metolachlor was applied might explain the limited removal of the metolachlor from the plant residues to the soil.

Other authors have studied the effect of sugarcane straw on herbicide interception. Cavenaghi et al. (2007) reported that in amounts of straw equal to or higher than 5 ton. ha^{-1}, the herbicide amicarbazone (solubility = 4,600 mg L^{-1} at pH 4-9) was almost completely intercepted. Therefore, the amount of the product capable of reaching the soil when amicarbazone was applied was nearly zero. However, a 20-mm rain that fell over the area removed a large portion of the herbicide. After applying a mixture of clomazone and hexazinone (solubility = 1,100 and 29,800 mg L^{-1} at 25°C, respectively), 2.5 mm of rain was enough to wash the mixture away from sugarcane straw at a rate of 5.0 ton. ha^{-1} (NEGRISOLI et al., 2011). These findings indicate that herbicides with higher solubility than that of S-metolachlor can be retained by a straw layer and that, depending on the solubility of the molecule, only small amounts rain or irrigation water is needed for the herbicide to be leached into the soil.

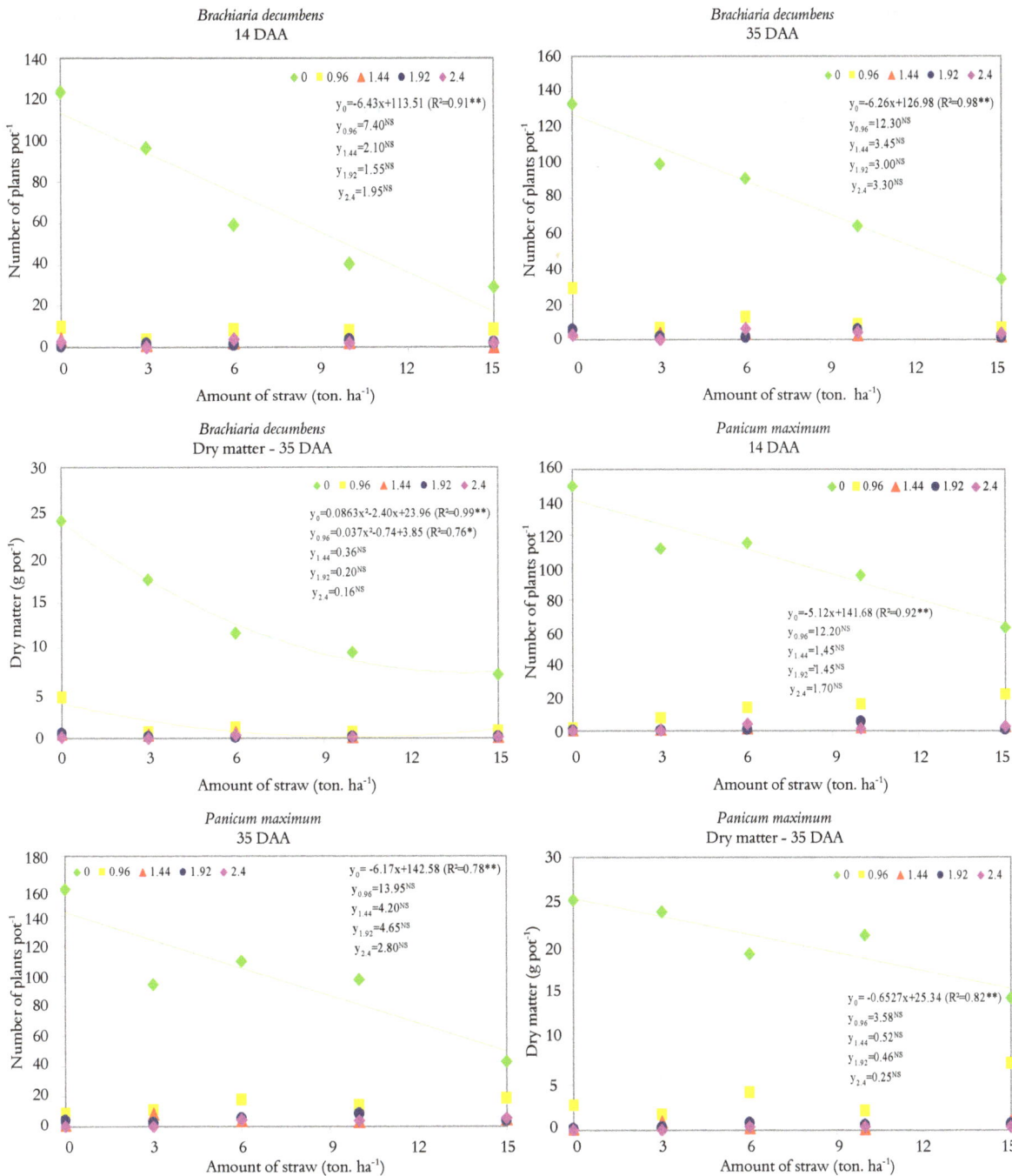

Figure 1. The number of *Brachiaria decumbens* and *Panicum maximum* plants at 14 and 35 days after the application of the herbicide S-metolachlor and the dry matter of the plants 35 DAA as functions of the amount of straw covering the soil surface and the doses of the herbicide.

The control of *Brachiaria decumbens* and *Panicum maximum* by the herbicide S-metolachlor as functions of the interval of time between herbicide application and the occurrence of rain and the amount of straw covering the soil surface (second experiment)

All the evaluated characteristics were significantly influenced by the interval of time between the herbicide application and the occurrence of simulated rain as well as by the interaction between rain and straw (Table 3). The presence or absence of straw on the soil surface had a significant influence on the emergence of *B. decumbens* at 21 and 42 DARS, the emergence of *P. maximum* at 21 DARS and on the dry matter of both species.

Without sugarcane straw, the number of plants and the dry matter of *B. decumbens* and *P. maximum* did not differ significantly among the time intervals between the application of S-metolachlor and the simulation of rain (Tables 4, 5, and 6). These results indicate that, when no straw covered the soil surface, the efficiency of S-metolachlor in controlling weeds is not hampered by rain regardless of whether the rain occurs one day before the application of the herbicide or 20 days later. However, when straw covered the soil surface, the time intervals had a significant effect on the herbicide efficacy. The

results thus show that the emergence and the dry matter of B. decumbens plants were lower when rain was simulated to occur just after the application of the herbicide and that these values were not significantly different from those obtained at 4, 8, and 12 DARS. Therefore, the control of B. decumbens by S-metolachlor was not affected by a 30-mm rain occurring up to 12 days after the herbicide application.

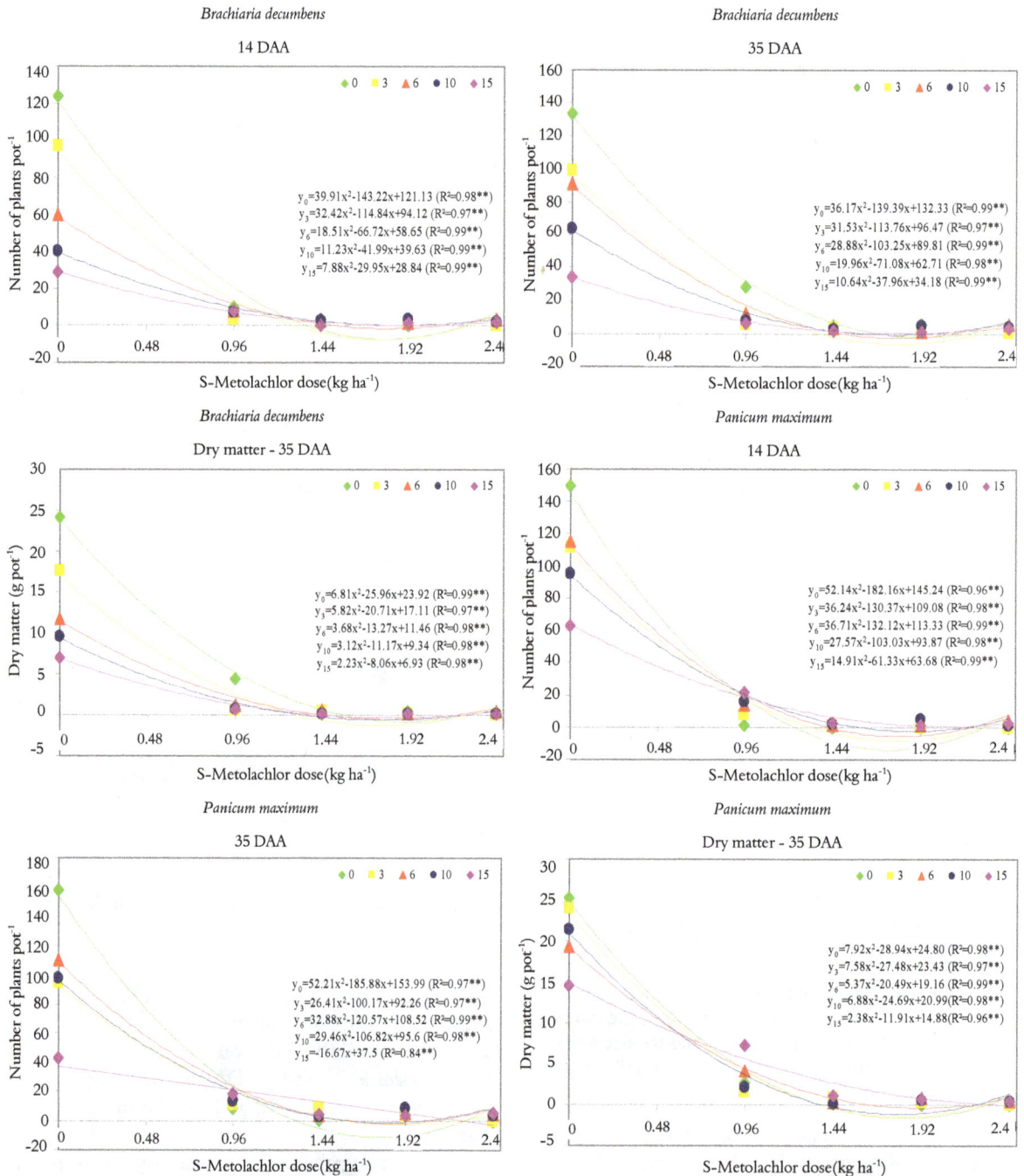

Figure 2. Number of *Brachiaria decumbens* and *Panicum maximum* plants at 14 and 35 days after the application (DAA) of the herbicide S-metolachlor and the dry matter of the plants 35 DAA as functions of the herbicide dose and the amount of straw on the soil surface.

Table 3. Results of an F test of the analysis of variance for number of plants of *Brachiaria decumbens* and *Panicum maximum* at 21 and 42 days after rain simulation (DARS) and dry matter of the aerial parts at 42 DARS.

Sources of variation	B. decumbens			P. maximum		
	Number of plants		Dry matter	Number of plants		Dry matter
	Days after herbicide application					
	21	42	42	21	42	42
Rain	101.99**	155.62**	231.89**	89.89**	103.68**	50.95**
Straw	4.81**	7.02*	0.26	9.85**	0.68	16.91**
Rain x Straw	64.26**	86.19**	130.28**	19.72**	19.38**	9.23**
CV (%)	28.09	24.31	6.26	34.31	34.06	49.24

**, * Significant at the levels of 1% and 5% of probability, respectively, by the F test of the analysis of variance.

Table 4. The number of plants per pot of *Brachiaria decumbens* at 21 and 42 days after rain simulation (DARS) as functions of the time interval between the application of S-metolachlor and rain simulation and of presence or absence of straw covering the soil surface.

Rain simulation - days after the herbicide application	21 DARS		42 DARS	
	Straw (ton. ha⁻¹)			
	0	10	0	10
One day before	3.50 a A[1]	36.25 c B	2.50 a A	34.00 cd B
Soon after	4.25 a A	5.25 a A	3.50 a A	4.50 a A
Four days later	8.00 a A	16.75 ab B	7.00 a A	14.75 ab B
Eight days later	4.25 a A	17.75 ab B	3.00 a A	14.00 ab B
Twelve days later	8.00 a A	15.00 ab A	9.00 a A	14.50 ab A
Sixteen days later	9.50 a A	26.00 bc B	7.75 a A	24.00 bc B
Twenty days later	8.75 a A	32.50 c B	6.25 a A	30.25 cd B
Control treatment[2]	111.00 b B	34.00 c A	106.00 b B	35.25 d A
LSD (by row)	8.50		6.82	
LSD (by column)	13.41		10.75	

[1]Means within a column followed by the same lowercase letter and means in a row followed by the same uppercase letter are not significantly different at the 5% level of probability according to Tukey's test. [2]Rain simulation occurred on the same day as the herbicide application in the other treatments.

Table 5. The number of plants per pot of *Panicum maximum* at 21 and 42 days after rain simulation (DARS) as functions of the time interval between the application of S-metolachlor and rain simulation and of the presence or absence of straw covering the soil surface.

Rain simulation - days after the herbicide application	21 DARS		42 DARS	
	Straw (ton. ha⁻¹)			
	0	10	0	10
One day before	0.75 a A[1]	28.00 ab B	2.00 a A	17.25 ab B
Soon after	7.50 a A	12.25 a A	7.00 a A	8.75 a A
Four days later	8.50 a A	23.50 ab B	8.50 a A	16.00 ab A
Eight days later	6.75 a A	22.00 ab B	4.50 a A	17.25 ab B
Twelve days later	16.00 a A	20.75 ab A	15.50 a A	20.00 ab A
Sixteen days later	10.25 a A	33.75 b B	12.50 a A	26.50 ab B
Twenty days later	15.00 a A	33.50 b B	17.25 a A	31.50 b B
Control treatment[2]	139.50 b B	70.75 c A	143.00 b B	67.75 c A
LSD (by row)	12.90		13.37	
LSD (by column)	25.14		18.64	

[1]Means within a column followed by the same lowercase letter and means in a row followed by the same uppercase letter are not significantly different at the 5% level of probability according to Tukey's test. [2]Rain simulation occurred on the same day as the herbicide application in the other treatments.

Table 6. Dry matter (g pot⁻¹) of the aerial part of *Brachiaria decumbens* and *Panicum maximum* plants 42 days after rain simulation as functions of the time interval between the herbicide application and rain simulation and the presence or absence of straw covering the soil surface.

Rain simulation - days after the herbicide application	B. decumbens		P. maximum	
	Straw (ton. ha⁻¹)			
	0	10	0	10
One day before	0.39 a A[1]	4.84 e B	0.08 a A	2.02 ab A
Soon after	0.19 a A	0.23 a A	0.18 a A	0.21 a A
Four days later	0.29 a A	0.76 ab A	0.51 a A	2.59 ab B
Eight days later	0.17 a A	0.84 ab A	0.16 a A	2.36 ab B
Twelve days later	0.62 a A	0.83 ab A	0.42 a A	2.91 ab B
Sixteen days later	0.54 a A	1.77 bc B	0.44 a A	3.76 b B
Twenty days later	0.31 a A	2.31 c B	0.53 a A	5.05 b B
Control treatment[2]	12.97 b B	3.59 d A	14.16 a B	8.79 c A
LSD (by row)	0.71		2.01	
LSD (by column)	1.12		3.17	

[1]Means within a column followed by the same lowercase letter and means in a row followed by the same uppercase letter are not significantly different at the 5% level of probability according to Tukey's test. [2]Rain simulation occurred on the same day as the herbicide application in the other treatments.

The emergence of *P. maximum* plants was also reduced when rain was simulated to occur just after the application of the herbicide. This level of emergence was significantly different from that resulting from intervals of 16 days (at 21 DARS and for dry matter), 20 days and control treatment. Keeping 10 ton. ha⁻¹ of straw covering the soil surface and rainfall from 1 day before to 12 days after the application of the herbicide had no negative significant effects on the efficacy of S-metolachlor in controlling *P. maximum*.

Generally, S-metolachlor was more efficient in controlling *B. decumbens* when the soil surface was free of straw and when rain occurred 1 day before or 16 or 20 days after the herbicide application. In the other time intervals, the straw treatments did not differ among themselves. *P. maximum* plants were more efficiently controlled by S-metolachlor when no straw covered the soil surface and when the simulated rain occurred 4, 8, 12, 16, or 20 days after the herbicide was applied. However, the straw treatments did not differ among themselves when the rain took place one day before or soon after the herbicide was applied.

In the control treatment with straw, there was a reduction in plant emergence and dry matter accumulation in comparison with the control treatment without straw; the reduction in plant emergence of *B. decumbens* was 72% and that of *P. maximum* was 53%. These results are an indication that straw has an inhibitory effect on the emergence and development of those weeds due to physical, chemical or biological processes.

The inhibiting effect of sugarcane straw on the emergence of weeds was also reported by Correia and Durigan (2004). According to these authors, covering the soil with raw sugarcane straw reduced *B. decumbens* seed viability because the ungerminated seeds, after straw removal, were not capable of germinating even under favorable environmental conditions. The authors attributed the loss of *B. decumbens* seed viability to physical, chemical or biological factors inherent to the straw. Sugarcane straw maintained on the soil surface interferes with the germination of *B. decumbens* seeds and, consequently, the soil seed bank composition.

Conclusion

The response of *B. decumbens* and *P. maximum* to varied doses of S-metolachlor was not affected by the amount of sugarcane straw left on the soil surface.

When there was no straw covering the soil surface, simulated rain occurring before or up to 20 days after the application of the herbicide did not affect the efficacy of S-metolachlor in controlling *B. decumbens* and *P. maximum*. With 10 ton. ha^{-1} of straw, S-metolachlor was capable of

efficiently controlling the weeds even when it rained up to 12 days after application. *P. maximum* was controlled even when the rain fell one day before the herbicide application.

References

CARBONARI, C. A.; VELINI, E. D.; CORREA, M. R.; NEGRISOLI, E.; ROSSI, C. V.; OLIVEIRA, C. P. Efeitos de Período de permanência de clomazone + hexazinona no solo e na palha de cana-de-açúcar antes da ocorrência de chuvas na eficácia de controle de plantas daninhas. **Planta Daninha**, v. 28, n. 1, p. 197-205, 2010.

CAVENAGHI, A. L.; ROSSI, C. V. S.; NEGRISOLI, E.; COSTA, E. A. D.; VELINI, E. D.; TOLEDO, R. E. B. Dinâmica do herbicida amicarbazone (Dinamic) aplicado sobre palha de cana-de-açúcar (*Saccarum officinarum*), **Planta Daninha**, v. 25, n. 4, p. 831-837, 2007.

CORREIA, N. M.; DURIGAN, J. C. Emergência de plantas daninhas em solo coberto com palha de cana-de-açúcar. **Planta Daninha**, v. 22, n. 1, p. 11-17, 2004.

CORREIA, N. M.; DURIGAN, J. C.; MELO, W. J. Envelhecimento de resíduos vegetais sobre o solo e os reflexos na eficácia de herbicidas aplicados em pré-emergência. **Bragantia**, v. 66, n. 1, p. 101-110, 2007.

FONTES, J. R. A.; SILVA, A. A.; VIEIRA, R. F.; RAMOS, M. M. Lixiviação de herbicidas no solo aplicados com água de irrigação em plantio direto, **Planta Daninha**, v. 22, n. 4, p. 623-631, 2004.

KISSMANN, K. G. **Plantas infestantes nocivas**. 2. ed. São Paulo: BASF, 1997.

KURT-KARAKUS, P. B.; MUIR, D. C. G.; BIDLEMAN, T. F.; SMALL, F.; BACKUS, S.; DOVE, A. Metolachlor and atrazine in the great lakes. **Environmental Science and Technology**, v. 4, n. 12, p. 4678-4684, 2010.

MERSIE, W.; SEYBOLD, C. A.; WU, J.; MCNAMEE, C. Atrazine and metolachlor sorption to switchgrass residues. **Communications in Soil Science and Plant Analysis**, v. 37, n. 3-4, p. 465-472, 2006.

MUNOZ, A.; KOSKINEN, W. C.; COX, L.; SADOWSKY, M. J. Biodegradation and mineralization of metolachlor and alachlor by *Candida xestobii*. **Journal of Agricultural and Food Chemistry**, v. 59, n. 1, p. 619-627, 2011.

NEGRISOLI, E.; VELINE, E. D.; CORRÊA, M. R.; ROSSI, C. V. S.; CARBONARI, C. A.; COSTA, A. G. F.; PERIM, L. Influência da palha e da simulação de chuva sobre a eficácia da mistura formulada clomazone + hexazinone no controle de plantas daninhas em área de cana-crua. **Planta Daninha**, v. 29, n. 1, p. 169-177, 2011.

OLIVEIRA, M. F.; ALVARENGA, R. C.; OLIVEIRA, A. C.; CRUZ, J. C. Efeito da palha e da mistura atrazine e metolachlor no controle de plantas daninhas na cultura do milho, em sistema de plantio direto, **Pesquisa Agropecuária Brasileira**, v. 36, n. 1, p. 37-41, 2001.

RODRIGUES, B. N. Influência da cobertura morta no comportamento dos herbicidas imazaquin e clomazone. **Planta Daninha**, v. 11, n. 1/2, p. 21-28, 1993.

RODRIGUES, B. N.; ALMEIDA, F. L. S. **Guia de herbicidas**. 6. ed. Londrina: Edição dos autores, 2011.

TEASDALE, J. R.; SHELTON, D. R.; SADEGHI, A. M.; ISENSEE, A. R. Influence of hairy vetch residue on atrazine and metolachlor soil solution concentration and weed emergence. **Weed Science**, v. 51, n. 4, p. 628-234, 2003.

19

Ovicidal effect of the essential oils from 18 Brazilian *Piper* species: controlling *Anticarsia gemmatalis* (Lepidoptera, Erebidae) at the initial stage of development

Diones Krinski[1*], Luís Amilton Foerster[2] and Cicero Deschamps[3]

[1]Departamento de Ciências Biológicas, Faculdade de Ciências Agrárias, Biológicas, Engenharia e da Saúde, Universidade do Estado de Mato Grosso, Campus Universitário de Tangará da Serra, Rodovia MT 358, km 7, Jardim Aeroporto, Tangará da Serra, Mato Grosso, Brazil. [2]Departamento de Zoologia, Setor de Ciências Biológicas, Universidade Federal do Paraná, Curitiba, Paraná, Brazil. [3]Departamento de Agronomia, Setor de Ciências Agrárias, Universidade Federal do Paraná, Curitiba, Paraná, Brazil. *Author for correspondence. E-mail: diones.krinski@unemat.br

ABSTRACT. The toxicities of essential oils (EOs) from 18 species of Brazilian Piperaceae were assessed on eggs of the velvetbean caterpillar, *Anticarsia gemmatalis*. Oils were extracted using steam distillation, and dilutions were made for bioassays at concentrations of 0.25, 0.5, 1.0, 2.0, and 4.0%. All EOs reduced larval hatching. The lowest lethal concentrations were obtained from *Piper fuligineum* (SP), *Piper mollicomum* "chemotype 1" (SP), *Piper mosenii* (PR), *Piper aduncum* (PA) and *Piper marginatum* (PA). Ovicidal activity is related to the potential toxicity of several compounds, especially dilapiolle, myristicin, asaricine, spathulenol and piperitone. According to our results, EOs from 16 Brazilian *Piper* species have potential for use as biorational botanical insecticides.

Keywords: *Piper abutiloides*, *Piper fuligineum*, *Piper marginatum*, bioinseticides.

Introduction

The velvetbean caterpillar, *Anticarsia gemmatalis* Hübner (Lepidoptera: Eribidae), is the primary soybean defoliator in Brazil (Panizzi, Oliveira, & Silva, 2004). This species also damages other crops of economic importance (Rahman, Bridges, Chapin, & Thomas, 2007). It is mainly controlled using synthetic insecticides, but other control methods include the use of transgenic crop plants and the soil-dwelling bacterium, *Bacillus thuringiensis* Berliner (McPherson & Macrae, 2009; Castro et al., 2013).

However, new approaches are needed to reduce risks to the environment and natural enemies and to avoid or delay the onset of insecticide resistance (Loureiro, Moino-Junior, Arnosti, & Souza, 2002;

Petroski & Stanley, 2009; Rampelotti-Ferreira et al., 2010). Additionally, they may be safer and more environmentally acceptable. Thus, strategies for insect management should include alternatives to conventional insecticides. The use of plant-based insecticides is an alternative for the control of lepidopteran pests primarily by having low toxicity and short persistence in the environment (Costa, Silva, & Fiuza, 2004). In this context, plants of the family Piperaceae may be a promising alternative for the control of insect pests because they contain active principles with high insecticidal potential (Fazolin, Estrela, Catani, & Alécio, 2005; Fazolin, Estrela, Catani, Alécio, & Lima, 2007; Estrela, Fazolin, Catani, Alécio, & Lima, 2006; Barbosa et al., 2012).

The analysis and identification of chemical compounds of plant origin that are active against insects are important as they allow the discovery of new groups of plants with insecticidal potential, and they provide new perspectives for the synthesis and development of new bioactive compounds (Scott, Jensen, Philogène, & Arnason, 2008). Despite the widespread occurrence of the Piperaceae species in Brazil, research on its bioactivity against agricultural insect pests is incipient. In addition, pest management is performed during the phase in which the insects are causing damage to the crop, and in this case, A. gemmatalis is in the larval stage, where the larvae feed mainly on soybean leaves (Lourenção, Reco, Braga, Valle, & Pinheiro, 2010; Moscardi et al., 2012; Franco et al., 2014).

However, few studies have been conducted on the ovicidal effect of Piperaceae on any insect pest group (Laurent et al., 1997; Fazolin et al., 2005; Scott et al., 2008; Carneiro, Pereira, & Galbiati, 2011). Due to the importance of identifying alternative, environmentally sound methods for agricultural pest control, we determined the ovicidal action of essential oils (EOs) from leaves of Piperaceae species of various Brazilian regions against Anticarsia gemmatalis eggs.

Material and methods

The eggs used in the bioassays were obtained from a colony of A. gemmatalis maintained in the Laboratory of Integrated Control of Insects (LCII), and the oils were extracted by hydrodistillation in the Vegetable Ecophysiology Laboratory, both at the Federal University of Paraná (UFPR), Curitiba, Paraná State, Brazil.

Essential oils extraction - To obtain the EOs, we used dry leaves of Piper species collected from various Brazilian regions (Table 1, Figure 1).

Chromatographic analysis

Chromatographic analysis was performed in the Laboratory of Ecophysiology Vegetable and the Laboratory of Natural Products and Chemical Ecology (LAPEQ), both at the UFPR. The EOs were subjected to analysis by gas chromatography coupled to a flame ionization detector (HP- Agilent 7890A GC-FID) and by gas chromatography coupled to mass spectrometry (MS) (60–240°C at 3°C min. rate) using a fused-silica capillary column (30 m x 0.25 mm i.d. x 0.25 μm) coated with DB-5. The injector and detector temperatures were 280°C. Hydrogen was used as a carrier gas at a flow rate of 2.4 mL min.$^{-1}$; injection was in the split mode (1:20), and the injection volume was 1.0 μL. MS spectra were obtained using electron impact at 70 eV, with a scan interval of 0.5 s and mass range from 40 to 550 m/z. The initial identification of components of the EOs was carried out by comparison with previously reported values of retention indices, which was obtained by co-injection of oil samples and C11–C24 linear hydrocarbons and calculated according to the equation of Van den Dool and Kratz (1963). Subsequently, the MS acquired for each component was matched with those stored in the Wiley/NBS mass spectral library of the GC-MS system and with other published mass spectral data (Adams, 2007).

Table 1. *Piper* species evaluated and collection sites (geographic coordinates).

Brazilian regions	*Piper* species (State)	Geographic coordinates* - elevation (m)
North region	*Piper aduncum* L. (PA)	7°07'43.56"S 55°23'22.09"W - 231 m
	Piper malacophyllum Prels. (PA)	7°06'56.31"S 55°24'22.19"W - 210 m
	Piper marginatum L. (PA)	7°07'43.56"S 55°23'22.09"W - 231 m
South region	*Piper. caldense* C. DC. "chemotype 2" (PR)	25°29'41.6"S 49°00'50.6"W – 528 m
	Piper gaudichaudianum Kunth. "chemotype 1" (PR)	25°32'59.07"S 49°14'13.09"W - 919 m
	Piper gaudichaudianum "chemotype 2" (PR)	25°29'41.6"S 49°00'50.6"W – 528 m
	Piper mikanianum (Kunth) Steud. (PR)	25°33'21.53"S 49°13'51.34"W - 908 m
	Piper mosenii C. DC. (PR)	25°29'41.6"S 49°00'50.6"W – 528 m
Southeast region	*Piper abutiloides* Kunth. (SP)	24°12'11.27"S 48°33'42.36"W - 858 m
	Piper amalago L. (SP)	22°50'42.03"S 48°25'34.04"W - 763 m
	Piper arboreum Aubl. (SP)	22°51'37.19"S 48°26'14.32"W - 798 m
	Piper caldense "chemotype 1" (SP)	24°01'18.42"S 47°31'46.37"W - 722 m
	Piper crassinervium Kunth. (SP)	22°51'37.19"S 48°26'14.32"W - 798 m
	Piper mollicomum Kunth. "chemotype 1" (SP)	22°53'32.57"S 48°28'57.45"W - 853 m
	Piper mollicomum "chemotype 2" (SP)	22°53'21.40"S 48°28'12.23"W - 822 m
	Piper fuligineum Kunth. (SP)	22°53'32.57"S 48°28'57.45"W - 853 m
	Piper solmsianum C. DC. (SP)	24°01'16.05"S 47°31'48.41"W - 718 m
	Piper lhotzkyanum Kunth. (SP)	22°50'42.03"S 48°25'34.04"W - 763 m
Midwest region	*Piper tuberculatum* Jacq. (MT)	4°37'29.32"S 57°29'09.10"W - 385 m
	Piper hispidum Sw. (GO)	16°29'47.47"S 49°16'50.45"W - 825 m
	Piper umbellatum L. (GO)	16°31'00.40"S 49°16'15.16"W - 741 m

*Datum WGS84.

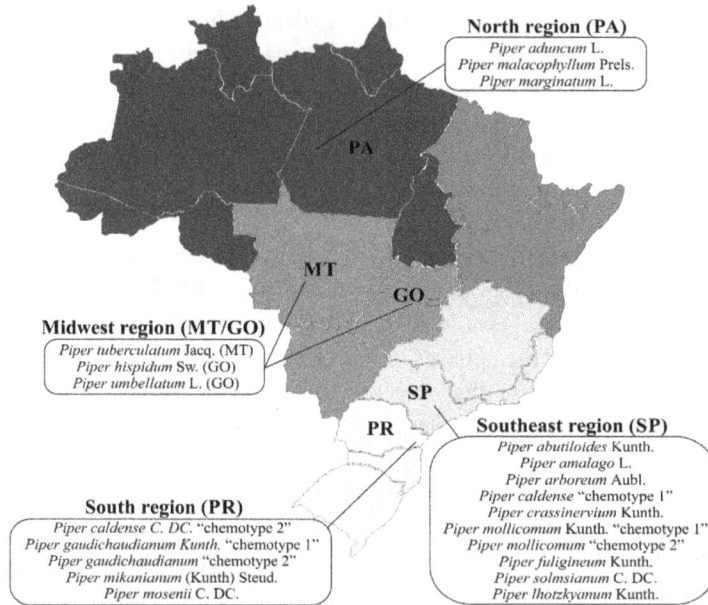

Figure 1. *Piper* species evaluated and collection regions (Brazilian states).

The species were identified, and herbarium specimens were deposited at Universidade do Estado de Mato Grosso, Campus Universitário de Tangará da Serra (UNEMAT/CUTS) in Herbarium Tangará (TANG). Collected leaves were stored in a greenhouse for 96h at 50°C to dry and then ground with a mill to obtain the leaf powder. The milled material was extracted by hydrodistillation. For each oil extraction, 50 g of vegetable powder were placed in a glass flask (2 L) containing 1 L of distilled water. The flask was heated in a heating mantle and boiled for four hours to obtain the EOs.

Ovicidal effects of essential oils on *Anticarsia gemmatalis* eggs

To evaluate the activity of EOs from Piperaceae leaves on *A. gemmatalis* eggs, the bioassays were performed by spraying eggs with an airbrush up to 24 hours after oviposition. Five oil concentrations (0.25, 0.5, 1.0, 2.0, and 4.0% diluted with acetone P.A.) and two control treatments (distilled in water and acetone P.A.) were evaluated, totalling seven treatments for each *Piper* species. The sprayings were performed with a calibrated airbrush at a pressure of 20 psi, which enabled the deposition of 1.5 mg cm^{-2} of each solution/concentration. Ten replicates containing ten eggs glued on blue paper were sprayed in each concentration, and the eggs were left to dry at room temperature. Each replicate was placed in glass tubes (10 cm x 1 cm), and all treatments were maintained in climatic chambers (Eletrolab Model EL 202) at 25 ± 1.0°C (SD), 70 ± 10% RH (SD), and a photoperiod of 12h L:D. Larval hatching was evaluated on the third day after oviposition, which is the normal time for

emergence of *A. gemmatalis* caterpillars at 25°C (Magrini, Botelho, & Silveira Neto, 1999).

Statistical analysis

For the statistical analysis, we used the results obtained three days after treatment in each EO. After this time, the hatching of *A. gemmatalis* caterpillars was recorded. The results were compared using an analysis of variance (ANOVA), and means were classified with the Scott Knott test (p < 0.05). Lethal concentrations causing 50, 75, and 90% mortality of the eggs (LC$_{50}$, LC$_{75}$, and LC$_{90}$) were calculated by Probit analysis (Finney, 1971) using Statistica software (version 7).

Results

All evaluated EOs reduced the larval hatching of *A. gemmatalis* compared with control treatments (water and acetone), except for *P. solmsianum* (SP) and *P. hispidum* (GO), which did not show significant differences in the average number of hatched larvae. The results showed that there were significant differences among *Piper* species on the ecclosion of *A. gemmatalis* eggs (Table 2). To consider a product as ovicidal, it should inhibit hatching by at least 75% (Picollo & Zerba, 1997). The *Piper* species showing ovicidal effects (mortality of eggs greater than 75%) at each concentration were *P. fuligineum* (SP) from the lowest concentration (0.25%); *P. aduncum* (PA), *P. mollicomum* 'chemotype 1' (SP) and *P. mosenii* (PR) from 0.5%; *P. caldense* 'chemotype 1' (SP) and *P. marginatum* (PA) at concentrations of 1.0%; *P. arboreum* (SP), *P. gaudichaudianum* (PR), *P. lhotzkyanum* (SP), *P. mikanianum* (PR), *P. mollicomum* 'chemotype 2' (SP), *P.*

caldense 'chemotype 2' (PR) and *P. tuberculatum* (MT) at 2.0%; and only *P. abutiloides* (SP) and *P. amalago* (SP) in the highest concentration (4.0%) (Table 2). Data concerning the lethal concentrations that inhibit hatching of 50, 75, and 90% of *A. gemmatalis* eggs (LC$_{50}$, LC$_{75}$, and LC$_{90}$) are reported in Table 3. The LC$_{50s}$ calculated after 72 hours ranged from 0.4% (*P. fuligineum* oil) to 1091.4% (*P. hispidum* oil). The lowest LC$_{50s}$ were observed for *P. fuligineum* (SP), *P.*

mollicomum 'chemotype 1' (SP), *P. mosenii* (PR), *P. aduncum* (PA) and *P. marginatum* (PA), in this order. Similar patterns were observed for the LC$_{75}$ and LC$_{90}$ with *P. caldense* 'chemotype 1' and *P. mollicomum* 'chemotype 2' being included among the species with smaller lethal concentrations (Table 3). The major chemical compounds found in each species of Piperaceae evaluated are shown in Table 4.

Table 2. Mean (± SD) number of caterpillars hatching (n= 10) after spraying of essential oils from dried leaves of different species of Piperaceae on *Anticarsia gemmatalis* eggs three days after treatment in each essential oil concentration.

Piper species (Brazilian state)	Treatments/Oil concentrations*							p	CV (%)
	Water	Acetone	0.25%	0.5%	1.0%	2.0%	4.0%		
P. abutiloides (SP)	9.6 ± 0.5 a	9.1 ± 0.9 a	8.0 ± 0.7 Bb	8.0 ± 0.9 Bb	8.0 ± 0.9 Bb	5.3 ± 3.2 Cc	0.0 ± 0.0 Dd	<0.0001	20.41
P. aduncum (PA)	9.6 ± 0.5 a	9.1 ± 0.9 a	3.3 ± 1.7 Ec	2.2 ± 1.4 Dd	5.8 ± 1.8 Cb	1.4 ± 1.1 De	0.0 ± 0.0 Df	<0.0001	16.04
P. amalago (SP)	9.6 ± 0.5 a	9.1 ± 0.9 a	9.0 ± 1.1 Aa	8.4 ± 1.1 Bb	8.5 ± 1.2 Bb	7.8 ± 1.4 Bb	0.0 ± 0.0 Dc	<0.0001	13.23
P. arboreum (SP)	9.6 ± 0.5 a	9.1 ± 0.9 a	8.9 ± 0.7 Aa	8.2 ± 1.0 Bb	8.0 ± 1.3 Bb	2.2 ± 2.2 Dc	0.0 ± 0.0 Dd	<0.0001	13.80
P. caldense 'chemotype 1' (SP)	9.6 ± 0.5 a	9.8 ± 0.4 a	9.5 ± 0.7 Aa	9.5 ± 3.0 Aa	0.0 ± 0.0 Fb	0.0 ± 0.0 Eb	0.0 ± 0.0 Db	<0.0001	8.89
P. caldense 'chemotype 2' (PR)	9.6 ± 0.5 a	9.1 ± 0.9 a	5.6 ± 1.6 Db	5.6 ± 1.7 Cb	4.3 ± 1.3 Dc	1.7 ± 1.1 Dd	0.0 ± 0.0 De	<0.0001	21.74
P. crassinervium (SP)	9.6 ± 0.5 a	9.1 ± 0.9 a	7.3 ± 1.1 Bb	6.5 ± 1.2 Cc	6.3 ± 0.9 Cc	0.2 ± 0.6 Ed	0.0 ± 0.0 Dd	<0.0001	15.27
P. fuligineum (SP)	9.6 ± 0.5 a	9.1 ± 0.9 a	2.3 ± 1.4 Fb	2.3 ± 1.6 Db	2.8 ± 1.8 Eb	0.0 ± 0.0 Ec	0.0 ± 0.0 Dc	<0.0001	30.05
P. gaudichaudianum 'chem. 1' (PR)	9.6 ± 0.5 a	9.1 ± 0.9 a	8.6 ± 1.2 Bb	8.4 ± 1.0 Bb	8.0 ± 1.2 Bb	0.1 ± 0.3 Ec	0.0 ± 0.0 Dc	<0.0001	13.80
P. gaudichaudianum 'chem. 2' (PR)	9.6 ± 0.5 a	9.1 ± 0.9 a	8.3 ± 0.7 Bb	8.0 ± 1.3 Bb	8.3 ± 1.2 Bb	7.8 ± 1.2 Bb	7.8 ± 2.0 Bb	0.0082	14.42
P. hispidum (GO)	9.6 ± 0.5 a	9.8 ± 0.4 a	9.7 ± 0.7 Aa	9.7 ± 0.5 Aa	9.7 ± 0.5 Aa	9.2 ± 0.9 Aa	9.8 ± 0.4 Aa	0.2907	6.06
P. lhotzkyanum (SP)	9.6 ± 0.5 a	9.1 ± 0.9 a	9.3 ± 0.7 Aa	9.5 ± 0.8 Aa	9.2 ± 0.8 Aa	0.0 ± 0.0 Eb	0.0 ± 0.0 Db	<0.0001	9.54
P. malacophyllum (PA)	9.6 ± 0.5 a	9.1 ± 0.9 a	5.4 ± 1.8 Dc	7.6 ± 2.0 Bb	7.7 ± 1.6 Bb	7.2 ± 1.8 Bb	4.4 ± 1.9 Cc	<0.0001	23.01
P. marginatum (PA)	9.6 ± 0.5 a	9.1 ± 0.9 a	6.4 ± 1.6 Cb	6.0 ± 1.9 Cb	2.1 ± 1.5 Ec	0.1 ± 0.3 Ed	0.0 ± 0.0 Dd	<0.0001	26.20
P. mikanianum (PR)	9.6 ± 0.5 a	9.1 ± 0.9 a	8.9 ± 1.4 Aa	7.7 ± 0.9 Bb	8.1 ± 1.3 Bb	0.8 ± 1.2 Ec	0.0 ± 0.0 Dc	<0.0001	16.06
P. mollicomum "chemotype 1" (SP)	9.6 ± 0.5 a	9.1 ± 0.9 a	4.0 ± 1.1 Eb	0.7 ± 1.0 Ed	2.5 ± 1.8 Ee	0.0 ± 0.0 Ed	0.0 ± 0.0 Dd	<0.0001	25.91
P. mollicomum "chemotype 2" (SP)	9.6 ± 0.5 a	9.1 ± 0.9 a	9.2 ± 0.8 Aa	8.2 ± 2.3 Bb	3.4 ± 2.0 Dc	0.0 ± 0.0 Ed	0.0 ± 0.0 Dd	<0.0001	18.80
P. mosenii (PR)	9.6 ± 0.5 a	9.1 ± 0.9 a	4.7 ± 1.8 Db	2.1 ± 1.1 Dd	3.6 ± 1.2 Dc	0.1 ± 0.3 Ee	0.1 ± 0.3 De	<0.0001	23.80
P. solmsianum (SP)	9.6 ± 0.5 a	9.8 ± 0.4 a	9.7 ± 0.5 Aa	9.7 ± 0.6 Aa	9.9 ± 0.3 Aa	9.7 ± 0.7 Aa	9.7 ± 0.5 Aa	0.919	5.39
P. tuberculatum (MT)	9.6 ± 0.5 a	9.1 ± 0.9 a	9.8 ± 0.6 Aa	9.1 ± 2.3 Aa	4.1 ± 2.8 Db	0.0 ± 0.0 Ec	0.0 ± 0.0 Dc	<0.0001	20.58
P. umbellatum (GO)	9.6 ± 0.5 a	9.1 ± 0.9 a	7.8 ± 1.5 Bb	8.0 ± 1.1 Bb	7.7 ± 0.9 Bb	7.9 ± 1.6 Bb	7.5 ± 1.4 Bb	0.0008	14.56
P	-	-	<0.0001	<0.0001	<0.0001	<0.0001	<0.0001		
CV (%)	-	-	16.11	17.33	22.92	40.93	37.10		

*Means followed by the same lowercase letter in the lines (comparing the concentrations for the same species) and capital letters in columns (comparing the same concentration for different species) do not differ by Scott Knott test (p < 0.05). Means highlighted in bold show concentrations with ovicidal activity (Picollo & Zerba, 1997).

Table 3. Lethal concentrations of essential oils from leaves of *Piper* species to make unviable 50, 75, and 90% (LC$_{50}$, LC$_{75}$, and LC$_{90}$) of *Anticarsia gemmatalis* eggs.

Piper species (Brazilian state)	Lethal Concentration (%)		
	LC$_{50}$ (ci[1])	LC$_{75}$ (ci)	LC$_{90}$ (ci)
P. abutiloides (SP)	1.9 (1.8 - 2.0)	2.9 (2.8 - 3.1)	3.6 (3.4 - 3.8)
P. aduncum (PA)	0.6 (0.3 - 0.9)	2.0 (1.6 - 2.3)	2.8 (2.4 - 3.3)
P. amalago (SP)	2.2 (2.1 - 2.3)	3.3 (3.1 - 3.4)	3.9 (3.7 - 4.1)
P. arboreum (SP)	1.7 (1.4 - 1.9)	2.6 (2.3 - 2.9)	3.19 (2.8 - 3.6)
P. caldense "chem. 1" (SP)	1.2 (0.6 - 1.8)	1.8 (1.1 - 2.4)	2.1 (1.4 - 2.9)
P. caldense "chem. 2" (PR)	1.1 (0.7 - 1.4)	2.2 (1.8 - 2.6)	2.8 (2.3 - 3.3)
P. crassinervium (SP)	1.3 (0.9 - 1.7)	2.2 (1.7 - 2.6)	2.7 (2.1 - 3.3)
P. fuligineum (SP)	0.4 (0.3 - 1.1)	1.5 (0.9 - 2.0)	2.1 (1.4 - 2.8)
P. gaudichaudianum "chem. 1" (PR)	1.5 (1.2 - 1.9)	2.3 (1.9 - 2.8)	2.8 (2.2 - 3.4)
P. gaudichaudianum "chem. 2" (PR)	12.9 (12.4 - 13.3)	21.6 (20.9 - 22.4)	26.9 (26.0 - 27.8)
P. hispidum (GO)	1091.4 (1095.2 -1087.7)	1683.2 (1689.0 - 1677.3)	2038.2 (2045.3 - 2031.1)
P. lhotzkyanum (SP)	1.6 (1.3 - 2.0)	2.4 (1.9 - 2.9)	2.9 (2.3 - 3.4)
P. malacophyllum (PA)	3.7 (3.6 - 3.8)	6.8 (6.6 - 6.9)	8.6 (8.3 - 8.8)
P. marginatum (PA)	1.0 (0.4 - 1.5)	1.8 (1.2 - 2.4)	2.3 (1.6 - 3.0)
P. mikanianum (PR)	1.6 (1.2 - 1.9)	2.4 (2.0 - 2.8)	2.9 (2.4 - 3.4)
P. mollicomum "chem. 1" (SP)	0.4 (0.3 - 1.2)	1.4 (0.8 - 2.0)	2.0 (1.3 - 2.7)
P. mollicomum "chem. 2" (SP)	1.3 (0.8 - 1.8)	2.0 (1.5 - 2.6)	2.5 (1.8 - 3.1)
P. mosenii (PR)	0.6 (0.0 - 1.1)	1.7 (1.2 - 2.2)	2.4 (1.8 - 3.0)
P. solmsianum (SP)	765.7 (771.1 - 760.3)	1172.2 (1180.5 -1164.0)	1416.1 (1426.1 - 1406.2)
P. tuberculatum (MT)	1.4 (1.0 - 1.9)	2.1 (1.6 - 2.7)	2.6 (1.9 - 3.2)
P. umbellatum (GO)	12.0 (11.6 - 12.3)	20.6 (20.0 - 21.2)	25.8 (25.1 - 26.6)

[1]*ci*: confidence interval (- 95% and + 95%). Values highlighted in bold show the minors lethal concentrations (LC$_{50}$ > 1%, LC$_{75}$ > 2%, and LC$_{90}$ > 2.5%).

Table 4. Main chemical compounds found in the *Piper* species tested against *Anticarsia gemmatalis* eggs.

RI*	Compounds	Piper species**/concentration of each compound (%)																				
		1	2	3	4	5	6	7	8	9	10	11	12	13	14	15	16	17	18	19	20	21
932	α-pinene	-	-	-	-	6.04	-	-	-	-	9.66	-	-	-	-	-	-	-	-	7.80	-	-
974	β-pinene	-	-	-	-	5.17	-	-	-	-	13.18	-	-	-	-	-	-	-	5.78	-	-	-
1014	α-terpinene	-	-	-	-	-	-	-	-	-	-	-	13.85	-	-	-	-	-	-	-	-	-
1022	o-cimene	-	-	-	-	-	-	-	-	-	-	-	6.18	-	-	-	-	-	-	-	-	-
1024	limonene	-	-	-	-	-	-	-	-	-	-	-	-	-	-	-	-	-	5.05	-	-	-
1025	β-phellandrene	-	-	-	-	-	-	-	-	-	-	-	-	-	-	-	-	-	-	-	-	-
1044	(E)-β-ocimene	-	8.16	-	-	-	-	-	-	-	-	-	-	6.14	-	4.39	-	-	-	-	-	-
1054	γ-terpinene	-	-	-	-	-	-	-	-	-	-	-	21.46	-	-	-	-	-	-	-	-	-
1086	terpinolene	-	-	-	-	-	-	-	-	-	-	-	8.18	-	-	-	-	11.34	-	-	-	-
1249	piperitone	-	7.82	-	-	-	-	-	-	-	-	-	-	-	-	-	-	-	-	-	-	-
1374	α-copaene	-	-	-	-	-	-	-	-	-	-	-	-	-	-	-	-	-	-	-	5.36	-
1417	(E)-caryophyllene	-	-	-	9.5	8.72	-	7.21	-	4.31	17.84	-	8.82	-	-	-	-	-	16.82	9.16	7.87	5.39
1457	allo-aromadendrene	5.53	-	-	-	-	-	-	-	-	-	-	-	-	-	-	-	-	-	-	-	-
1484	germacrene B	-	-	-	20.42	6.79	-	-	-	-	-	6.08	7.46	-	6.11	-	6.38	-	-	-	8.38	-
1489	β-selinene	-	-	-	-	-	-	-	-	-	-	-	-	5.93	-	-	-	-	-	-	-	16.12
1495	asaricine	-	11.86	-	-	-	-	-	10.80	-	-	-	-	-	-	7.21	-	-	-	-	-	-
1500	bicyclogermacrene	6.33	-	15.78	21.31	-	-	8.89	5.64	5.12	6.39	11.49	21.71	5.39	-	-	20.41	-	-	32.47	-	10.64
1505	(E, E)-α-farnesene	5.47	-	-	-	-	-	-	-	-	-	-	-	-	-	-	-	-	-	-	-	-
1517	myristicin	22.7	12.61	-	-	-	66.75	-	-	-	-	-	-	15.75	-	-	-	-	-	-	-	-
1518	δ-cadinene	-	-	-	-	6.11	-	-	-	6.69	6.86	-	-	-	-	-	-	-	9.12	-	-	-
1537	α-cadinene	-	-	-	-	-	-	-	-	-	-	-	7.98	-	-	-	-	-	-	-	-	-
1555	elemicina	-	-	-	-	-	-	-	-	-	-	-	-	-	-	13.21	-	-	-	-	-	-
1562	(E)-nerolidol	-	-	-	-	-	-	11.74	-	4.02	5.03	-	-	-	-	-	-	-	-	-	-	-
1574	germacrene D-ol	-	-	5.13	-	-	-	-	-	-	-	-	-	-	-	-	-	-	-	-	-	-
1577	spathulenol	-	-	-	-	-	-	-	4.14	-	-	-	7.22	-	-	-	-	-	21.08	-	7.54	-
1582	caryophyllene oxide	-	-	-	-	-	-	-	-	-	-	-	13.97	-	-	-	-	-	9.82	-	6.35	-
1586	Thujopsan-2-α-ol	-	-	-	-	-	-	-	4.32	-	-	-	-	-	-	-	-	-	-	-	-	-
1608	epoxy II humulene	-	-	-	-	-	-	-	-	-	-	-	-	-	-	-	-	-	-	-	-	-
1620	dill apiole	-	6.54	-	-	-	-	-	3.82	-	-	-	-	-	-	-	-	-	-	-	-	-
1624	1-epi-cubenol	-	-	-	-	-	-	-	-	24.22	-	-	-	-	-	-	-	-	-	-	-	-
1652	α-cadinol	-	-	5.56	-	-	-	-	-	-	-	-	-	-	-	-	-	-	-	-	-	-
1654	4,6-dimethyl-5-vinyl-1,2-benzodioxide	-	-	-	-	-	-	-	-	-	-	-	-	-	-	13.9	-	-	-	-	-	-
1683	cadalene	-	-	-	-	-	-	-	-	33.73	-	-	-	-	-	-	-	-	-	-	-	-
1700	eudesm-7(11)-en-4-ol	-	-	-	-	-	-	-	-	-	-	-	-	-	-	26.4	-	-	-	-	-	-
1759	ciclocolorenone	-	-	-	-	-	-	-	-	-	-	-	-	-	-	-	-	9.58	-	-	-	-

*RI= retention index.**Piper species: 1- *P. abutiloides* (SP); 2- *P. aduncum* (PA); 3- *P. amalago* (SP); 4- *P. arboreum* (SP); 5- *P. caldense* (PR); 6- *P. caldense* (SP); 7- *P. crassinervium* (SP); 8- *P. fuligineum* (SP); 9- *P. gaudichaudianum* "chemotype 1" (PR); 10- *P. gaudichaudianum* "chemotype 2" (PR); 11- *P. hispidum* (MT); 12- *P. lhotzkyanum* (SP); 13- *P. malacophyllum* (PA); 14- *P. marginatum* (PA); 15- *P. mikanianum* (PR); 16- *P. mollicomum* "chemotype 1" (SP); 17- *P. mollicomum* "chemotype 2" (SP); 18- *P. mosenii* (PR); 19- *P. solmsianum* (SP); 20- *P. tuberculatum* (MT); 21- *P. umbellatum* (SP).

Discussion

The toxicity of some species such as *P. aduncum* against different orders of insects of medical and agricultural importance has already been reported (Estrela et al., 2006; Silva, Ribeiro, Souza, & Correa, 2007; Scott et al., 2008; Misni, Othman, & Sulaiman, 2011; Souto, Harada, Andrade, & Maia, 2012; Piton, Turchen, Butnariu, & Pereira, 2014; Turchen, Piton, Dall'Oglio, Butnariu, & Pereira, 2016a). However, the egg stage is probably the least studied in terms of susceptibility to chemicals, and the few studies that assessed the vulnerability of eggs were often reported by accident, arising from the application of insecticides to other developmental phases of insects (Smith & Salkeld, 1966). Past studies report that mineral oils act primarily as ovicides by depressing the respiratory rate when applied directly to moth eggs (Riedl, Halaj, Kreowski, Hilton, & Westigard, 1995). The length of respiratory expression and the dose of oil in contact with the egg are critical to provoke mortality (Smith

& Pearce, 1948). Other studies have reported that the ovicidal nature of the oils is due to the natural tendency of oils to block the oxygen supplied to the developing embryo or due to the toxicity of some inherent chemical constituents of the oil (Rajapakse & Senanayake, 1997).

In general, ovicidal activity is directly proportional to the increase in the concentration tested. Considering previous information, the ovicidal effects of the EOs evaluated in our study can be attributed either to physical or chemical properties. Physically, when the oils come into contact with the surface of the eggs, they can cover the areas of gas exchange between the embryo and the external environment (corium and micropyles), thus interfering in the normal embryo development. Chemically, the compounds present in each oil may exhibit different toxicity rates and can operate concurrently with the physical properties of the oils, thus causing the death of the eggs.

This characteristic was observed for some of the Piperaceae oils tested in our study. We observed that some compounds in particular (or perhaps synergistically) may be acting on *A. gemmatalis* eggs and causing their death. The EOs that showed bioactivity contained among the most abundant compounds, mainly asaricine (*P. aduncum, P. fuligineum* and *P. mollicomum* 'chemotype 1'), myristicin (*P. aduncum, P. abutiloides, P. Caldense*, and *P. marginatum*), spathulenol (*P. fuligineum, P. lhotzkyanum, P. mollicomum* SP, and *P. solmsianum*), (E)-caryophyllene and germacrene B (*P. arboreum* and *P. tuberculatum*), dillapiol (*P. aduncum* e *P. fuligineum*), (E)-β-ocimene (*P. aduncum, P. Marginatum*, and *P. mollicomum* 'chemotype 1'), limonene (only in *P. mosenii*), (E)-nerolidol (only in *P. crassinervium*), piperitone (*P. aduncum*), 1-epi-cubenol and cadalene (only in *P. gaudichaudianum* 'chemotype 1'), 4,6-dimethyl-5-vinyl-1,2-benzodioxide, eudesm-7 (11)-en-4-ol and ciclocolorenone (only in *P. mikanianum*), α-copaene (only in *P. tuberculatum*), and (E, E)-α-farnesene and allo-aromadendrene (only in *P. abutiloides*) (Table 4).

Some compounds, such as (E)-caryophyllene and bicyclogermacrene, were discarded as possible agents of egg mortality, even when they were in large quantities in some oils, as recorded for *P. caldense* 'chemotype 2', *P. crassinervium, P. gaudichaudianum* 'chemotype 2', *P. lhotzkyanum, P. solmsianum, P. amalago, P. Hispidum*, and *P. umbellatum* (Table 4). Apparently, these compounds are common among Piperaceae, as observed in other chemical constitutions of several *Piper* species (Santos, Moreira, Guimarães, & Kaplan, 2001; Mundina et al., 2001; Cruz, Cáceres, Álvarez, Apel, & Henriques, 2011; Santana et al., 2015).

Among the 21 chemotypes of Piperaceae tested, *P. aduncum, P. caldense, P. fuligineum, P. marginatum, P. mollicomum* chemotype 1, and *P. mosenii* were the most effective against *A. gemmatalis* eggs. *Piper aduncum* and *P. marginatum* presented compounds already reported as insecticides in previous works, such as dillapiol, asaricine (sarisan), piperitone and myristicin (Bizzo et al., 2001; Morais et al., 2007; Qin, Huang, Li, Chen, & Peng, 2010; Souto et al., 2012; Santana et al., 2015; Ribeiro, Camara, & Ramos, 2016; Krinski & Foerster, 2016). For *P. caldense, P. fuligineum, P. Mollicomum*, and *P. mosenii*, there are no studies regarding the insecticidal activity, and our work is the first to test the ovicidal activity against a lepidopteran pest of importance in Brazilian agriculture.

It was shown that the chemical compounds present in these Piperaceae can be further evaluated among the isolation of active molecules to develop new botanical formulations, especially when we consider that the majority of these substances has been reported as potential insecticides in many other studies conducted worldwide (Santos et al., 2010; 2011; 2013; Cáceres & Kato, 2014, Brito, Baldin, Silva, Ribeiro, & Vendramim, 2015; Krinski & Foerster, 2016, Sanini et al. 2017, Turchen, Hunhoff, Paulo,Souza, Pereira, 2016b) and other plant families (Isman, 2000; Koul, Walia, & Dhaliwal, 2008; Tripathi, Upadhyay, Bhuiyan, & Bhattacharya, 2009; Coitinho, Oliveira, Gondim-Júnior, & Câmara, 2010; Ntalli & Menkissoglu-Spiroudi, 2011, Zoubiri & Baaliouamer, 2011; Baskar & Ignacimuthu, 2012; Baskar, Muthu, Raj, Kingsley, & Ignacimuthu, 2012; Krishnappa & Elumalai, 2012; Cáceres & Kato, 2014, Krinski & Massaroli, 2014; Krinski, Massaroli, & Machado, 2014; Backiyaraj et al., 2015; Massaroli, Pereira, & Foerster; 2016; Costa, Santana, Oliveira, & Serrão, 2017).

The effects of EOs from *Piper* species on egg hatching of *A. gemmatalis* were evaluated, and their respective major compounds were reported as alternative plant products with effectiveness to control insect pests. The literature reports that there are many phytochemical studies on several species of *Piper*, demonstrating the presence of a variety of secondary metabolites, including alkaloids, amides, propenilfenóis, lignins, neoligninas, terpenes, steroids, kawapirenos, chalcones, dihydrochalcones, flavones, and flavonones (Dyer & Palmer, 2004; Tchoumbougnang et al., 2009).

Thus, the use of these oils can affect different functions in the insects, possibly due to the synergism that occurs in unique, natural and complex mixtures that may decrease the resistance of these organisms (Ntalli & Menkissoglu-Spiroudi, 2011).

Field and semi-field studies must be conducted to assess whether the same pattern of results obtained in laboratory studies is maintained in the field, both for *A. gemmatalis* and for other insects and food crops (Albuquerque, 1993; Krinski & Pelissari, 2012; Krinski, Favetti, & Butnariu, 2012; Favetti, Krinski, & Butnariu, 2013; Krinski, 2013; Krinski, 2015; Krinski & Godoy, 2015; Krinski, Foerster, & Grazia, 2015; Krinski & Foerster, 2017; Martins & Krinski, 2016).

Therefore, more studies are needed to search for new plant species with bioactive principles as well as for the chemical synthesis of effective ingredients and identification of the target sites of the toxic molecule. There is a lack of major commitment by the chemical industry to promote research and development of new biopesticides, based on the results obtained for plants considered bio-insecticides. Such studies have increased in recent years, however, without the concomitant development of new compounds for use in agriculture (Isman & Grieneisen, 2014).

Conclusion

The ovicidal activity observed in our study may indicate the potential toxicity of the main chemical components found in 16 *Piper* species tested and the possible synergistic action among these compounds. Based on the ovicidal effect and lethal concentrations, the species *P. aduncum* (PA), *P. caldense* (SP), *P. fuligineum* (SP), *P. marginatum* (PA), *P. mollicomum* (SP), and *P. mosenii* (PR) were the most promising for the management of *A. gemmatalis* in egg stage.

Thus, we noticed that the Piperaceae species showed to be toxic to *A. gemmatalis* eggs by reducing or inhibiting larval hatching of treated eggs.

Acknowledgements

The authors acknowledge the Coordenação de Aperfeiçoamento de Pessoal de Nível Superior (CAPES) for providing scholarship to the first author (CAPES – Proc.: 1468981/2015). We also thank the Dra. Micheline Carvalho-Silva of University of Brasilia (UnB) for the identification of the Piperaceae used in this research, Dra. Beatriz Helena Lameiro de Noronha Sales Maia of Federal University of Paraná (UFPR) for chromatographic analysis of the essential oils in the Laboratory of Natural Products and Chemical Ecology (LAPEQ), and Dr. Wanderley do Amaral for the collection of *P. gaudichaudianum* 'chemotype 2' and *P. mosenii* species (PR).

References

Adams, R. P. (2007). *Identification of essential oil components by Gas Chromatography/Mass Spectroscopy*. Gruver, TX: Allured Publishing Corporation.

Albuquerque, G. S. (1993). Planting time as a tactic to manage the small rice stink bug, *Oebalus poecilus* (Hemiptera, Pentatomidae), in Rio Grande do Sul, Brazil. *Crop Protection, 12*(8), 627-630. doi: 10.1016/0261-2194(93)90128-6

Backiyaraj, M., Elumalai, A., Kasinathan, D., Mathivanan, T., Krishnappa K., & Elumalai K. (2015). Bioefficacy of *Caesalpinia bonducella* extracts against tobacco cutworm, *Helicoverpa armigera* (Hub.) (Lepidoptera: Noctuidae). *Journal of Coastal Life Medicine, 3*(5), 382-388. doi: 10.12980/JCLM.3.2015JCLM-2014-0058

Barbosa, Q. P. S., Câmara, C. A. G., Ramos, C. S., Nascimento, D. C. O., Lima-Filho, J. V., & Guimarães, E. F. (2012). Chemical composition, circadian rhythm and antibacterial activity of essential oils of *Piper divaricatum*: a new source of safrole. *Química Nova, 35*(9), 1806-1808. doi: 10.1590/S0100-40422012000900019

Baskar, K., & Ignacimuthu S. (2012). Ovicidal activity of *Atalantia monophylla* (L) Correa against *Helicoverpa armigera* Hubner (Lepidoptera: Noctuidae). *Journal of Agricultural Technology, 8*(3), 861-868.

Baskar, K., Muthu, C., Raj, G. A. Kingsley, S., & Ignacimuthu., S. (2012). Ovicidal activity of *Atalantia monophylla* (L) Correa against *Spodoptera litura* Fab. (Lepidoptera: Noctuidae). *Asian Pacific Journal of Tropical Biomedicine, 2*(12), 987-991. doi: 10.1016/S2221-1691(13)60011-8

Bizzo, H. R., Lopes, D., Abdala, R. V., Pimentel, F. A., Souza, J. A., Pereira, M. V. G., ... Guimarães, E. F. (2001). Sarisan from leaves of *Piper hispidinervum* C. DC (Long pepper). *Flavour and Fragrance Journal, 16*(2), 113-115. doi: 10.1002/ffj.957

Brito, E. F., Baldin, E. L. L., Silva, R. C. M., Ribeiro, L. P., & Vendramim, J. D. (2015). Bioactivity of Piper extracts on *Tuta absoluta* (Lepidoptera: Gelechiidae) in tomato. *Pesquisa Agropecuária Brasileira, 50*(3), 196-202. doi: 10.1590/S0100-204X2015000300002

Cáceres, A., & Kato, M. J. (2014). Importance of a multidisciplinary evaluation of *Piper* genus for development of new natural products in Latin America. *International Journal of Phytocosmetics and Natural Ingredients, 1*(3), 1-7. doi: 10.15171/ijpni.2014.04

Carneiro, A. P., Pereira, M. J. B., & Galbiati, C. (2011). Biocidal effect of the *Annona coriacea* Mart 1841 on eggs and nymphs of the vector *Rhodnius neglectus* Lent 1954. *Neotropical Biology and Conservation, 6*(2), 131-136. doi: 10.4013/nbc.2011.62.08

Castro, A. A., Corrêa, A. S., Legaspi, J. C., Guedes, R. N. C., Serrão, J. E., & Zanuncio, J. C. (2013). Survival and behavior of the insecticide-exposed predators *Podisus nigrispinus* and *Supputius cincticeps* (Heteroptera: Pentatomidae). *Chemosphere, 93*(6), 1043-1050. doi: 10.1016/j.chemosphere.2013.05.075

Coitinho, R. L. B. C., Oliveira, J. V., Gondim-Júnior, M. G. C., & Câmara, C. A. G. (2010). Persistence of essential oils in stored maize submitted to infestation of maize weevil. *Ciência Rural, 40*(7), 1492-1496. doi: 10.1590/S0103-84782010005000109

Costa, E. L. N., Silva, R. F. P., & Fiuza, L. M. (2004). Efeitos, aplicações e limitações de extratos de plantas inseticidas. *Acta Biologica Leopoldensia, 26*(2), 173-185.

Costa, M. S., Santana, A. E. G., Oliveira, L. L., & Serrão, J. E. (2017). Toxicity of squamocin on larvae, its predators and human cells. *Pest Management Science, 73*(3), 636-640. doi: 10.1002/ps.4350

Cruz, S. M., Cáceres, A., Álvarez, L. E., Apel, M. A., & Henriques, A. T. (2011). Chemical diversity of essential oils of 15 *Piper* species from Guatemala. *Acta Horticulturae, 964*(1), 39-46. doi: 10.17660/ActaHortic.2012.964.4

Dyer, L. A., & Palmer, A. D. N. (2004). *Piper*: A model genus for studies of Pytochemistry, Ecology, and Evolution. New York, US: Kluwer Academic Publisher.

Estrela, J. L. V., Fazolin, M., Catani, V., Alécio, M. R., & Lima, M. S. (2006). Toxicity of essential oils of *Piper aduncum* and *Piper hispidinervum* against *Sitophilus zeamais*. *Pesquisa Agropecuária Brasileira*, *41*(2), 217-222. doi: 10.1590/S0100-204X2006000200005

Favetti, B. M., Krinski, D., & Butnariu, A. R. (2013). Egg parasitoids of *Edessa meditabunda* (Fabricius) (Pentatomidae) in lettuce crop. *Revista Brasileira de Entomologia*, *57*(2), 236-237. doi: 10.1590/S0085-56262013005000014

Fazolin, M., Estrela, J. L. S., Catani, V., & Alécio, M. R. (2005). Toxicity of *Piper aduncum* oil to adults of *Cerotoma tingomarianus* Bechyné (Coleoptera: Chrysomelidae). *Neotropical Entomology*, *34*(3), 485-489. doi: 10.1590/S1519-566X2005000300018

Fazolin, M., Estrela, J. L. S., Catani, V., Alécio, M. R., & Lima, M. S. (2007). Insecticidal properties of essential oils of *Piper hispidinervum* C. DC.; *Piper aduncum* L. and *Tanaecium nocturnum* (Barb. Rodr.) Bur. & K. Shum against *Tenebrio molitor* L., 1758. *Ciência e Agrotecnologia*, *31*(1), 113-120. doi: 10.1590/S1413-70542007000100017

Finney, D. J. (1971). *Probit analysis*. Cambridge, UK: University Press.

Franco, A. A., Queiroz, M. S., Peres, A. R., Rosa, M. E, Campos, A. R., & Campos, Z. R. (2014). Preferência alimentar de *Anticarsia gemmatalis* Hübner (Lepidoptera: Noctuidae) por cultivares de soja. *Científica*, *42*(1), 32-38. doi: 10.15361/1984-5529.2014v42n1p32-38

Isman, M. B. (2000). Plant essential oils for pest and disease management. *Crop Protection*, *19*(8/10), 603-608. doi: 10.1016/S0261-2194(00)00079-X

Isman, M. B., & Grieneisen, M. L. (2014). Botanical insecticide research: many publications, limited useful data. *Trends in Plant Science*, *19*(3), 140-145. doi: 10.1016/j.tplants.2013.11.005

Koul, O., Walia, S., & Dhaliwal, G. S. (2008). Essential oils as green pesticides: Potential and constraints. *Biopesticides International*, *4*(1), 63-84.

Krinski, D. (2013). First report of phytophagous stink bug in chicory crop. *Ciência Rural*, *43*(1), 42-44.

Krinski, D. (2015). First report of Squash Vine Borer, *Melittia cucurbitae* (Harris, 1828) (Lepidoptera, Sessidae) in Brazil and South America: distribution extension and geographic distribution map. *Check List*, *11*(3), 1625-1626. doi: 10.15560/ 11.3.1625

Krinski, D., & Foerster, L. A. (2016). Toxicity of essential oils from leaves of five Piperaceae species in rice stalk stink bug eggs, *Tibraca limbativentris* (Hemiptera: Pentatomidae). *Ciência e Agrotecnologia*, *40*(6), 155-167. doi: 10.1590/1413-70542016406021616

Krinski, D., & Foerster, L. A. (2017). Damage by *Tibraca limbativentris* Stål (Pentatomidae) to upland rice cultivated in Amazon rainforest region (Brazil) at different growth stages. *Neotropical Entomology*, *46*(1), 107-114. doi: 10.1007/s13744-016-0435-5

Krinski, D., & Massaroli, A. (2014). Nymphicidal effect of vegetal extracts of *Annona mucosa* and *Annona crassiflora* (Magnoliales, Annonaceae) against rice stalk stink bug,

Tibraca limbativentris (Hemiptera, Pentatomidae). *Revista Brasileira de Fruticultura*, *36*(spe1), 217-224. doi: 10.1590/S0100-29452014000500026

Krinski, D., & Pelissari, T. D. (2012). Occurrence of the stinkbug *Edessa meditabunda* F. (Pentatomidae) in differents cultivars of lettuce *Lactuca sativa* L. (Asteraceae). *Bioscience Journal*, *28*(4), 654-659.

Krinski, D., Favetti, B. M., & Butnariu, A. R. (2012). First record of *Edessa meditabunda* (F.) on lettuce in Mato Grosso State, Brazil. *Neotropical Entomology*, *41*(1), 79-80. doi: 10.1007/s13744-011-0012-x

Krinski, D., Foerster, L. A., & Grazia, J. (2015). *Hypatropis inermis* (Hemiptera, Pentatomidae): First report on rice crop. *Revista Brasileira de Entomologia*, *59*(1), 12-13. doi: 10.1016/j.rbe.2014.11.001

Krinski, D., & Godoy, A. F. (2015). First record of *Helicoverpa armigera* (Lepidoptera: Noctuidae) feeding on *Plectranthus neochilus* (Lamiales: Lamiaceae) in Brazil. *Florida Entomologist*, *98*(4), 1238-1240. doi: 10.1653/024.098.0434

Krinski, D., Massaroli, A., & Machado, M. (2014). Insecticidal potential of the Annonaceae family plants. *Revista Brasileira de Fruticultura*, *36*(spe1), 225-242. doi: 10.1590/S0100-29452014000500027

Krishnappa, K., & Elumalai, K. (2012). Larvicidal and ovicidal activities of *Chloroxylon swietenia* (Rutaceae) essential oils against *Spodoptera litura* (Lepidoptera: Noctuidae) and their chemical compositions. *International Journal of Current Research in Life Sciences*, *1*(1), 3-7.

Laurent, D., Vilaseca, L. A., Chantraine, J. M., Ballivian, C., Saavedra. G., & Ibañez, R. (1997). Insecticidal activity of essential oils on *Triatoma infestans*. *Phytotherapy Research*, *11*(4), 283-290.

Loureiro, E. S., Moino-Junior, A., Arnosti, A., & Souza, G. C. (2002). Effect of chemical products used in lettuce and chrysanthemum on entomopathogenic fungi. *Neotropical Entomology*, *31*(2), 263-269. doi: 10.1590/S1519-566X2002000200014

Lourenção, A. L., Reco, P. C., Braga, N. R., Valle, G. E., & Pinheiro, J. B. (2010). Yield of soybean genotypes under infestation of the velvetbean caterpillar and stink bugs. *Neotropical Entomology*, *39*(2), 275-281. doi: 10.1590/S1519-566X2010000200020

Magrini, E. A., Botelho, P. S. M., & Silveira Neto, S. (1999). Biologia de *Anticarsia gemmatalis* Hübner, 1818 (Lepidoptera: Noctuidae) na cultura da soja. *Scientia Agricola*, *56*(3), 547-555. https://dx.doi.org/10.1590/S0103-90161999000300006

Martins, A. L., & Krinski, D. (2016). First record of the parasitoid *Gonatopus flavipes* Olmi, 1984 (Hymenoptera, Dryinidae) in Brazil's Amazon forest. *Journal of Hymenoptera Research*, *50*(3), 191-196. doi: 10.3897/JHR.50.8897

Massarolli, A., Pereira, M. J. B., & Foerster, L. A. (2016). *Annona mucosa* Jacq. (Annonaceae): A promising phytoinsecticide for the control of *Chrysodeixis includens* (Walker) (Lepidoptera: Noctuidae). *Journal of Entomology*, *13*(4), 132-140. doi: 10.3923/ je.2016.132.140

McPherson, R. M., & Macrae, T. C. (2009). Evaluation of transgenic soybean exhibiting high expression of a synthetic *Bacillus thuringiensis* cry1A transgene for suppressing lepidopteran population densities and crop injury. *Journal of Economic Entomology*, *102*(4), 1640-1648. doi: 10.1603/029.102.0431

Misni, N., Othman, H., & Sulaiman, S. (2011). The effect of *Piper aduncum* Linn. (Family: Piperaceae) essential oil as aerosol spray against *Aedes aegypti* (L.) and *Aedes albopictus* Skuse. *Tropical Biomedicine*, *28*(2), 249-258.

Morais, S. M., Facundo, V. A., Bertini, L. M., Cavalcanti, E. S. B., Anjos-Junior, J. F., Ferreira, S. A., ... Souza-Neto, M. A. (2007). Chemical composition and larvicidal activity of essential oils from *Piper* species. *Biochemical Systematics and Ecology*, *35*(10), 670-675. doi: 10.1016/j.bse.2007.05.002

Moscardi, F., Bueno, A. F., Sosa-Gómez, D. R., Roggia, S., Hoffman-Campo, C. B., Pomari, A. F., ... Yano, S. A. C. (2012). Artrópodes que atacam as folhas da soja. In C. B. Hoffman-Campo, B. S. Corrêa-Ferreira, & F. Moscardi (Eds.), *Soja*: manejo integrado de insetos e outros artrópodes-praga (p. 213-309). Brasília, DF: Embrapa.

Mundina, M., Vila, R., Tomi, F., Tomas, X., Ciccio, J. F., Adzet, T., ... Canigueral, S. (2001). Composition and chemical polymorphism of the essential oils from *Piper lanceaefolium*. *Biochemical Systematics and Ecology*, *29*(7), 739-748.

Ntalli, N.G., & Menkissoglu-Spiroudi, U. (2011). Pesticides of botanical origin: a promising tool in plant protection. In M. Stoytcheva (Ed.), *Pesticides-formulation, effects fate* (p. 3-24). Croatia: InTech. doi: 10.5772/13776

Panizzi, A. R., Oliveira, L. J., & Silva, J. J. (2004). Survivorship, larval development and pupal weight of *Anticarsia gemmatalis* (Hübner) (Lepidoptera: Noctuidae) feeding on potential leguminous host plants. *Neotropical Entomology*, *33*(5), 563-567. doi: 10.1590/S1519-566X2004000500004

Petroski, R. J., & Stanley, D. W. (2009). Natural compounds for pest and weed control. *Journal of Agricultural and Food Chemistry*, *57*(18), 8171–8179. doi:http://doi.org/10.1021/jf803828w

Picollo, M. I., & Zerba, E. (1997). Embryogenesis. In R. U. Carcavallo, G. L. Galíndez, J. Jurberg, & H. Lent (Eds.), *Atlas dos vetores da Doença de Chagas nas Américas*. (p. 265-270). Rio de Janeiro, RJ: Fiocruz.

Piton, L. P., Turchen, L. M., Butnariu, A. R., & Pereira, M. J. B. (2014). Natural insecticide based-leaves extract of *Piper aduncum* (Piperaceae) in the control of stink bug brown soybean. *Ciência Rural*, *44*(11), 1915-1920. doi: 10.1590/0103-8478cr20131277

Qin, W., Huang, S., Li, C., Chen, S., & Peng, Z. (2010). Biological activity of the essential oil from the leaves of *Piper sarmentosum* Roxb. (Piperaceae) and its chemical constituents on *Brontispa longissima* (Gestro) (Coleoptera: Hispidae). *Pesticide Biochemistry and Physiology*, *96*(3), 132-139. doi: 10.1016/j.pestbp.2009.10.006

Rahman, K., Bridges, W. C., Chapin, J. W., & Thomas, J. S. (2007). Three cornered alfalfa hopper (Hemiptera: Membracidae): seasonal occurrence, girdle distribution, and response to insecticide treatment on peanut in South Carolina. *Journal of Economic Entomology*, *100*(4), 1229-1240. doi: 10.1603/0022-0493(2007)100[1229:TAHHMS]2.0.CO;2

Rajapakse, R., & Senanayake, S. (1997). Effectiveness of seven vegetable oils against *Callosobruchus chinensis* L. in pigeon pea *Cajanus cajan* L. *Entomon*, *22*(3/4), 179-183.

Rampelotti-Ferreira, F. T., Ferreira, A., Prando, H. F., Tcacenco, F. A., Grützmacher, A. D., & Martins, J. F. S. (2010). Selectivity of chemical pesticides used in rice irrigated crop at fungus *Metarhizium anisopliae*, microbial control agent of *Tibraca limbativentris*. *Ciência Rural*, *40*(4), 745-751. doi: 10.1590/S0103-84782010005000062

Ribeiro, N. E., Camara, C., & Ramos, C. (2016). Toxicity of essential oils of *Piper marginatum* Jacq. against *Tetranychus urticae* Koch and *Neoseiulus californicus* (McGregor). *Chilean Journal of Agricultural Research*, *76*(1), 71-76. doi: 10.4067/S0718-58392016000100010

Riedl, H., Halaj, J., Kreowski, W., Hilton, R., & Westigard, P. (1995). Laboratory evaluation of mineral oils for control of *Codling moth* (Lepidoptera: Tortricidae). *Journal of Economic Entomology*, *88*(1), 140-147. doi: 10.1093/jee/88.1.140

Sanini, C., Massarolli, A., Krinski, D., Butnariu, A.R., 2017. Essential oil of spiked pepper, *Piper aduncum* L. (Piperaceae) for the control of caterpillar soybean looper, *Chrysodeixis includens* Walker (Lepidoptera: Noctuidae). *Brazilian Journal of Botany*, *40*(2), 399-404. doi: 10.1007/s40415-017-0363-6

Santana, H. T., Trindade, F. T. T, Stabeli, R. G., Silva, A. A. E, Militão, J. S. L. T., & Facundo, V. A. (2015). Essential oils of leaves of *Piper* species display larvicidal activity against the dengue vector, *Aedes aegypti* (Diptera: Culicidae). *Revista Brasileira de Plantas Medicinais*, *17*(1), 105-111. doi: 10.1590/1983-084X/13_052

Santos, M. R. A., Lima, R. A., Silva, A. G., Teixeira, C. A. D., Alpirez, I. P. V., Facundo, & V. A. (2013). Chemical constituents and insecticidal activity of the crude acetonic extract of *Piper alatabaccum* Trel & Yuncker (Piperaceae) on *Hypothenemus hampei* Ferrari. *Revista Brasileira de Plantas Medicinais*, *15*(3), 332-336. doi: 10.1590/S1516-05722013000300004

Santos, M. R. A., Lima, R. A., Silva, A. G., Teixeira, C. A. D., Lima, D. K. S., & Polli, A. R. (2011). Facundo, V.A. Atividade inseticida do extrato de raiz de *Piper hispidum* H.B.K. (Piperaceae) sobre *Hypothenemus hampei* Ferrari. *Revista Saúde e Pesquisa*, *4*(3), 335-340.

Santos, M. R. A., Silva, A. G., Lima, R. A., Lima, D. K. S., Sallet, L. A. P., Teixeira, C. A. D., ... Facundo, V. A. (2010). Inseticidal activity of *Piper hispidum* (Piperaceae) leaves extract on (*Hypothenemus hampei*). *Brazilian Journal of Botany*, *33*(2), 319-324. doi: 10.1590/S0100-84042010000200012

Santos, P. R. D., Moreira, D. L., Guimarães, E. F., & Kaplan, M. A. (2001). Essential oil analysis of 10

Piperaceae species from the Brazilian Atlantic forest. *Phytochemistry, 58*(4), 547-551.

Scott, I. M., Jensen, H. R., Philogène, B. J. R., & Arnason, J.T. (2008). A review of *Piper* spp. (Piperaceae) phytochemistry, insecticidal activity and mode of action. *Phytochemistry Reviews, 7*(1), 65-75. doi: 10.1007/s11101-006-9058-5

Silva, W. C., Ribeiro, J. D., Souza, H. E. M., & Correa, R. S. (2007). Insecticidal activity of *Piper aduncum* L. (Piperaceae) on *Aetalion* sp. (Hemiptera: Aetalionidae), plague of economic importance in Amazon. *Acta Amazonica, 37*(2), 293-298. doi: 10.1590/S0044-59672007000200017

Smith, E. H., & Salkeld, E. H. (1966). The use and action of ovicides. *Annual Review of Entomology, 11*(1), 331-368. doi: 10.1146/annurev.en. 11.010166.001555

Smith, E., & Pearce, G. (1948). The mode of action of petroleoum oils as ovicides. *Journal of Economic Entomology, 41*(2), 173-180. doi: 10.1093/jee/41.2.173

Souto, R. N., Harada, A. Y., Andrade, E. H., & Maia, J. G. (2012). Insecticidal activity of *Piper* essential oils from the Amazon against the fire ant *Solenopsis saevissima* (Smith) (Hymenoptera: Formicidae). *Neotropical Entomology, 41*(6), 510-517. doi: 10.1007/ s13744-012-0080-6

Tchoumbougnang, F., Jazet, D. P. M., Sameza, M. L., Fombotioh, N., Wouatsa, N. A. V., Amvam, Z. P. H., & Menut, C. (2009). Comparative essential oils composition and insecticidal effect of different tissues of *Piper capense* L., *Piper guineense* Schum. et Thonn., *Piper nigrum* L. and *Piper umbellatum* L. grown in Cameroon. *African Journal of Biotechnology, 8*(3), 424-431.

Tripathi, A. K., Upadhyay, S., Bhuiyan, M., & Bhattacharya., P. R. (2009). A review on prospects of essential oils as biopesticide in insect-pest management. *Journal of Pharmacognosy and Phytotherapy, 1*(5), 52-63.

Turchen, L. M., Piton, L. P., Dall'Oglio, E. L., Butnariu, A. R., & Pereira, M. J. B. (2016a). Toxicity of *Piper aduncum* (Piperaceae) essential oil against *Euschistus heros* (F.) (Hemiptera: Pentatomidae) and non-effect on egg parasitoids. *Neotropical Entomology, 45*(5), 604-611. doi: 10.1007/s13744-016-0409-7

Turchen, L. M., Hunhoff, L. M., Paulo, M. V., Souza, C. P. R., & Pereira, M. J. B. (2016b). Potential phytoinsecticide of *Annona mucosa* (Jacq) (Annonaceae) in the control of brown stink bug. *Bioscience Journal, 32*(3), 581-587. doi: 10.14393/BJ-v32n3a2016-32803

Van den Dool, H., & Kratz, P. D. (1963). A generalization of the retention index system including liner temperature programmed gas-liquid partition chromatography. *Journal of Chromatography A, 11*(1), 463-467. doi: 10.1016/S0021-9673(01)80947-X

Zoubiri, S., & Baaliouamer, A. (2011). Potentiality of plants as source of insecticide principles. *Journal of Saudi Chemical Society, 18*(6), 925-938. doi: 10.1016/j.jscs.2011.11.015

20

Periods of weed interference in chickpea grown under different doses of nitrogen fertilizer topdressing

Carita Liberato Amaral, Guilherme Bacarin Pavan, Fernanda Campos Mastrotti Pereira and Pedro Luis da Costa Aguiar Alves[*]

*Faculdade de Ciências Agrárias e Veterinárias, Universidade Estadual Paulista "Júlio de Mesquita Filho", Rod. de Acesso Prof. Paulo Donato Castellane, Km 05, Zona rural, 14884-900, Jaboticabal, São Paulo. Brazil. *Author for correspondence. E-mail: plalves@fcav.unesp.br*

ABSTRACT. Weed interference can reduce chickpea growth and, therefore, productivity depending on the period of coexistence and the nutritional status of the crop, among other factors. A study was performed over two crop years to estimate the critical period of weed interference (CPWI) during chickpea production under three doses of nitrogen (N) fertilizer topdressing (0, 50, and 75 kg N ha^{-1}). The experiments were conducted at 0, 7, 14, 21, 28, 35, 42, 56, 63, and 140 days after emergence (DAE) of chickpea/weed coexistence under two conditions: initially weed-free and initially weed-infested. The presence of weeds negatively affected chickpea production and reduced yields by 70% on average regardless of N rate, rendering the crop economically unfeasible. The CPWI ranged from 5 to 76 DAE and was not affected by N topdressing up to 75 kg N ha^{-1} in both crop years, assuming an acceptable production loss of 5%. Although the CPWI without fertilization (0 kg N ha^{-1}) was similar to that when fertilized with 50 and 75 kg N ha^{-1}, the two topdressing doses increased chickpea productivity by 37% and 51%, on average, respectively.

Keywords: *Cicer arietinum* L.; coexistence periods; critical period of interference; productivity.

Introduction

Chickpea (*Cicer arietinum* L.) is a very important crop that is mainly used for human and animal food (Mohammadi, Javanshir, Khooie, Mohammadi, & Salmasi, 2005; Hossain, Hasan, Sultana, & Bari, 2016), and it is the second most widely grown legume worldwide (Pang et al., 2017) after soybean (Varshney et al., 2014). This crop can be grown in many areas, including marginal land and low-fertility areas (Esfahani et al., 2014), and its cultivation plays a key role in maintaining soil fertility, especially in tropical regions (Varshney et al., 2009), thus representing an important component of crop rotation. Current global chickpea production is approximately 13 million tons (Mt) (FAO, 2014), with an expected increase to 17 Mt in 2020 (Abate et al., 2012).

Chickpea plays important roles in the human diet (Ulukan, Bayraktar, & Koçak, 2012) and agricultural systems (Varshney et al., 2014). The seeds are rich in fiber, vitamins, carbohydrates, mineral salts (Ulukan et al., 2012), unsaturated fatty acids and β-carotene (Gaur, Jukanti, & Varshney, 2012) and are a good source of protein, with a content of approximately 21% (Esfahani et al., 2014). Therefore, this crop plays a key role in the food security of developing countries and is an important component of subsistence agriculture (Varshney et al., 2014).

Weeds represent a great barrier to the productivity of several agricultural crops (Kaushik, Rai, Sirothia, Sharma, & Shukla, 2014), and similar to other crops, chickpea can be threatenedby both direct and indirect weed interference, which can quantitatively and qualitatively reduce production depending on the severity (Amaral, Pavan, Souza, Martins, & Alves, 2013; Amaral, Souza, Pavan, Gavassi, & Alves, 2015). Singh and Bajpai (1996) and Ratnam, Rao, and Reddy (2011) found that weed interference can decrease chickpea productivity by more than 85%, and Kaushik et al. (2014) observed a loss in productivity of more than 65% that reduced financial gains by 42%. However, weeds do not interfere equally during all stages of crop growth; during some periods, chickpeas can tolerate the presence of weeds without any negative effects on productivity (Al-Thahabi, Yasin, Abu-Irmaileh, Haddad, & Saxena, 1994).

There are three different weed interference periods: the period prior to weed interference (PPWI) that begins with crop emergence and during which the crop may coexist with weeds without decreased productivity; the total period of weed interference prevention (TPIP) that starts with crop emergence and during which weeds should be controlled to enable the crop to reach its productivity potential; and the critical period of weed interference (CPWI), which is the interval between these two periods. Periods of weed interference in agricultural crops can be used to optimize the weed control period (Amaral, Souza, Pavan, Gavassi, & Alves, 2013), thus enabling a reduction in the use of pesticides and/or weeding through the development of bioeconomic models for use in integrated weed management systems (Mohammadi et al., 2005; Amaral et al., 2015) and avoiding crop losses or damages, thereby resulting in an economically viable yield (Tepe, Erman, Yergin, & Bükün, 2011).

Interference periods vary widely depending on factors including environmental conditions and the characteristics of the soil, weed community and the crop itself (Tepe et al., 2011). Chickpeas are very sensitive to weed interference due to their slow growth rate and limited leaf area during the early stages of growth and establishment (Kaushik et al., 2014).

The yield gap in chickpea culture can be narrowed by adopting advanced production technologies that balance nutrition, weed management and the use of high-yielding varieties (Rani & Krishna, 2016). Soil nutrient availability, especially of nitrogen (N), phosphorus (P) and potassium (K), is among the most important factors that affect the competitive relationships between the crop and weeds (Tang et al., 2014).

As a leguminous crop, chickpea fixes N from the atmosphere, but there is strong evidence that it may be inferior to other grain legumes in terms of this function. Therefore, there is a need to determine the level of N needed to obtain higher yieldsof good quality (Rani & Krishna, 2016).

The chickpea is a robust plant, but mineral nutrient limitation is considered a major environmental stressor that contributes to yield loss (Valenciano, Boto, & Marcelo, 2011). Understanding plant responses to fertilizer aids in the development of fertilization strategies and is a key component of integrated weed management programs (Blackshaw et al., 2003).

The aim of this study was to assess the effects of three N topdressing regimes on the critical periods of interference (the PPWI, TPIP and CPWI) for the natural weed community and chickpea productivity. The study attempted to answer the following questions: a) Does N topdressing alter the critical periods of interference? b) Does N topdressing improve chickpea productivity and interfere with the weed community?

Material and methods

During 2011 and 2012, three experiments were performed under field conditions at 21° 14' S latitude and 48° 17' W longitude in the municipality of Jaboticabal, São Paulo State, Brazil at an average altitude of 615 meters above sea level. The climate of the region is defined as tropical and is classified as Cwa.

Climatological data (Table 1) show that the experiments were conducted under similar environmental conditions. Average temperature was similar in both years, although rainfall was much higher in 2012, with a cumulative rainfall of 110.5 mm compared with 44.9 mm in 2011. These moisture variations were mitigated using supplemental irrigation, and sprinklers were used whenever considered necessary based on visual inspection. By the end of the experiment, 400 mm of water had accumulated in both areas following the recommendations of EMBRAPA (2010).

Table 1. Maximum, average and minimum temperatures and monthly rainfall between May and September 2011 and 2012 in Jaboticabal, São Paulo State, Brazil.

Year	Months	Maximum temperature ($^{\circ}$C)	Average temperature ($^{\circ}$C)	Minimum temperature ($^{\circ}$C)	Rainfall (mm)
2011	May	31.60	20.20	8.80	6.50
	June	30.10	16.75	3.40	26.70
	July	31.60	19.10	6.60	0.00
	August	35.40	19.80	4.20	9.40
	September	37.30	22.85	8.40	2.30
2012	May	30.30	19.55	8.80	20.40
	June	30.20	21.00	11.80	48.60
	July	31.70	18.75	5.80	9.00
	August	31.60	21.15	10.70	0.00
	September	36.60	21.40	6.20	32.50

The experiments were in different fields in the different years. The experimental areas were previously used to grow soybean (*Glycine max*) in 2011 and maize (*Zea mays*) in 2012. A composite soil sample was collected seventy days before the sowing of chickpeas, and chemical and physical analyses of the sample are presented in Table 2. Based on the results, dolomitic limestone was applied to raise the base saturation (V%) to 70% in both years.

Table 2. Physical and chemical properties of the soil in which chickpeas were sown in 2011 and 2012 in Jaboticabal, São Paulo State, Brazil.

Soil parameter	Year 2011	Year 2012
pH (CaCl$_2$)	5.5	4.5
Organic matter (g dm^{-3})	20.0	19.0
P$_{residue}$ (mg dm^{-3})	69.0	24.0
K$^+$ (mMol$_c$ dm^{-3})	3.0	3.4
Ca^{2+} (mMol$_c$ dm^{-3})	40.0	11.0
Mg^{2+} (mMol$_c$ dm^{-3})	18.0	6.0
H+Al^{3+} (mMol$_c$ dm^{-3})	34.0	42.0
Base saturation (mMol$_c$ dm^{-3})	61.0	20.4
Cation exchange capacity (mMol$_c$ dm^{-3})	95.0	62.4
Fertility rate [V%]	64	33
Clay	546	380
Silt	241	37
Sand	130	245
Grit	83	338
Texture	Clayey	Clayey

Chickpeas (BRS Cícero, Kabuli group) were sown during the first half of May, in both years, under a conventional tillage system and the rate of 14 seeds per meter at a 45-cm inter-row spacing. Seeds were previously treated with thiamethoxam and carboxin + thiram, and fertilization at sowing comprised 150 kg ha^{-1} of formulated fertilizer (04-14-08). Thinning was performed after emergence, leaving 12 plants per meter.

Three doses of N topdressing were used in both years, corresponding to three experiments: I – 0 kg

N ha^{-1}; II - 50 kg N ha^{-1}; and III - 75 kg N ha^{-1}. Fertilization was performed at 40 days after sowing when the plants were at the "vegetative growth" phenological stage before flowering.

At each N dose, the treatments consisted of increasing periods of coexistence and weed control; the treatments were analyzed from crop emergence and divided into two groups. In the first group, weeds were allowed to coexist from crop emergence to the end of the respective coexistence period (infested with weeds - IWW): IWW until 0 (IWW$_0$), 7 (IWW$_7$), 14 (IWW$_{14}$), 21 (IWW$_{21}$), 28 (IWW$_{28}$), 35 (IWW$_{35}$), 42 (IWW$_{42}$), 56 (IWW$_{56}$), 63 (IWW$_{63}$), and 140 (IWW$_{140}$) days after emergence (DAE), after which weeds were controlled and the plots were kept clean until harvest. In the second group, the crop was maintained free of weeds from emergence to the end of the respective control period (free of weeds - FOW): 0 (FOW$_0$), 7 (FOW$_7$), 14 (FOW$_{14}$), 21 (FOW$_{21}$), 28 (FOW$_{28}$), 35 (FOW$_{35}$), 42 (FOW$_{42}$), 56 (FOW$_{56}$), 63 (FOW$_{63}$), and 140 (FOW$_{140}$); after the indicated periods, weeds were allowed to grow freely in the plots, coexisting with the crop until harvest. Hand weeding was performed to maintain the plots "clean" of weeds.

The experiments were conducted using a randomized block design with four replicates. Each experimental plot comprised five six-meter-long sowing rows, resulting in a total area of 13.5 m^2. The three central rows were samples and evaluated, and one meter was discarded at each end, resulting in a final useful area of 5.4 m^2.

In the treatments corresponding to the weed-infested periods, the weed community was sampled at the end of each predetermined period by collecting weed samples from 0.75 m^2 of the useful area of the respective plots, corresponding to three subsamples of 0.25 m^2, using frames that were randomly placed in the plot row and inter-row areas. The weed species were identified, separated and dried in a convection drying oven at 70°C for 96h for subsequent measurement of shoot dry mass (DM). In the treatments corresponding to weed-free periods, the weed community was evaluated at 70 DAE (before harvest).

The chickpea crops were harvested at 140 DAE, when the grain moisture contents ranged from 13 to 15%. The chickpea productivity data were fitted to a Boltzmann sigmoidal regression model to estimate the PPWI, TPIP, and CPWI, as performed by Kuva, Pitelli, Christoffoleti, and Alves (2000) and Cardoso, Alves, Severino, and Vale (2011): y = [(P1 - P2) / (1 + e (x - xi) / dx))] + P2, where y is chickpea yield (t ha^{-1}) according to the period of coexistence or control; P1 is the maximum production (t ha⁻¹) obtained in plants without weed interference throughout the cycle; P2 is the minimum production (t ha⁻¹) obtained in plants coexisting with weeds during the period; (P1 - P2) is the yield losses (t ha⁻¹); x is the upper limit of the coexistence control period (days); xi is the upper limit of the interaction or control period, which corresponds to the intermediate value between the maximum and minimum output (days); and dx indicates the rate of production loss of due to the duration of coexistence [(t ha⁻¹) dia⁻¹]. The periods of interference were determined by accepting productivity losses of 2.5, 5, and 10% compared with those obtained in plots that were maintained free of weeds throughout the crop cycle.

The percentage losses compared with the weed-free plots were calculated based on grain productivity data as follows: PL (%) = ((Ra - Rb)/Ra) x 100, where PL is the percent loss of chickpea productivity; Ra is the chickpea yield in the coexistence periods; and Rb is the chickpea yield in the weed control period. These data were correlated with the accumulated weed dry mass using a linear or quadratic regression model.

Result

In 2011, the weed community comprised 15 species belonging to 11 botanical families; Amaranthaceae and Poaceae were the most represented, at three species each. However, Amaranthaceae and Brassicaceae had the highest density values, with 36.49 and 33.29% of the total number of individuals sampled (10,430), respectively. Conversely, the Poaceae family only accounted for 5.81% of the individuals sampled, less than Portulacaceae (10.25%). The family Brassicaceae was only represented by *Raphanus raphanistrum* in 2011. During the periods of infestation in 2011, *R. raphanistrum* and *Amaranthus viridis* had the highest total weed dry mass, corresponding to 72.58 and 12.38% of the total weed dry mass (36.16 kg accumulated in 180 m² sampled), respectively.

In 2012, the weed community comprised 23 weed species belonging to 13 families, yielding a total dry mass accumulation of 18.46 kg over 180 m². The families with the greatest species densities were Brassicaceae (35.33%), Asteraceae (17.09%) and Portulacaceae (14.92%). In 2012, the family Amaranthaceae only accounted for 7.99% of 11,602 individuals sampled, a value that was lower than that

for Cyperaceae (8.46%) and Solanaceae (8.20%). The family Asteraceae was represented by five species in 2012 but was represented only by *Parthenium hysterophorus* in 2011. The family Brassicaceae had the highest abundance in terms of the number of individuals and was represented by *R. raphanistrum*, *Lepidium virginicum* and *Coronopus didymus*, which accounted for 63.19, 26.06, and 10.76% of the total number of individuals of the family, respectively. The species *R. raphanistrum* and *Bidens pilosa* were the most dominant among the species with the highest densities (*A. viridis*, *B. pilosa*, *C. didymus*, *Cyperus rotundus*, *L. virginicum*, *Nicandra physaloides*, *P. hysterophorus*, *Portulaca oleracea*, and *R. raphanistrum*) in the analyzed agro-ecosystem and exhibited the highest accumulations of dry mass (67.72 and 10.21%, respectively).

Higher N doses caused reduced species diversity and plant density in both years, whereas the opposite was true of biomass; N fertilization led to increased biomass accumulation. *R. raphanistrum* was among the most important species regardless of N dose applied, both in 2011 and 2012.

Figure 1 presents curves fitted to Boltzmann's equation; according to this model, productivity declined from 2,450.34 to 670.02 kg ha⁻¹ (0 kg N ha⁻¹), from 3,876.75 to 925.59 kg ha⁻¹ (50 kg N ha⁻¹) and from 3,302.40 to 1,205.85 kg ha⁻¹ (75 kg N ha⁻¹) in 2011 (based on maximum and minimum yields), representing decreases of 72.66, 76.12, and 63.49%, respectively, with increasing N doses. Productivity decreases from 1,779.58 to 687.79 kg ha⁻¹ (0 kg N ha⁻¹), from 2,786.67 to 758.55 kg ha⁻¹ (50 kg N ha⁻¹) and from 2,865.24 to 846.84 kg ha⁻¹ (75 kg N ha⁻¹) were obtained in 2012, representing decreases of 61.35, 72.78, and 70.44%, respectively. These results demonstrate the high susceptibility of chickpeas to weed interference.

Considering 2.5, 5, and 10% chickpea productivity losses as acceptable, the presence of weeds affected the crop at 6, 7, and 11 DAE (PPI), respectively, in 2011 when using 0 kg N ha⁻¹ (Figure 1a; Table 3). The results also showed that weeds should be controlled until 82, 76, and 61 DAE to achieve maximum production losses of 2.5, 5, and 10%, respectively (TPIP). Thus, the periods during which weeds should be controlled (CPWI) are as follows: from 6 to 82 DAE (2.5% acceptable productivity loss), from 7 to 76 DAE (5% acceptable productivity loss) and from 11 to 61 DAE (10% acceptable productivity loss). Following the TPIP, weed control did not increase chickpea productivity.

Coexistence with the weed community began to affect crop productivity at 4, 5, and 9 DAE (PPI) in the experiment using 50 kg N ha^{-1} in 2011 (Figure 1b), assuming chickpea productivity losses of 2.5, 5, and 10% (Table 3), respectively. Furthermore, weed control for maximum losses of 2.5, 5, and 10% should be performed until 69, 65, and 57 DAE, respectively (TPIP). Thus, the periods during which weeds should be controlled (CPWI) are as follows: from 4 to 69 DAE (2.5% acceptable productivity loss), from 5 to 65 DAE (5%

acceptable productivity loss) and from 9 to 57 DAE (10% acceptable productivity loss).

Also in 2011, PPWIs of 9, 10, and 14 DAE and TPIPs of 45, 40 and 31 were obtained using 75 kg N ha^{-1} (Figure 1c) for chickpea productivity losses of 2.5, 5, and 10%, respectively (Table 3). Thus, the periods during which weeds should be controlled (CPWI) are as follows: from 9 to 45 DAE (2.5% acceptable productivity loss), from 10 to 40 DAE (5% acceptable productivity loss) and from 14 to 31 DAE (10% acceptable productivity loss).

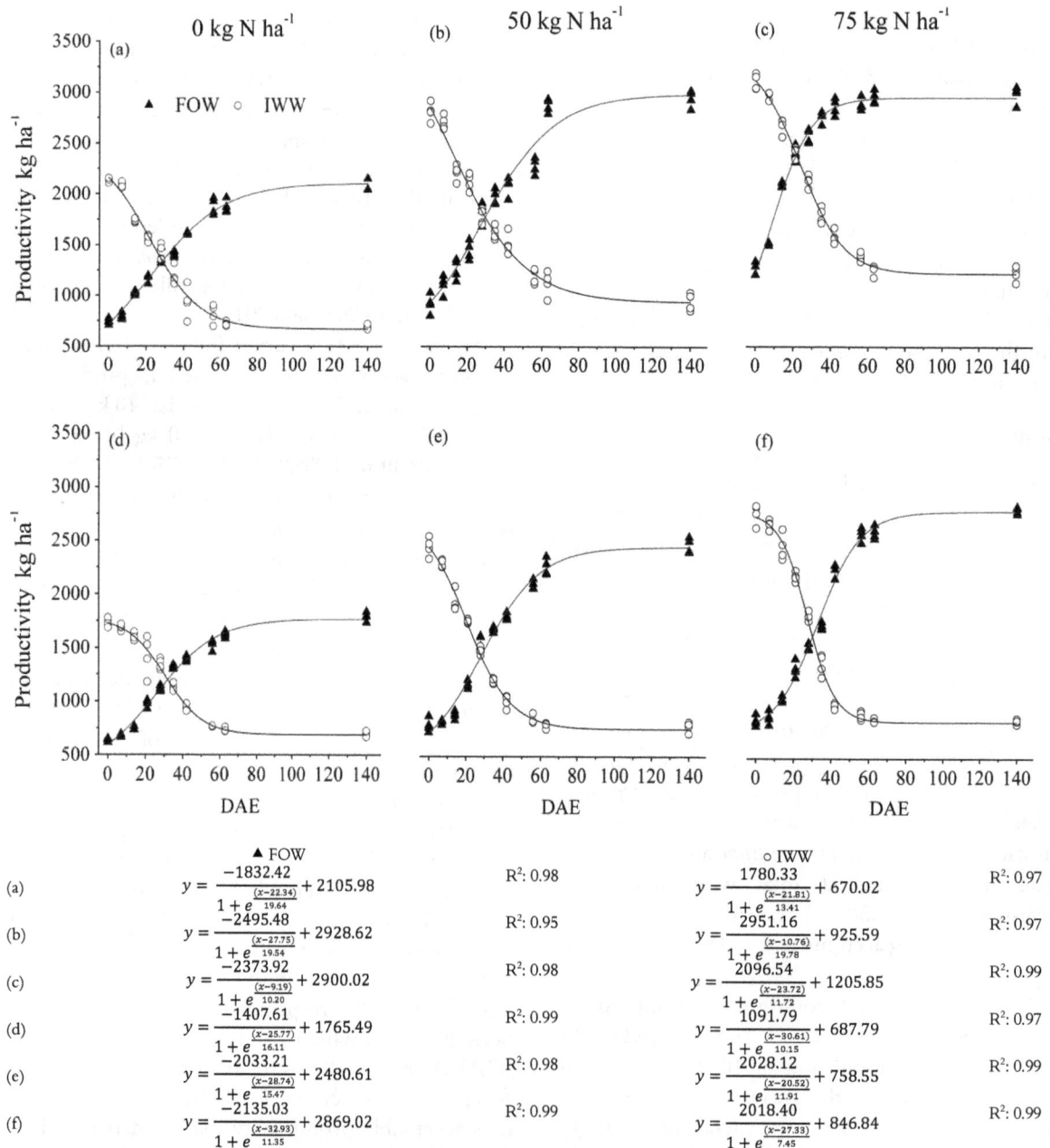

	▲ FOW		○ IWW	
(a)	$y = \dfrac{-1832.42}{1+e^{\frac{(x-22.34)}{19.64}}} + 2105.98$	R^2: 0.98	$y = \dfrac{1780.33}{1+e^{\frac{(x-21.81)}{13.41}}} + 670.02$	R^2: 0.97
(b)	$y = \dfrac{-2495.48}{1+e^{\frac{(x-27.75)}{19.54}}} + 2928.62$	R^2: 0.95	$y = \dfrac{2951.16}{1+e^{\frac{(x-10.76)}{19.78}}} + 925.59$	R^2: 0.97
(c)	$y = \dfrac{-2373.92}{1+e^{\frac{(x-9.19)}{10.20}}} + 2900.02$	R^2: 0.98	$y = \dfrac{2096.54}{1+e^{\frac{(x-23.72)}{11.72}}} + 1205.85$	R^2: 0.99
(d)	$y = \dfrac{-1407.61}{1+e^{\frac{(x-25.77)}{16.11}}} + 1765.49$	R^2: 0.99	$y = \dfrac{1091.79}{1+e^{\frac{(x-30.61)}{10.15}}} + 687.79$	R^2: 0.97
(e)	$y = \dfrac{-2033.21}{1+e^{\frac{(x-28.74)}{15.47}}} + 2480.61$	R^2: 0.98	$y = \dfrac{2028.12}{1+e^{\frac{(x-20.52)}{11.91}}} + 758.55$	R^2: 0.99
(f)	$y = \dfrac{-2135.03}{1+e^{\frac{(x-32.93)}{11.35}}} + 2869.02$	R^2: 0.99	$y = \dfrac{2018.40}{1+e^{\frac{(x-27.33)}{7.45}}} + 846.84$	R^2: 0.99

Figure 1. Chickpea productivity as a function of the periods of weed coexistence and absence in 2011 (a, b, c) and 2012 (d, e, f) with topdressings of 0 kg N ha^{-1} (a, d), 50 kg N ha^{-1} (b, e) and 75 kg N ha^{-1} (c, f). One curve represents the productivity of initially weed-infested chickpeas (○ IWW), and the second represents that of initially weed-free chickpeas (▲ FOW).

Table 3. Period prior to weed interference (PPWI), total period of weed interference prevention (TPIP) and critical period of weed interference (CPWI) as a function of the tolerated reductions in yield for the experiments conducted in 2011 and 2012 with topdressings of 0, 50, and 75 kg N ha^{-1}.

Year	Topdressing	Tolerated reduction								
		2.5%			5%			10%		
		Period (days after emergence)								
		PPWI	TPIP	CPWI	PPWI	TPIP	CPWI	PPWI	TPIP	CPWI
2011	0 kg N ha^{-1}	6	82	6-82	7	76	7-76	11	61	11-61
	50 kg N ha^{-1}	4	69	4-69	5	65	5-65	9	57	9-57
	75 kg N ha^{-1}	9	45	9-45	10	40	10-40	14	31	14-31
2012	0 kg N ha^{-1}	8	59	8-59	11	64	11-64	17	55	17-55
	50 kg N ha^{-1}	9	59	9-59	10	57	10-57	13	50	13-50
	75 kg N ha^{-1}	7	63	7-63	10	61	10-61	15	53	15-53

In 2012, in the experiment using 0 kg N ha^{-1} (Figure 1d) and accepting 2.5, 5, and 10% crop productivity losses (Table 3), the PPWIs were 8, 11, and 17 DAE, respectively, and the TPIPs were 59, 64, and 55 DAE. Consequently, the CPWIs were from 8 to 59 DAE (51 days), from 11 to 64 DAE (53 days) and from 17 to 55 DAE (38 days), respectively.

Also during 2012, in the experiment using 50 kg N ha^{-1} (Figure 1e) and accepting productivity losses of 2.5, 5, and 10%, the PPWIs were 9, 10, and 13 DAE, respectively, and the TPIPs were 59, 57, and 50 DAE. Consequently, the CPWIs obtained in this experiment were from 9 to 59 DAE (50 days), from 10 to 57 DAE (47 days) and from 13 to 50 DAE (47 days), respectively (Table 3).

Also in 2012, PPWIs of 7, 10, and 15 DAE, TPIPs of 63, 61 and 53 DAE and CPWIs from 7 to 63 DAE (56 days), from 10 to 61 DAE (51 days) and from 15 to 53 DAE (38 days) were obtained for productivity losses of 2.5, 5, and 10%, respectively, (Table 3) when using 75 kg N ha^{-1} (Figure 1f).

Comparing treatments in the total absence of weeds (FOW$_{140}$) with the treatments in the presence of weeds throughout the entire crop cycle (IWW$_{140}$), the data for 2011 demonstrate grain productivity decreases of 67.16, 67.65, and 58.69% when using 0, 50, and 75 kg N ha^{-1}, respectively, and the maximum productivities obtained were 2,101, 2,890, and 2,936 kg ha^{-1} (FOW$_{140}$), respectively; these results demonstrate the high susceptibility of chickpeas to weed interference.

Chickpea yields in the IWW$_0$ periods for the 2011 harvest were 2,138, 2,755, and 3,051 kg ha^{-1} when fertilized with 0, 50, and 75 kg N ha^{-1}, respectively. The 2012 harvest showed lower yields than those obtained in 2011; total yields were 1,733, 2,482, and 2,846 kg ha^{-1} for the treatments fertilized with 0, 50, and 75 kg N ha^{-1}, representing decreases of 18.94, 9.91, and 6.72%, respectively, when compared with the maximum yields of the previous year. The productivities measured for the FOW$_{140}$ periods were 2,101, 2,890, and 2,936 kg ha^{-1} in 2011 and 1,792, 2,501, and 2,883 in 2012, representing decreases of 14.71, 13.46, and 1.80% in the treatments fertilized with 0, 50, and 75 kg N ha^{-1}, respectively.

Decreased grain productivity was positively correlated with weed dry mass accumulation (p < 0.0001) in the three experiments conducted in 2011 (Figure 2a, b, and c), and the productivity losses were directly proportional to the dry mass of the weed community in all experiments, regardless of N fertilization dose. In 2011, the relationship between weed dry weight and seed yield loss could be described by a linear regression model. In contrast, the increase in weed dry weight also led to seed yield loss in 2012, but the weed dry weight associated with maximum crop yield was identified by polynomial regression analysis, which indicated a quadratic polynomial response of seed yield loss to weed dry weight.

Discussion

The weed community in the chickpea crops differed between 2011 and 2012 in terms of species composition, density and dry mass accumulation. These differences might be explained by the different agricultural practice histories used in preceding crops, which might have benefited some species over others by creating the appropriate characteristics for the occupation and/or establishment of specific species in the niche. In addition to the previous agricultural practices, the N fertilization might have affected the weed flora because the treatments with higher N doses exhibited decreased numbers and diversity of species in both years (data not shown). For some plant species, especially grasses, the use of high doses of fertilizers might negatively affect species diversity (Lorenzo, Michelea, Sebastian, Johannes, & Angelo, 2007); this suggests an inverse relationship between soil nutrient availability and plant species diversity. The species *R. raphanistrum*, *A. viridis*, and *B. pilosa* had a higher incidence in both years. Amaral et al. (2015), while studying the interference of *A. viridis*, *B. pilosa*, *R. raphanistrum*, *C. rotundus*, *Digitaria nuda*, and *Eleusine indica* on the vegetative growth of chickpea, observed that *A. viridis*, *D. nuda*, and *E. indica* were the most competitive and aggressive species; these species compromised crop growth until 90 days after emergence.

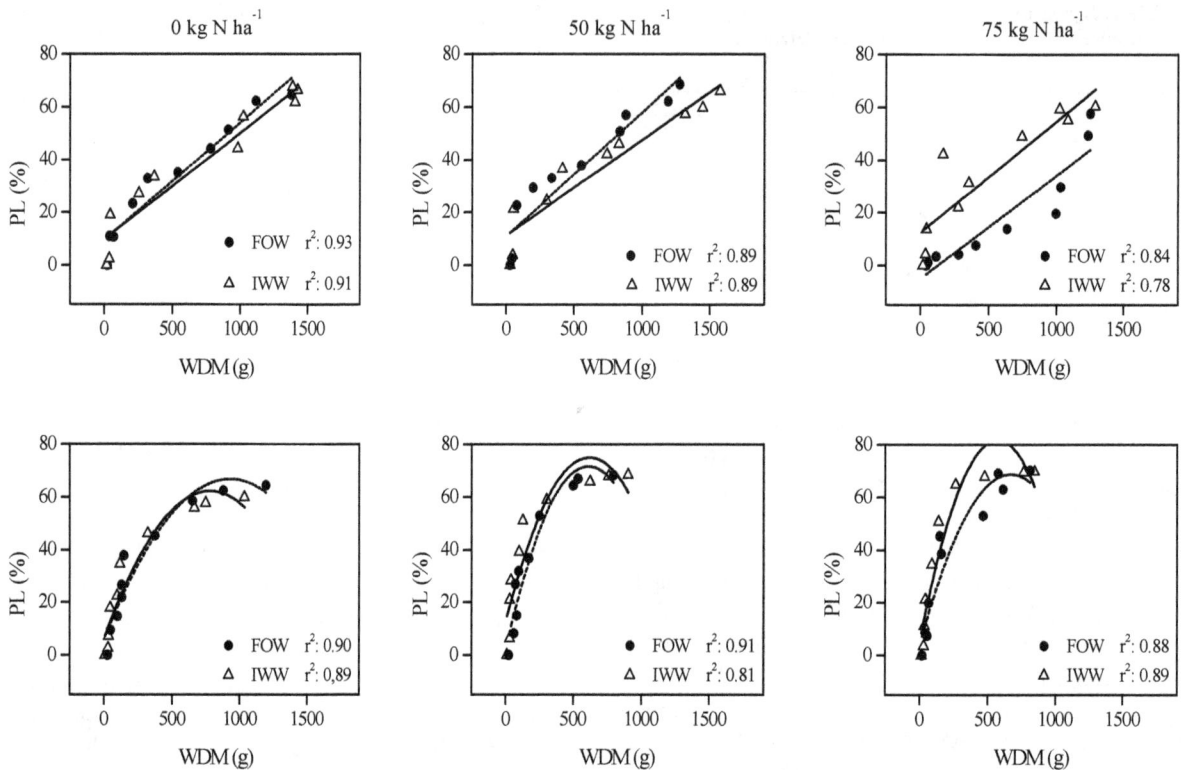

Figure 2. Percent loss (PL) of chickpea productivity as a function of accumulated weed dry mass (WDM) in the treatments subjected to different periods of coexistence in 2011 (a, b, c) and 2012 (d, e, f) with topdressing fertilization at 0 kg N ha-1 (a, d), 50 kg N ha-1 (b, e), and 75 kg N ha-1 (c and f). FOW: initially free of weeds; IWW: initially infested with weeds, p < 0.0001.

In both years, longer duration of weed pressure on the crop caused higher chickpea productivity losses, thereby indicating the crop sensitivity to coexistence with a weed community, regardless of the use of N topdressing or the composition of that community. Weed interference has effects on crops that are often irreversible, and the recovery of growth, development or productivity might not occur after removing the stress caused by the coexistence of weeds (Bressanin, Nepomuceno, Martins, Carvalho, & Alves, 2013). The advantages of topdressing fertilization for plants grown in the absence of weeds are indisputable, but the presence of weeds generates great uncertainty about the effectiveness of fertilization. Nitrogen is the nutrient that weeds and crop species most compete for (Shafiq, Hassan, Ahmad, & Rashid, 1994). Nitrogen fertilization might increase the competitive ability of the crop species, thereby decreasing the competitive pressure and weed suppression in the crop (Shafiq et al., 1994). The decrease in species diversity might also be related to crop growth, and higher N doses result in shorter critical weed-free periods due to the resulting rapid shoot growth (Yamauti, Alves, & Bianco, 2012).

The lower productivity observed in the experiments conducted in 2012 compared with those conducted in 2011 might be due in part to the weed diversity (15 species in 2011 vs. 23 in 2012), the previous crop (soybean in 2011 vs. maize in 2012), environmental conditions and soil characteristics. However, the yields in the FOW_0 of the plots (mean: 1,732.8 kg ha^{-1}) remained within the expected productivity range for the cultivar (from 1,600 to 2,700 kg ha^{-1}; EMBRAPA, 2010), even in the most unfavorable soil without topdressing (the experiment with the lowest productivity - 2012, experiment I).

The experimental PPWIs of both years ranged from 4 to 7 DAE if a productivity loss from 2.5 to 10% was considered acceptable, and they ranged from 5 to 11 DAE when considering a 5% productivity loss. Similar PPWI values have been estimated in other species belonging to the family Fabaceae, including the common bean (*Phaseolus vulgaris*) at 4 DAE (Borchartt, Jakelaitis, Valadão, Venturoso, & Santos, 2011), soybean (*Glycine max*) with values ranging from 11 to 17 DAE (Silva et al., 2009) for 5% losses only, and the Jam chickpea cultivar (also from the Kabuli group) in Tabriz, northwestern Iran, for which a PPWI of 17 DAE and a TPIP of 60 DAE were obtained in 2003 (Mohammadi et al., 2005). Al-Thahabi et al. (1994) observed a chickpea CPWI in Jordan ranging from

35 to 49 DAE. The low PPI values of the chickpea crop might be explained by its slow growth, open canopy and short stature, which reduce the ability of the crop to compete with weeds (Mohammadi et al., 2005).

In 2011, the increased use of N shortened the CPWI, but this result was not observed in 2012.Therefore, under the experimental conditions in this study, there was no relationship between the amount of N topdressing applied and CPWI.

The increase in the period of crop coexistence with the weed community drastically reduced the observed productivity, regardless of the absence or presence of fertilization. Yield decreases of greater than 50% were observed compared with plants that remained weed-free throughout the crop cycle (FOW_{140}) in all experiments. Most weeds exhibit faster initial growth than chickpeas, providing a great competitive advantage for weed populations, thereby inhibiting crop growth and reducing the incidence of light, which might affect photosynthesis and crop yield (Tepe et al., 2011). Al-Thahabi et al. (1994) observed that chickpea production was decreased by 81% due to weed interference, thus confirming the sensitivity of the crop to this factor. The cited authors noted a significant negative correlation between weed dry mass accumulation and crop seed production, a finding that was primarily attributed to a decrease in the number of pods per plant and the 100-seed weight.

The difference observed between the periods of interference from one harvest to the other suggests that critical periods can depend on several factors, including temperature and soil moisture, weed density, weed species composition (Tepe et al., 2011), time to weed emergence (Scholten, Parreira, & Alves, 2011), climate (Mohammadi et al., 2005; Tepe et al., 2011), drought (Parreira, Barroso, Fernandes, & Alves, 2015), light intensity (Retta, Vanaderlip, Higgins, Moshier, & Feyerherm, 1991), soil fertility (Mohammadi et al., 2005; Yamauti et al., 2012), and the characteristics of the crop itself, including species and cultivar used (Parreira, Alves, Lemos, & Portugal, 2014), crop sowing density and distribution patterns (Scholten et al., 2011), and sowing season (Mohammadi et al., 2005).

The need for fertilizer when growing chickpeas is not very well known and requires further study. Although chickpea is a legume, it responded positively to topdressing with N fertilizer. Fertilization might therefore give the crop a competitive advantage over the weed community. The comparison between the weed-infested and weed-free treatments during 2011 in the experiments with N topdressing showed increases of 37.55 and 28.86% (50 kg N ha^{-1}) and 39.74 and

42.70% (75 kg N ha^{-1}) compared with the experiment in the absence of fertilization for the periods FOW and IWW, respectively. In 2012, the observed productivity gains were even larger at 39.56 and 43.22% (50 kg N ha^{-1}) and 60.88 and 64.22% (75 kg N ha^{-1}) for the periods FOW and IWW, respectively. Methods such as topdressing with N-based fertilizer have been advocated for bean crops due to the resulting increases in grain yield (Gomes Jr. et al., 2005). Bressanin et al. (2013) noted that topdressing with N fertilizer increased the productivity of the 'Rubi' bean, even in the presence of weeds, and favored the crop competitively, thereby increasing the period prior to weed interference (PPI).

Topdressing with N fertilizer led to productivity gains of 617 and 913 kg ha^{-1} in 2011 and to gains of 749 and 1,113 kg ha^{-1} in 2012 when applied at the rates of 50 and 70 kg ha^{-1} N, respectively. The average annual price of the fertilizer used in the topdressing (urea) (CONAB, 2012) was U\$D 0.61 per kilogram (base year 2012), and between 112 and 167 kg urea is necessary for application at the rates of 50 and 75 kg N ha^{-1}. Furthermore, the technical coefficient for chickpea fertilization (EMBRAPA, 2010) and the machine-hour cost of the tractor and fertilizer spreader for June 2012 (CONAB, 2012) should be added to the cost. Finally, the international market price of chickpeas is approximately U\$D 1 per kilogram (FAO, 2015; Where Food Comes From, 2014). Based on these numbers, topdressing with N fertilizer at the rates of 50 to 75 kg N ha^{-1} would cost between U\$D 222.40 to 256.81 ha^{-1} and generate a return between approximately U\$D 617.00 to 1113.00 ha^{-1}.

Conclusion

In conclusion, the presence of weeds negatively affects chickpea production and can cause considerable yield losses, rendering the crop economically unfeasible. The CRWI for chickpea production during both years was affected regardless of the N topdressing dose and ranged from 5 to 76 DAE in 2011 and 10 to 64 DAE in 2012, assuming an acceptable productivity loss of 5%. Although the critical periods of interference obtained in the 50 and 75 kg N ha^{-1} treatments were similar to those obtained in the treatments without fertilization (0 kg N ha^{-1}), N fertilization increased chickpea productivity, leading to economic gains.

Acknowledgements

CLA acknowledges the Coordenação de Aperfeiçoamento de Pessoal de Nível Superior

(CAPES) and the Fundação de Amparo à Pesquisa do Estado de São Paulo (FAPESP) for PhD fellow ships (grants #2010/14018-0 and grants #2010/07809-1). PLCAA acknowledges the Conselho Nacional de Desenvolvimento Científico e Tecnológico (CNPq) for research productivity fellowships.

References

Abate, T., Alene, A. D., Bergvinson, D., Shiferaw, B., Silim, S., Orr, A., & Asfaw, S. (2012). *Tropical grain legumes in Africa and south Asia: knowledge and opportunities*. Nairobi, KE: International Crops Research Institute for the Semi Arid Tropics.

Al-Thahabi, S. A., Yasin, J. Z., Abu-Irmaileh, B. E., Haddad, N. I., & Saxena, M. C. (1994). Effect of weed removal on productivity of chickpea (*Cicer arietinum* L.) and lentil (*Lens culinaris* Med.) in a mediterranean environment. *Journal of Agronomy and Crop Science*, *172*(5), 333-341. doi: 10.1111/j.1439-037X.1994.tb00184.x

Amaral, C. L., Pavan, G. B., Souza, M. C., Martins, J. V. F., & Alves, P. L. C. A. (2015). Relações de interferência entre plantas daninhas e a cultura do grão-de-bico. *Bioscience Journal*, *31*(1), 37-46. doi: 10.14393/BJ-v31n1a2015-17971

Amaral, C. L., Souza, M. C., Pavan, G. B., Gavassi, M. A., & Alves, P. L. C. A. (2013) High sourgrass threshold interfere on chick-peas development in tropical conditions. *African Journal of Agricultural Research*, *8*(2), 167-172. doi: 10.5897/AJAR12.1875

Blackshaw, R. E., Brandt, R. N., Janzen, H. H., Entz, T., Grant, C. A., & Derksen, D. A. (2003). Differential response of weed species to added nitrogen. *Weed Science*, *51*(4), 532-539. doi: 10.1614/0043-1745(2003)051[0532:DROWST]2.0.CO;2

Borchartt, L., Jakelaitis, A., Valadão, F. C. A., Venturoso, L. A. C., & Santos C. L. (2011). Períodos de interferência de plantas daninhas na cultura do feijoeiro-comum (*Phaseolus vulgaris* L.). *Revista Ciência Agronômica*, *42*(3), 725-734.

Bressanin, F. N., Nepomuceno, M. P., Martins, J. V. F., Carvalho, L. B., & Alves, P. L. C. A. (2013). Influência da adubação nitrogenada sobre a interferência de plantas daninhas em feijoeiro. *Revista Ceres*, *60*(1), 43-52. doi: 10.1590/S0034-737X2013000100007

Cardoso, G. D., Alves, P. L. C. A., Severino, L. S., & Vale, L. S. (2011). Critical periods of weed control in naturally green colored cotton BRS Verde. *Industrial Crops and Products*, *34*(1), 1198-1202. doi: 10.1016/j.indcrop.2011.04.014

Companhia Nacional de Abastecimento [CONAB]. (2012). *Preços dos Insumos Agropecuários*. Retrieved on Dec. 12, 2012 from http://www.conab.gov.br/detalhe.php?a=1303&t=2

Empresa Brasileira de Pesquisa Agropecuária [EMBRAPA]. (2010). *Grão-de-bico Cícero*. Retrieved on Aug. 23, 2010 from http://www.cnph.embrapa.br/paginas/produtos/cultivares/grao_de_bico_cicero.htm

Esfahani, M. N., Sulieman, S., Schulze, J., Yamaguchi-Shinozaki, K., Shinozaki, K, & Tran L. S. P. (2014). Mechanisms of physiological adjustment of N_2 fixation in *Cicer arietinum* L. (chickpea) during early stages of water deficit: single or multi-factor controls. *The Plant Journal*, *79*(6), 964-980. doi: 10.1111/tpj.12599

Food and Agriculture Organization of the United Nations [FAO]. (2015). Base de dados FAOSTAT. Retrieved on June 7, 2017 from http://faostat.fao.org

Gaur, P. M., Jukanti, A.K., & Varshney, R. K. (2012). Impact of genomic technologies on chickpea breeding strategies. *Agronomy*, *2*(3), 199-221. doi: 10.3390/agronomy2030199

Gomes Jr., F. G., Lima, E. R., Leal, A. J. F., Matos, F. A., Sá, M. E., & Haga, K. I. (2005). Teor de proteína em grãos de feijão em diferentes épocas e doses de cobertura nitrogenada. *Acta Scientiarum. Agronomy*, *27*(3), 455-459. doi: 10.4025/ actasciagron.v27i3.1409

Hossain, M. B., Hasan, M. M., Sultana, R., & Bari, A. K. M. A. (2016). Growth and yield response of chickpea to different levels of boron and zinc. *Fundamental and Applied Agriculture*, *1*(2), 82-86.

Kaushik, S. S., Rai, A. K., Sirothia, P., Sharma, A. K., & Shukla, A. K. (2014). Growth, yield and economics of rain fed chickpea (*Cicer arietinum* L.) as influenced by integrated weed management. *Indian Journal of Natural Products and Resources*, *5*(2), 282-285.

Kuva, M. A., Pitelli, R. A., Christoffoleti, P. J., & Alves, P. L. C. A. (2000). Períodos de interferência das plantas daninhas na cultura da cana-de-açúcar. I - Tiririca. *Planta Daninha*, *18*(2), 241-251.

Lorenzo, M., Michela, S., Sebastian, K., Johannes, I., & Angelo, P. (2007). Effects of local factors on plant species richness and composition of Alpine meadows. *Agriculture, Ecosystems and Environment*, *119*(3-4), 281-288. doi: 10.1016/j.agee.2006.07.015

Mohammadi, G., Javanshir, A., Khooie, F. R., Mohammadi, S. A., & Salmasi, S. Z. (2005). Critical period of weed interference in chickpea. *Weed Research*, *45*(1), 57-63. doi: 10.1111/j.1365-3180.2004.00431.x

Pang, J., Turner, N. C., Khan, T., Du, Y. L., Xiong, J. L, Colmer, T. D., Devilla, R., Stefanova, K., & Siddique, K. H. M. (2017). Response of chickpea (*Cicer arietinum* L.) to terminal drought: leaf stomatal conductance, pod abscisic acid concentration, and seed set. *Journal of Experimental Botany*, *68*(8), 1973-1985. doi: 10.1093/jxb/erw153

Parreira, M. C., Alves, P. L. C. A., Lemos, L. B., & Portugal, J. (2014). Comparação entre métodos para determinar o período anterior à interferência de plantas daninhas em feijoeiros com distintos tipos de hábitos de crescimento. *Planta Daninha*, *32*(4), 727-738. doi: 10.1590/S0100-83582014000400007

Parreira, M. C., Barroso, A. A. M., Fernandes, J. M. P. E. V., & Alves, P. L. C. A. (2015). Effect of drought stress on periods prior of weed interference (PPWI) in bean crop using arbitrary and tolerance estimation. *Australian Journal of Crop Science*, *9*(12), 1249-1256.

Rani, B. S., & Krishna, T. G. (2016). Response of chickpea (*Cicer arietinum* L.) varieties to nitrogen on a calcareous vertisols. *Indian Journal of Agricultural Research*, *50*(3), 278-281. doi: 10.18805/ijare.v50i3.10749

Ratnam, M., Rao, A. S., & Reddy, T. Y. (2011). Integrated weed management in chickpea (*Cicer arietinum* L.). *Indian Journal of Weed Science*, *43*(1 & 2), 70-72.

Retta, A., Vanaderlip, R. L., Higgins, R. A., Moshier, L. J., & Feyerherm, A. M. (1991). Suit-ability of corn growth models for incorporation of weed and insect stresses. *Agronomy Journal*, *83*(4), 757-765. doi: 10.2134/agronj1991.00021962008300040021x

Scholten, R., Parreira, M. C., & Alves, P. L. C. A. (2011). Período anterior à interferência das plantas daninhas para a cultivar de feijoeiro 'Rubi' em função do espaçamento e da densidade de semeadura. *Acta Scientiarum. Agronomy*, *33*(2), 313-320. doi: 10.4025/actasciagron.v33i2.5646

Shafiq, M., Hassan, A., Ahmad, N., & Rashid, A. (1994). Crop yields and nutrient uptake by rainfed wheat and mungbean as affected by tillage, fertilization, and weeding. *Journal of Plant Nutrition*, *17*(4), 561-577. doi: 10.1080/01904169409364750

Silva, A. F., Concenço, G., Aspiazú, I., Ferreira, E. A., Galon, L., Freitas, M. A. M., ... Ferreira, A. F. (2009). Período anterior à interferência na cultura da Soja-RR em condições de baixa, média e alta infestação. *Planta Daninha*, *27*(1), 57-66. doi: 10.1590/S0100-83582009000100009

Singh, V. K., & Bajpai R. P. (1996). Effects of crop production inputs on gram (*Cicer arietinum*) in North-eastern hills zone of Madhya Pradesh. *Indian Journal of Agronomy*, *41*(3), 655-656.

Tang, L., Cheng, C., Wan, K., Li, R., Wang, D., Tao, Y., ...Chen, F. (2014). Impact of fertilizing pattern on the biodiversity of a weed community and wheat growth. *Plos one*, *9*(1), 1-11. doi: 10.1371/journal.pone.0084370

Tepe, I., Erman, M., Yergin, R., & Bükün, B. (2011). Critical period of weed control in chickpea under non-irrigated conditions. *Turkish Journal of Agriculture and Forestry*, *35*(5), 525-534. doi: 10.3906/tar-1007-956

Ulukan, H., Bayraktar, N., & Koçak, N. (2012). Agronomic importance of first development of chickpea (*Cicer arietinum* L.) under semi-arid conditions: I. effect of powder humic acid. *Pakistan Journal of Biological Sciences*, *15*(4), 203-207.

Valenciano, J. B., Boto, J. A., & Marcelo, V. (2011). Chickpea (Cicer arietinum L.) response to zinc, boron and molybdenum application under field conditions. *New Zealand Journal of Crop and Horticultural Science*, *39*(4), 217-229. doi: 10.1080/01140671.2011.577079

Varshney, R. K., Hiremath, P. J., Lekha, P., Kashiwagi, J., Balaji, J., Deokar, A. A., ... Hoisington, D. A. (2009). A comprehensive resource of drought - and salinity - responsive ESTs for gene discovery and marker development in chickpea (*Cicer arietinum* L.). *BMC Genomics*, *10*, 1-18. doi: 10.1186/1471-2164-10-523

Varshney, R. K., Thudi, M., Nayak, S. N., Gaur, P. M., Kashiwagi, J., Krishnamurthy, L., ... Viswanatha, K. P. (2014). Genetic dissection of drought tolerance in chickpea (*Cicer arietinum* L.) *Theoretical and Applied Genetics*, *127*(2), 445-462. doi: 10.1007/s00122-013-2230-6

Where Food Comes From. (2014) *Chickpeas: Cost of Production and Marketing Loan Rate*. Retrieved on Dec. 22, 2014 from http://wherefoodcomesfrom.com/article/7044/Chickpeas-Cost-of-Production-and-Marketing-Loan-Rate

Yamauti, M. S., Alves, P. L. C. A., & Bianco, S. (2012). Effects of mineral nutrition on inter- and intraspecific interference of peanut (*Arachis hypogaea* L.) and hairy beggarticks (*Bidens pilosa* L.). *Interciencia*, *37*(1), 65-69.

21

NPK and flavonoids affecting insect populations in *Dimorphandra mollis* seedlings

Germano Leão Demolin Leite[*], Farley William Souza Silva, Rafael Eugênio Maia Guanabens, Luiz Arnaldo Fernandes, Lourdes Silva Figueiredo and Leonardo Ferreira Silva

Laboratório de Entomologia Universitário, Instituto de Ciências Agrárias, Universidade Federal de Minas Gerais, Av. Universitária, 1000, Cx. Postal 125, 39404-006, Montes Claros, Minas Gerais, Brazil. *Author for correspondence. E-mail: gldleite@ig.com.br

ABSTRACT. The study evaluated the influence of different levels of nitrogen (N), phosphorus (P) and potassium (K), and flavonoids on the population of insects in *Dimorphandra mollis* Benth (Leguminosae) seedlings. The treatments associated with the highest level of attacks by *Frankliniella schulzei* (Trybon) (Thysanoptera: Thripidae) were 600 mg dm^{-3} of P and 50 mg dm^{-3} of K. The highest level of attacks by Coccidae occurred for 300 of P and 150 and 250 mg dm^{-3} of K. The last two treatments also exhibited the highest level of attacks by Pseudococcidae. On the other hand, the control exhibited higher levels of flavonoids and a lower level of insect attacks. We observed a small positive effect of N levels on attack by *F. schulzei*. The levels of N, P and K negatively affected the levels of flavonoids in the leaves of *D. mollis*. We detected no significant effects of flavonoid levels on the populations of Coccidae, Pseudoccocidae and *F. schulzei*. Higher numbers of Coccidae and Pseudococcidae were observed in the abaxial face of apical leaves. However, higher numbers of *F. schulzei* were observed on the adaxial face at lower heights in the canopy. The preferred treatment for the production of *D. mollis* seedlings is the control (without fertilization) because it showed higher flavonoid levels than other treatments and did not result in higher insect numbers.

Keywords: "fava d'anta", Coccidae, Pseudococcidae, *Frankliniella schulzei*.

Introduction

The use of medicinal plants to cure or to prevent disease increases each year. The cultivation of these plants and/or the extraction of substances from plants that possess therapeutic properties has become an attractive alternative for agriculture and is interesting and viable at the national level of the industry (SIMÕES et al., 2000). However, some species that show great potential for the pharmaceutical industry are not receiving sufficient attention and are little studied by the scientific community.

Among the species of the Brazilian flora that deserve more extensive study is *Dimorphandra mollis* Benth (Leguminosae), commonly known as "fava d'anta", a native species of the Brazilian Savanna, found in Minas Gerais, São Paulo, Goiás, Mato Grosso and Mato Grosso do Sul (LORENZI; MATOS, 2002; PACHECO et al., 2010). The *D. mollis* is a deciduous pioneer tree of medium size. The bioflavonoid rutin or rutoside is extracted from its fruits in the prematuration stage (LORENZI; MATOS, 2002; LUCCI; MAZZAFERA, 2009). The proportion of rutin occurring in the dry matter can vary from 6 to 10% (HUBINGER et al., 2010; SOUZA et al., 1991). Other substances, such as rhamnose and quercetin, can also be extracted from the fruits (PETACCI et al., 2010).

Species of plants such as *D. mollis* that are subject to intensive extraction activities are thereby at risk for extinction. Accordingly, such species require management strategies and cultivation procedures (GONÇALVES et al., 2010; LEITE et al., 2006; VIANA E SOUZA; LOVATO, 2010). An alternative is the production of seedlings for colonization of the areas in which extractive activity occurs or for commercial plantations. This approach requires knowledge of the nutritional demands of the plants, primarily in regard to such macronutrients as nitrogen (N), potassium (K) and phosphorus (P). Consequently, the domestication of the medicinal plants is an alternative approach for the production of Phytotherapic substances for the industry. However, the seedlings can be attacked by insects that can affect the quantity and the quality of the product (LEITE et al., 2008).

Moreover, the fertilization of plants can influence vulnerability to insect attack (LEITE et al., 2003). Indeed, different levels of nutrients can produce physiological and morphological alterations in plants (COELHO et al., 1999; LARA, 1991). Food has very significant effects on the distribution and abundance of insects. It has direct influences on insect populations and affects biological, morphological and behavioral processes (GALLO et al., 2002). The nutrients N, P and K have important functions in plants and interfere with insect population dynamics. Previous studies of insects associated with *D. mollis* have investigated ants and bees in adult plants (CINTRA et al., 2002, 2005a and b; ROTHER et al., 2009).

The aim of this work is to evaluate the influence of N, P and K levels and flavonoids on the insects associated with *D. mollis* seedlings.

Material and methods

These studies were conducted under greenhouse conditions in the "Instituto de Ciências Agrárias da Universidade Federal de Minas Gerais (ICA-UFMG)", Montes Claros, Minas Gerais State, from July to November 2005.

Seeds of *D. mollis* were subjected to scarification on the side opposite the hilum and were immersed in water for 24h. Soon thereafter, six seeds were planted in a three dm^3 vase. The soil used was a dystrophic Red Latosol collected from a layer 0-20 cm deep. Soil physical and chemical properties, determined according to the methodology of Embrapa (1997), were as follows: pH in water 4.6, P = 0.6 mg dm^{-3}, Ca = 11.0 $mmol_c$ dm^{-3}, Mg = 4.0 $mmol_c$ dm^{-3}, K = 1.0 $mmol_c$ dm^{-3}, Al = 37 $mmol_c$ dm^{-3}, H + Al = 140 $mmol_c$ dm^{-3}, S = 16.0 $mmol_c$ dm^{-3}, t = 53.0, m = 70%, T = 156 $mmol_c$ dm^{-3}, V = 10%, organic matter = 24 g kg^{-1}, sand = 500 g kg^{-1}, silt = 80 g kg^{-1} and clay = 420 g kg^{-1}.

After germination, the two largest plants were chosen from each pot and the other plants eliminated. The experimental design was entirely randomized with three repetitions and 19 treatments: six levels of nitrogen (50, 100, 150, 200, 250 and 300 mg dm^{-3}) in the form of NH_4NO_3, six levels of phosphorus (100, 200, 300, 400, 500, 600 mg dm^{-3}) in the form of H_3PO_4 six levels of potassium (50, 100, 150, 200, 250 and 300 mg dm^{-3}) in the form of KNO_3, and the control (without fertilization).

To complement the fertilization, the nutrients Mg, S, B, Cu, Zn and Ca were used (60, 40, 0.35, 1.5, 0.5, and 200 mg dm^{-3}, respectively). Fertilization was performed one week before the seeds were planted. Irrigation was conducted with distilled water in order to avoid contamination and to ensure the effects of the fertilization.

The aerial part of each plant was collected after 150 days, at the beginning of December 2005. The aerial part was placed in Kraft paper bags, dried in a forced-circulation oven at 60°C for three days until constant weight was achieved, and used for determination of dry matter. Subsequently, this material was ground in a Wiley mill (20 mesh), and the extraction of total flavonoids was performed according to Mendes et al. (2005).

The evaluations of insect occurrence (natural infestation) and of the level of damage (percent) were performed by visual inspection weekly in the morning on one leaf from the apical, medium and basal parts of the canopy of each plant, on both leaf faces. Defoliation was determined visually by estimating the percentage of leaf area lost on a scale from 0-100% by increments of 5% of the total area removed (SASTAWA et al., 2004, MIZUMACHI et al., 2006).

Regression analysis (p < 0.05) was applied to relate data on insects to N, P and K and flavonoids and to relate flavonoids to soil attributes. Data were transformed using the square root function $\sqrt{x+0.5}$ and subjected to an analysis of variance and to the Tukey or Scott-Knott tests (p < 0.05).

Results and discussion

Scale insects of the families Coccidae and Pseudococcidae (Hemiptera) and thrips *Frankliniella schulzei* (Trybon) (Thysanoptera: Thripidae) were observed in *D. mollis* seedlings. The attack of scale insects produced honeydew on the *D. mollis* leaves. These families of scale insects include the pest species *Coccus viridis* (Green) in citrus culture and *Pseudococcus adonidum* (L.) in horticultural plants. *Frankliniella schulzei* produced scratches on the leaves of *D. mollis* seedlings and thereby caused leaf chlorosis. This insect is a pest species on several crops including cotton and horticultural plants (GALLO et al., 2002).

The treatments associated with the highest level of attacks by *F. schulzei* were 600 mg dm^{-3} of P and 50 mg dm^{-3} of K (Table 1). The highest level of attacks by Coccidae occurred for 300 mg dm^{-3} of P and 150 and 250 mg dm^{-3} of K. The last two treatments also exhibited the highest level of attacks by Pseudococcidae (Table 1). The control exhibited higher levels of flavonoids and a lower level of insect attacks (Table 1).

Table 1. Effect of nitrogen (N), potassium (K) and phosphorus (P) on the leaf damage (%) produced by thrips *Frankliniella schulzei* and on the number of Pseudococcidae and Coccidae leaf^{-1} face and flavonoid levels (% dry matter) in *Dimorphandra mollis* seedlings.

Treatments	Leaf damage (%)	Coccidae	Pseudo-coccidae	Flavonoids (%)
Without fertilization	0.43 b	0 b	0 b	4.59 a
50 mg dm^{-3} of N	0.92 b	0.01 b	0 b	1.78 b
100 mg dm^{-3} of N	1.03 b	0.03 b	0 b	1.88 b
150 mg dm^{-3} of N	1.47 b	0 b	0 b	1.74 b
200 mg dm^{-3} of N	0.79 b	0 b	0 b	1.64 b
250 mg dm^{-3} of N	1.36 b	0.01 b	0 b	1.53 b
300 mg dm^{-3} of N	0.33 b	0 b	0 b	1.83 b
100 mg dm^{-3} of P	0.86 b	0 b	0 b	1.67 b
200 mg dm^{-3} of P	1.00 b	0.09 b	0.010 b	1.40 b
300 mg dm^{-3} of P	0.94 b	0.22 a	0.006 b	1.83 b
400 mg dm^{-3} of P	1.49 b	0 b	0 b	1.19 b
500 mg dm^{-3} of P	1.17 b	0 b	0 b	1.56 b
600 mg dm^{-3} of P	3.06 a	0.05 b	0 b	1.39 b
50 mg dm^{-3} of K	2.73 a	0.12 b	0.006 b	1.98 b
100 mg dm^{-3} of K	1.09 b	0.07 b	0.006 b	2.04 b
150 mg dm^{-3} of K	1.16 b	0.30 a	0.078 a	1.60 b
200 mg dm^{-3} of K	1.31 b	0.02 b	0 b	1.75 b
250 mg dm^{-3} of K	1.26 b	0.21 a	0.073 a	1.99 b
300 mg dm^{-3} of K	1.08 b	0 b	0.006 b	1.78 b

The following averages for the same letter in the line do not differ for the Scott-Knott test to 5% of probability.

Excess of N or deficiency of K can cause accumulation of free amino acids and can consequently increase the populations of sucking insects on plants (JANSSON; EKBOM, 2002). Species of sucking insects such as aphids, scale, leafhoppers, whitefly and thrips and several species of phytophagous mites are not able to extract amino

acids from proteins for their own use. Consequently, these insects depend on free amino acids in the plants (GALLO et al., 2002). Phosphorus can participate in cell wall synthesis. It can accordingly influence the function of the cell wall as a barrier to disease. Phosphorus can also occur in toxic compounds and can affect metabolic routes that counteract insects or diseases (MALAVOLTA, 2004). We noted a small positive effect of N levels on attack by *F. schulzei* (Figure 1). However, the effects of N, P and K on attacks by Coccidae and Pseudococcidae as well as the effects of P and K in relation to *F. schulzei* were not significant (data not shown). In other words, we did not observe a substantial effect of the levels of N, P and K on the insects. This finding perhaps results from the low population densities of insects on *D. mollis* seedlings (Table 1).

Dimorphandra mollis participates in a symbiosis with soil bacteria that assimilate atmospheric N (PEREIRA; OLIVEIRA, 2005). Probably for this reason, the N levels did not affect the growth of the seedlings. According to Pinto and Lameira (2001), the synthesis of compounds originating from secondary metabolism can be induced in several ways; for example, by nutritional stress. Mendes et al. (2005) also verified a reduction in the levels of total flavonoids as a function of increasing doses of P in *D. mollis* seedlings cultivated in nutrient-rich solution. In the present experiment, the stressed seedlings (without fertilization) produced larger quantities of secondary metabolites, including flavonoids (Figure 2). These seedlings were not preferred by the insects (Table 1). Gazzoni et al. (1997) observed higher mortality in caterpillars given a diet of mixed flavonoids (rutin and quercetin) at the highest doses.

We detected no significant effect of flavonoid levels on the scale insects (Coccidae + Pseudoccocidae) and *F. schulzei* populations (Figure 1). The flavonoids are probably important as an initial barrier against insects (antixenosis). Once the insects were established on the plants, they may have experienced deleterious effects on their life cycles (antibiosis). This process would explain their low density in *D. mollis* seedlings. In other words, the lowest level of flavonoids observed in *D. mollis* leaves (1.02%) could have been enough to affect these insects negatively. Macedo et al. (2002) observed a strong negative effect on *Callosobruchus maculates* F. (Coleoptera: Bruchidae) larvae of 1% trypsin inhibitor isolated from *D. mollis* seeds. According to Fernandes et al. (2004), an increase in the concentration of nitrogen in the plant reduces the concentration of substances related to defense.

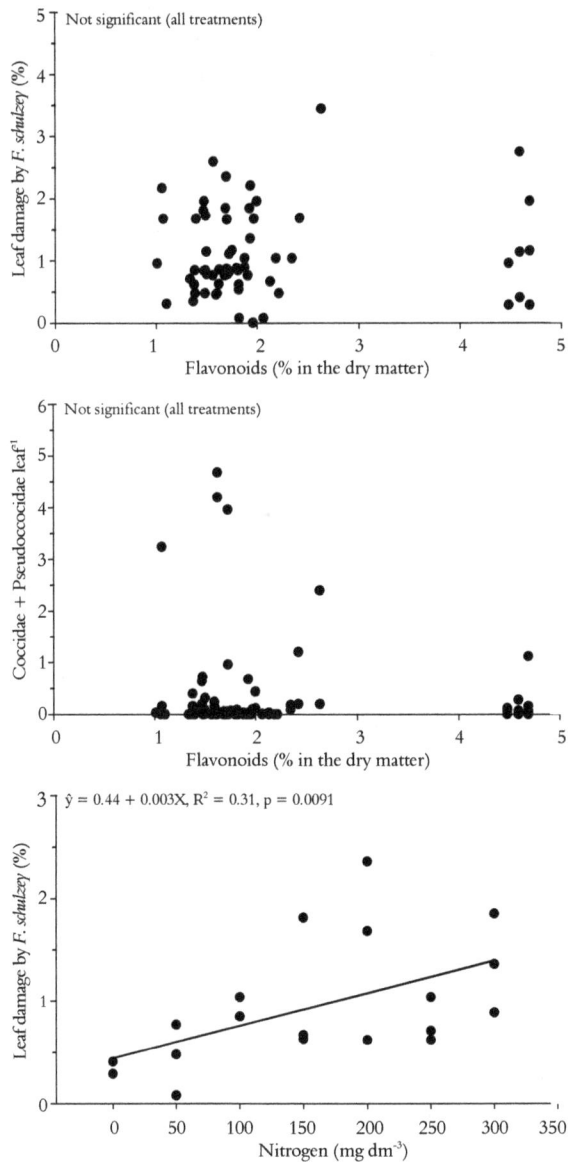

Figura 1. Effects of flavonoids levels on the attack of *Frankliniella schulzei* and Coccidae + Pseudococcidae populations, and nitrogen levels on the attack of *F. schulzein* in leaves of *Dimorphandra mollis* seedilings. Each symbol represent one repetition.

These substances include carbon-rich tannins and terpenes. The quantity and quality of the soluble compounds of nitrogen produced depend on the source of nitrogen used. These compounds could induce a larger or a smaller degree of pest resistance in the plant (BORTOLI; MAIA, 1994). Therefore, the effect of NPK on the insects is both direct (nutritional value of the plant) and indirect (defense of the plant), as observed in this work.

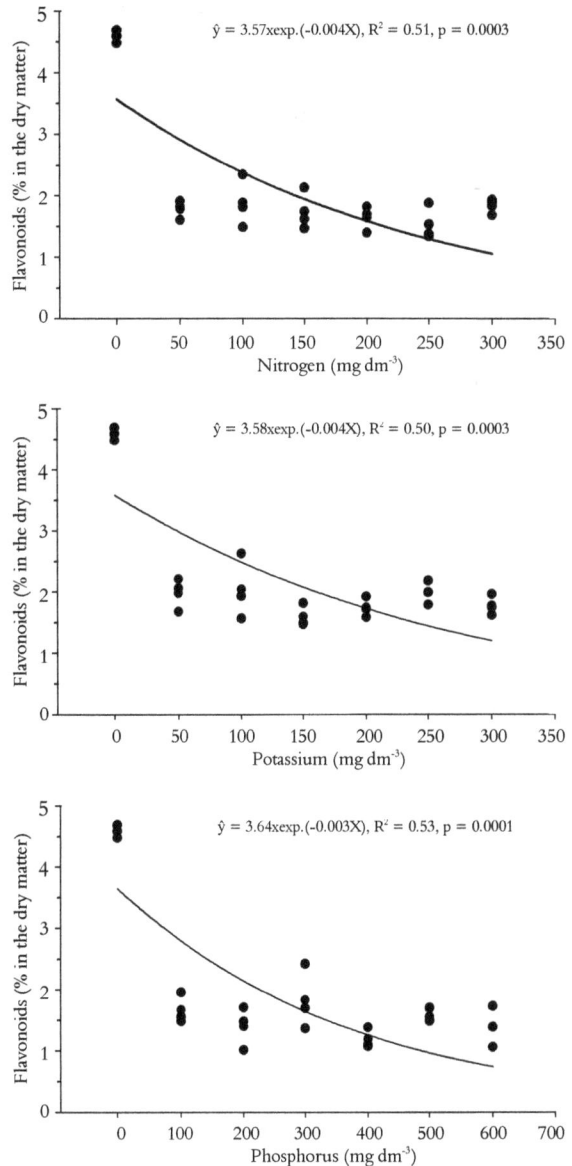

Figura 2. Effects of levels of nitrogen, potassium and phosphorus on the flavonoids levels in leaves of *Dimorphandra mollis* seedlings. Each symbol represent one repetition.

In all the treatments, the numbers of Coccidae and Pseudococcidae observed on the abaxial face on apical leaves were higher than the numbers observed on plants in the medium and basal heights of the canopy (Table 2). However, higher numbers of *F. schulzei* were observed on the adaxial face of the leaves of the inferior third of the canopy (Table 2). This distribution probably resulted from competition for space and food among species. Insects, particularly sucking insects, attack the abaxial face of apical leaves because these parts of the plant are more tender (the quantities of calcium and fiber are smaller) and are of higher nutritional value (higher level of N) (SILVA et al., 1998; LEITE et al., 2002; CHAU et al., 2005; SANTOS et al., 2003).

Table 2. Effect of canopy height and leaf face on leaf damage (%) and on the numbers of thrips *Frankliniella schulzei*, Pseudococcidae and Coccidae leaf^{-1} face in *Dimorphandra mollis* seedlings under different concentrations of nitrogen (N), phosphorus (P) and potassium (K) in the soil.

Insects	Canopy height of plants under different N levels		
	Apical	Medium	Bottom
Thrips damage (%)	0.00 b	0.24 b	3.87 a
Thrips	0.00 b	0.01 ab	0.02 a
Coccidae	0.033 a	0.012 ab	0.006 b
Pseudococcidae	0.027 a	0.004 b	0.001 b
	Plants under different P levels		
Thrips damage (%)	0.00 b	0.41 b	5.21 a
Pseudococcidae	0.017 a	0.010 ab	0.000 b
	Plants under different K levels		
Thrips damage (%)	0.00 b	0.31 b	5.06 a
Coccidae	0.30 a	0.17 ab	0.03 b
Pseudococcidae	0.11 a	0.10 ab	0.01 b
Insects	Leaf face of plants under different N levels		
	Adaxial	Abaxial	
Thrips	0.02 a	0.01b	
Coccidae	0.00 b	0.03 a	
Pseudococcidae	0.00 b	0.02 a	
	Plants under different P levels		
Coccidae	0.00 b	0.14 a	
Pseudococcidae	0.00 b	0.02 a	
	Plants under different K levels		
Thrips	0.01 a	0.00 b	
Coccidae	0.00 b	0.33 a	
Pseudococcidae	0.00 b	0.14 a	

The following averages for the same letter in the line do not differ for the test of Tukey to 5% of probability.

Conclusion

Scale insects can be pests of *D. mollis* seedlings because they suck the sap and coat the leaf with honeydew, thereby facilitating the growth of soot mold. Thrips can also be a pest in this plant because they produce scratches on the leaves of *D. mollis* seedlings and therefore cause chlorosis and premature fall of leaves. The preferred treatment for the production of *D. mollis* seedlings is the control condition (without fertilization) because it showed higher levels of flavonoids than did other treatments and because it did not result in higher insect numbers.

References

BORTOLI, S. A.; MAIA, I. G. Influência da aplicação de fertilizantes na ocorrência de pragas. In: SÁ, M. E.; BUZZETI, S. (Ed). **Importância da adubação na qualidade dos produtos agrícolas**. São Paulo: Ícone, 1994. p. 53-63.

CHAU, A.; HEINZ, M.; DAVIES, F. T. Influences of fertilization on *Aphis gossypii* and insecticide usage. **Journal of Applied Entomology**, v. 129, n. 2, p. 89-97, 2005.

CINTRA, P.; MALASPINA, O.; PETACCI, F. Toxicity of *Dimorphandra mollis* to workers of *Apis mellifera*. **Journal of Brazilian Chemistry Society**, v. 13, n. 1, p. 115-118, 2002.

CINTRA, P.; BUENO, F. C.; BUENO, O. C.; MALASPINA, O.; PETACCI, F.; FERNANDES, J. B. Astilbin toxicity to leaf-cutting ant *Atta sexdens rubropilosa* (Hymenoptera: Formicidae). **Sociobiology**, v. 45, n. 2, p. 347-353, 2005a.

CINTRA, P.; MALASPINA, O.; BUENO, O. C.; PETACCI, F.; FERNANDES, J. B.; VIEIRA, P. C.; DA SILVA, M. F. G. F. Oral toxicity of chemical substances found in *Dimorphandra mollis* (Caesalpiniaceae) against honeybees (*Apis mellifera*) (Hymenoptera: Apidae). **Sociobiology**, v. 45, n. 1, p. 141-149, 2005b.

COELHO, S. A. M. P.; OLIVEIRA, D. M.; BUENO, A. F.; CALAFIORI, M. H. Efeito de potássio sobre a população de mosca-branca, *Bemisia tabaci* (GENN, 1889) em feijoeiro, *Phaseolus vulgaris* L. **Ecossistema**, v. 24, n. 1, p. 25-27, 1999.

EMBRAPA-Empresa Brasileira de Pesquisa Agropecuária. **Centro Nacional de Pesquisa de Solos**. Manual de Métodos de Análises de Solos. 2. ed. Rio de Janeiro: Embrapa, 1997.

FERNANDES, L. C.; FAGUNDES, M.; SANTOS, G. A.; SILVA, G. M. Abundance of herbivore insects associated to pequizeiro (*Caryocar brasiliense*). **Revista Árvore**, v. 28, n. 6, p. 919-924, 2004.

GALLO, D.; NAKANO, O.; SILVEIRA NETO, S.; CARVALHO, R. P. L.; BATISTA, G. C.; BERTI FILHO, E.; PARRA, J. R. P.; ZUCCHI, R. A.; ALVES, S. B.; VENDRAMIM, J. D.; MARCHINI, L. C.; LOPES, J. R. S.; OMOTO, C. **Manual de Entomologia Agrícola**. Piracicaba: Fealq, 2002.

GAZZONI, D. L.; HÜLSMEYER, A.; HOFFMANN-CAMPO, C. B. Efeito de diferentes doses de rutina e de quercetina na biologia de *Anticarsia gemmatalis* Hübner, 1818 (Lep., Noctuidae). **Pesquisa Agropecuária Brasileira**, v. 32, n. 7, p. 673-681, 1997.

GONÇALVES, A. C.; VIEIRA, F. A.; FIORAVANTE REIS, C. A.; CARVALHO, D. Conservation of *Dimorphandra mollis* Benth. (Fabaceae) based on the genetic structure of natural populations. **Revista Árvore**, v. 34, n. 1, p. 95-101, 2010.

HUBINGER, S. Z.; CEFALI, L. C.; VELLOSA, J. C. R.; SALGADO, H. R. N.; ISAAC, V. L. B.; MOREIRA, R. R. D. *Dimorphandra mollis*: an alternative as a source of flavonoids with antioxidant action. **Latin American Journal of Pharmacy**, v. 29, n. 2, p. 271-274, 2010.

JANSSON, J.; EKBOM, B. The effect of different plant nutrient regimes on the aphid *Macrosiphum euphorbiae* growing on petunia. **Entomologia Experimentalis et Applicata**, v. 104, n. 1, p. 109-116, 2002.

LARA, F. M. **Princípios de resistência de plantas a insetos**. São Paulo: Ícone, 1991.

LEITE, G. L. D.; COSTA, C. A.; ALMEIDA, C. I. M.; PICANÇO, M. Efeito da adubação sobre a incidência de traça-do-tomateiro e alternaria em plantas de tomate. **Horticultura Brasileira**, v. 21, n. 3, p. 448-451, 2003.

LEITE, G. L. D.; PICANÇO, M.; JHAM, G. N.; ECOLE, C. C. Effect of leaf characteristics, natural enemies and climatic conditions on the intensities of *Myzus persicae* and *Frankliniella schulzei* attacks on *Lycopersicon esculentum*. **Arquivos do Instituto Biológico**, v. 69, n. 4, p. 71-82, 2002.

LEITE, G. L. D.; PIMENTA, M.; FERNANDES, P. L.; VELOSO, R. V. S.; MARTINS, E. R. Factors affecting arthropods associated with five accessions of Brazilian ginseng (*Pfaffia glomerata*) in Montes Claros, Brazil. **Acta Scientiarum. Agronomy**, v. 30, n. 1, p. 7-11, 2008.

LEITE, G. L. D.; VELOSO, R. V. S.; ZANUNCIO, J. C.; FERNANDES, L. A.; ALMEIDA, C. I. M. Phenology of *Caryocar brasiliense* in the Brazilian cerrado region. **Forest Ecology and Management**, v. 236, n. 2-3, p. 286-294, 2006.

LORENZI, H.; MATOS, F. J. A. **Plantas medicinais no Brasil**: nativas e exóticas cultivadas. Nova Odessa: Instituto Plantarum, 2002.

LUCCI, N.; MAZZAFERA, P. Rutin synthase in fava d'anta: Purification and influence of stressors. **Canadian Journal of Plant Science**, v. 89, n. 5, p. 895-902, 2009.

MACEDO, M. L. R.; MELLO, G. C.; FREIRE, M. G. M.; NOVELLO, J. C.; MARANGONI, S.; MATOS, D. G. G. Effect of a trypsin inhibitor from *Dimorphandra mollis* seeds on the development of*Callosobruchus maculatus*. **Plant Physiology and Biochemistry**, v. 40, n. 10, p. 891-898, 2002.

MALAVOLTA, E. **O fósforo na planta e interação com outros elementos**. São Pedro: Potafós, 2004.

MENDES, A. D. R.; MARTINS, E. R.; FERNANDES, L. A.; MARQUES, C. C. L. Produção de biomassa e flavonóides totais por fava d'anta (*Dimorphandra mollis* Benth) sob diferentes níveis de fósforo em solução nutritiva. **Revista Brasileira de Plantas Medicinais**, v. 7, n. 2, p. 7-11, 2005.

MIZUMACHI, E.; MORI, A.; OSAWA, N.; AKIYAMA, R.; TOKUCHI, N. Shoot development and extension of *Quercus serrata* saplings in response to insect damage and nutrient conditions. **Annals of Botany**, v. 98, n. 1, p. 219-226, 2006.

PACHECO, M. V.; MATTEI, V. L.; MATOS, V. P.; SENA, L. H. Moura. Germination and vigor of *Dimorphandra mollis* Benth. seeds under different temperatures and substrates. **Revista Árvore**, v. 34, n. 2, p. 205-213, 2010.

PEREIRA, J. A. A.; OLIVEIRA, C. A. Efeito do *Eucalyptus camaldulensis* sobre a colonização micorrítica e a nodulação em *Dimorphandra mollis* e *Stryphnodendron adstringens*, em Brasilândia, Minas Gerais. **Cerne**, v. 11, n. 4, p. 409-415, 2005.

PETACCI, F.; FREITAS, S. S.; BRUNETTI, I. L.; KHALIL, N. M. Inhibition of peroxidase activity and scavenging of reactive oxygen species by astilbin isolated from *Dimorphandra mollis* (Fabaceae, Caesalpinioideae). **Biological Research**, v. 43, n. 1, p. 63-74, 2010.

PINTO, J. E. B. P.; LAMEIRA, O. A. **Micropropagação e metabólicos secundários *in vitro* de plantas medicinais**. Lavras: UFLA/FAEP, 2001.

ROTHER, D. C.; SOUZA, T. F.; MALASPINA, O.; BUENO, O. C.; SILVA, M. F. G. F.; VIEIRA, P. C.; FERNANDES, J. B. Susceptibility of workers and larvae of social bees in relation to ricinine. **Iheringia Serie Zoologia**, v. 99, n. 1, p. 61-65, 2009.

SANTOS, T. M.; BOIÇA JÚNIOR, A. L.; SOARES, J. J. Influência de tricomas do algodoeiro sobre os aspectos biológicos e capacidade predatória de *Chrysoperla externa* (Hagen) alimentada com *Aphis gossypii* Glover. **Bragantia**, v. 62, n. 2, p. 243-254, 2003.

SASTAWA, B. M.; LAWAN, M.; MAINA, Y. T. Management of insect pests of soybean: effects of sowing date and intercropping on damage and grain yield in the Nigerian Sudan savanna. **Crop Protection**, v. 23, n. 2, p. 155-161, 2004.

SILVA, C. C.; JHAM, G. N.; PICANÇO, M.; LEITE, G. L. D. Comparison of leaf chemical composition and attack patterns of *Tuta absoluta* (Meyrick) (Lepidóptera: Gelechiidae) in three tomato species. **Agronomia Lusitana**, v. 46, n. 2-4, p. 61-71, 1998.

SIMÕES, C. M. O.; SCHENKEL, E. P.; GOSMANN, G.; MELLO, J. C. P.; MENTZ, L. A.; PETROVICK, P. R. **Farmacognosia**: da planta ao medicamento. Porto Alegre: UFRGS; Florianópolis: UFSC, 2000.

SOUZA, M. P.; MATOS, M. E. O.; MATOS, F. J. A.; MACHADO, M. I. L.; CRAVEIRO, A. A. **Constituintes químicos ativos de plantas medicinais brasileiras**. Fortaleza: EUFC, 1991.

VIANA E SOUZA, H. A.; LOVATO, M. B.. Genetic diversity and structure of the critically endangered tree *Dimorphandra wilsonii* and of the widespread in the Brazilian Cerrado *Dimorphandra mollis*: Implications for conservation. **Biochemical Systematics and Ecology**, v. 38, n. 1, p. 49-56, 2010.

22

Isolation and characterization of *Bacillus thuringiensis* strains active against *Elasmopalpus lignosellus* (Zeller, 1848) (Lepidoptera, Pyralidae)

Janaina Zorzetti[1*], Ana Paula Scaramal Ricietto[1], Fernanda Aparecida Pires Fazion[1], Ana Maria Meneguim[2], Pedro Manuel Oliveira Janeiro Neves[1] and Gislayne Trindade Vilas-Bôas[1]

[1]*Universidade Estadual de Londrina, Rodovia Celso Garcia Cid, Km 445, 86047-902, Londrina, Paraná, Brazil. [2]Instituto Agronômico do Paraná, Londrina, Paraná, Brazil. *Author for correspondence. E-mail: zorzettijanaina@gmail.com*

ABSTRACT. *Elasmopalpus lignosellus* (Zeller, 1848) (Lepidoptera, Pyralidae) is an insect pest of 60 economically important crops, including sugarcane, wheat, soybean, rice, beans, sorghum, peanuts, and cotton. The aim of this work was to select and characterize *Bacillus thuringiensis* isolates with insecticidal activity against *E. Lignosellus* that could be used as an alternative method of control. Selective bioassays were done to evaluate the toxicity of 47 isolates against first instar larvae of *E. lignosellus*. For the most toxic bacterial strains, the lethal concentration (LC_{50}) was estimated and morphological, biochemical and molecular methods were used to characterize the isolates. Among the 47 isolates tested, 12 caused mortality above 85% and showed LC_{50} values from 0.038E+8 to 0.855E+8 spores mL^{-1}. Isolates BR83, BR145, BR09, BR78, S1534, and S1302 had the lowest LC_{50} values and did not differ from the standard HD-1 strain; the exception was BR83. The protein profiles produced bands with molecular masses of 60-130 kDa. The genes *cry1*, *cry2*, *cry3*, and *cry11* were identified in the molecular characterization. The morphological analysis identified three different crystal inclusions: bipyramidal, spherical and cuboidal. Among the tested isolates, 12 isolates have potential for biotechnological control of *E. Lignosellus* by development of new biopesticides or genetically modified plants.

Keywords: biological control, *cry* genes, entomopathogenic bacteria, lesser cornstalk borer.

Introduction

The lesser cornstalk borer *Elasmopalpus lignosellus* (Zeller, 1848) (Lepidoptera, Pyralidae) is a polyphagous pest, and larvae feed on more than 60 species of cultivated plants (Viana, 2004). These plants include crops of high economic value, such as corn, beans, wheat, soy, peanuts, and sugarcane, which suffer extensive losses by attack from this pest. The larvae damage newly germinated plants and reduces the number of seedlings per planting area. Larvae penetrate

the stalk of a recently sprouted plant, make galleries toward the central core and then feed inside the stem causing new leaves to dry up and die, resulting in the so-called "dead heart" (Gallo et al., 2002).

The lesser cornstalk borer is a difficult pest to control because it can remain close to the plant stem, within the stem, or in silken web habitats and land shelters they build on the soil surface. In experiments conducted with different control methods, pest management using pheromone and light traps and soil cover with *Crotalaria jucea* resulted in a small reduction in the pest population (Jham, Silva, Lima, & Viana, 2007; Gill, McSorley, Goyal, & Webb, 2010). Thus, preventive chemical control and seed treatment remain the most widely used methods to control *E. lignosellus*. However, when chemicals are indiscriminately applied, human contamination and environmental imbalance can result, leading to an increase in the pest population.

Entomopathogenic bacteria, such as *Bacillus thuringiensis*, are among the alternatives to reduce the use of insecticides for pest control. The insecticidal characteristics of these bacteria are caused by the formation of parasporal crystals in the early sporulation phase. These crystals are composed of Cry proteins, which are toxic to a variety of insects that attack crops of high economic value (Vilas-Bôas, Peruca, & Arantes, 2007; Vidal-Quist, Castañera, & González-Cabrera, 2009).

The toxic activity of these proteins against insect pests led to the formulation of bioinsecticides and the selection of genes encoding insecticidal proteins to produce transgenic plants resistant to different species of insects. Several *Bacillus thuringiensis* isolates specific to insects of the orders Lepidoptera, Coleoptera, and Diptera have been investigated (Pardo-López, Soberón, & Bravo, 2013). These isolates typically harbour one or more *cry* genes, and the isolates containing a wider range of genes are the most targeted. Thus, further studies are required to select these isolates, identify the *cry* genes, and assess isolate toxicity (Sun, Fu, Ding, & Xia, 2008).

Although the search for isolates of *B. thuringiensis* that are effective against *E. lignosellus* is of great significance for the management of this insect pest, studies remain limited. Therefore, the aim of this work was to select and characterize native isolates of *B. thuringiensis* toxic to *E. lignosellus*, with the goal to conduct further studies focused on new formulations of bioinsecticides and development of genetically modified plants.

Material and methods

Insect rearing

Larvae of *E. lignosellus* were reared on an artificial diet according to the methodology described by Greene, Leppla, and Dickerson (1976). The adults were maintained at 27 ± 2°C and 60 ± 10% RH with a 14h photoperiod in plastic cages (10 cm diameter, 20 cm height) coated with filter paper and closed on the upper end with tissue and on the lower end with a petri dish (14.3 cm diameter) and fed a 10% aqueous honey solution. The eggs

obtained were transferred to petri dishes at 25°C for incubation; the first instar larvae were used in bioassays.

Bacterial isolates

Forty-seven native isolates of *B. thuringiensis* were examined from the Collection of Entomopathogenic Microorganisms of Londrina State University (Universidade Estadual de Londrina, UEL) and the Brazilian Agricultural Research Corporation, Embrapa Genetic Resources and Biotechnology (Empresa Brasileira de Pesquisa Agropecuária – Embrapa Recursos Genéticos e Biotecnologia). The HD-1 strain of *B. thuringiensis* subsp. *kurstaki* was obtained from the Collection of *B. thuringiensis* at the Institut Pasteur, Paris, France.

Selective bioassay to choose the most toxic isolates

Suspensions of each *B. thuringiensis* isolate were prepared by adding 1.0 mg of lyophilized material to 1.0 mL of sterile distilled water. The artificial diet was prepared according to Greene et al. (1976) and distributed (3 mL) when still liquid into glass tubes (2 cm diameter x 3 cm height). After the diet solidified, 50 μL of a mixture of spores and crystals were applied on the diet surface. The glass tubes were kept in a laminar flow hood until the complete absorption of the suspension by the diet. Subsequently, five first-instar larvae were released inside each glass tube, which were sealed with a plastic lid. The bioassay consisted of three replicates with four glass tubes for each *B. thuringiensis* isolate. The standard strain *B. thuringiensis* subsp. *kurstaki* HD-1 (Btk) and water were used as positive and negative controls, respectively. The insects were maintained in an incubator (27 ± 2°C, 60 ± 10% RH and a 14h photoperiod) for six days after which mortality was assessed. The corrected mortality was calculated using Abbott's control adjusted mortality (Abbott, 1925). The data were subjected to analysis of variance (ANOVA), and the means were compared using Tukey's test at 5% probability. The most toxic isolates, those that caused a reliable mortality (above 85%) and therefore had potential for further testing, were used in dose-response bioassays and evaluated according to their molecular, protein and morphological profiles. The mortality rate was selected based on the minimum efficacy threshold (80% efficacy) required for pesticide registration in Brazil (MAPA, 1995).

Estimation of the lethal concentration (LC$_{50}$) of *B. thuringiensis* isolates

Bioassays for dose estimation were performed with the 12 isolates that showed the greatest toxicity in the selective bioassays and with the HD-1 standard strain. Seven suspensions of spores and crystals of *B. thuringiensis* were prepared to estimate the concentration of each isolate that would cause 50% mortality in *E. lignosellus* larvae (LC$_{50}$). The suspensions were prepared with 5.0 mg of lyophilized material that was diluted in 5 mL of sterile distilled water. Dilutions were performed using the initial suspension to obtain the seven concentrations used in the study (1.0, 0.2, 0.1, 0.05, 0.025, 0.008, and 0.0025 mg mL^{-1} in sterile distilled water), and the number of spores per mL of water in each dilution was counted using a Neubauer chamber. The bioassay was conducted in the same way as previously described. For each concentration evaluated, three replicates with four tubes were used, for a total of 20 larvae per replicate and 60 per concentration. The mortality data were subjected to Probit analysis (Finney, 1971) to determine the lethal concentration. The LC$_{50}$ bioassay results were analysed by checking for the overlap of the 95% confidence intervals according to Probit analysis.

Protein and molecular characterization of *B. thuringiensis* isolates pathogenic to *E. lignosellus*

Genomic DNA samples of the *B. thuringiensis* strains were isolated according to the method described by Ricieto, Fazion, Carvalho Filho, Vilas-Boas, and Vilas-Bôas (2013). The isolates were cultivated for 15h at 30°C on plates containing Luria-Bertani (LB) medium (Bertani, 1951). With the aid of a sterile toothpick, a colony of approximately 2 mm in diameter was transferred to microtubes containing 200 μL of TE (10 mMTris; 1 mM EDTA; pH 8.0). The suspension was homogenized and incubated in boiling water for 10 min. Then, the suspension was centrifuged at 12,000 xg for 3 min, and the supernatant was transferred to a new tube and used as DNA template in the PCR reactions. The presence of the genes *cry1*, *cry2*, *cry3*, *cry4A*, *cry4B*, *cry10*, and *cry11* was analysed using specific primers and amplification conditions (Céron, Ortí, Quintero, Güereca, & Bravo, 1995, Bravo et al., 1998; Ibarra et al., 2003; Vidal-Quist et al., 2009). DNA amplification was performed using an Endurance TC-412 thermocycler. For each amplification reaction, a total reaction volume of 20 μL was prepared that contained 1 U *Taq* DNA polymerase (Invitrogen, Brazil), 2.0 μL of Buffer 10 (200 mM Tris-HCl, pH 8.0, 500 mM KCl), 1.5 mM MgCl$_2$, 0.25 mM dNTP, 0.5 μM each primer, 2 μL of DNA and sterile Milli-Q water. The same reaction was used for all the primers. The PCR products were visualized by electrophoresis

on 1.2% agarose gel in TBE buffer (89 mM Tris Borate, 2 mM EDTA, pH 8.0) stained with Syber Safe (Invitrogen, UK) using a 100 bp DNA ladder (Invitrogen, UK). After electrophoresis, the gel images were captured using a Sony Cyber-shot 8.1 digital camera.

The *B. thuringiensis* isolates were characterized by the protein profile of their crystals using protein electrophoresis on 10% polyacrylamide gel (SDS-PAGE). Initially, the crystals were obtained according to the protocol described by Lecadet, Chaufaux, Ribier, and Lereclus. (1992). Each isolate was cultivated in NB medium (Downes & Ito, 2001) at 30°C for 72h at 200 rpm, until complete sporulation. The *B. thuringiensis* subsp. *kurstaki* HD-1 standard strain was used as the reference.

Morphological characterization of *B. thuringiensis* isolates

The morphological characterization of the isolates was initially performed by optical microscopy using a microscope (Model CHS; Olympus Optical Co. Ltd., Tokyo, Japan) with a 100x phase contrast lens. For electron microscopy, the lyophilized material of each isolate used in the previous bioassays was directly deposited over metal supports and coated with gold for 180 s, using a 40 mA current under vacuum (10-1 mbar) in a BAL-TEC model SCD–050 Sputter Coater (Santos et al., 2009). Subsequently, the material was analysed using a scanning electron microscope Philips QUANTA 200 (FEI) in high vacuum under 20 kV tension with a working distance of 10.2 mm. The selected images were captured and stored for later analysis.

Results

Selective bioassay and determination of the Median Lethal Concentration (LC$_{50}$) of *Bacillus thuringiensis* isolates

Among the 47 isolates of *B. thuringiensis* tested, 12 isolates (25.53%), in addition to *B. thuringiensis* subsp. *kurstaki* (Btk) HD-1 strain, caused mortality of *E. lignosellus* above 85%, for a total of 13 isolates, which were selected for all other conclusive tests. Dose-response bioassays were used to evaluate these 13 isolates. The X^2 values related to the LC$_{50}$ were no significant for 12 of the isolates, which indicated that the data were homogeneous for those strains and fit the Probit analysis model (Table 1) (Finney, 1971).

Table 1. Mortality (%) and LC_{50} of *Bacillus thuringiensis* isolates against first instar larvae of *Elasmopalpus lignosellus* (Zeller, 1848) (Lepidoptera, Pyralidae) (n = 60) on the sixth day of the dose-response bioassay.

Isolate	[a]Mortality (%)	[b]LC_{50} (Spores mL^{-1})	CI $_{(95\%)}$ Lower	Upper	Slope ± SE	X^2
BR83	100.00 a	0.038E+8 a	0.025E+8	0.053E+8	1.92 ± 0.26	2.004 ns
BR145	100.00 a	0.059E+8 ab	0.020E+8	0.122E+8	1.25 ± 0.13	11.009 ns
BR09	100.00a	0.063E+8 ab	0.037E+8	0.102E+8	1.47 ± 0.13	11.070 ns
BR78	100.00 a	0.086E+8 ab	0.030E+8	0.179E+8	1.19 ± 0.14	10.806 ns
S1534	100.00 a	0.090E+8 ab	0.043E+8	0.189E+8	1.46 ± 0.18	7.098 ns
S1302	93.33 ab	0.097E+8 b	0.071E+8	0.134E+8	1.30 ± 0.13	3.724 ns
HD1	100.00 a	0.114E+8 b	0.084E+8	0.151E+8	1.38 ± 0.15	3.384 ns
BR52	100.00 a	0.141E+8 bc	0.098E+8	0.250E+8	1.38 ± 0.24	6.702 ns
BR38	100.00 a	0.346E+8 c	0.233E+8	0.445E+8	2.17 ± 0.31	1.637 ns
BR53	100.00 a	0.378E+8 c	0.239E+8	0.593E+8	1.69 ± 0.23	5.969 ns
S545	91.23 b	0.392E+8 c	0.218E+8	0.767E+8	1.39 ± 0.20	8.546 ns
S1269	94.92 ab	0.855E+8 d	1.39E+8	0.543E+8	1.43 ± 0.30	3.978 ns
S1450	95.00 ab	(-)	(-)	(-)	0.96 ± 0.16	15.880 s★

[a]Mortality (%) from selective bioassay. Means followed by the same letter in the column are not significantly different (Tukey's test) at p< 0.05; [b]Means followed by the same letter in the column are not different from one another based on the overlap of 95% confidence intervals, according to Probit analysis; (-)significant X^2: Estimate of the LC_{50} was not possible. s★significant based on the Chi-squared test (p < 0.05). ns- not significant based on the Chi-squared test (p < 0.05).

The LC_{50} values of the selected isolates varied from 0.038E+8 to 0.855E+8 spores mL^{-1}. The lowest LC_{50} values were obtained for the group consisting of BR83, BR145, BR09, BR78, and S1534. The LC_{50} values within that group did not differ statistically according to the Probit analysis, as shown by the overlap of the 95% confidence intervals (Table 1). Only theLC$_{50}$ value of isolate BR83 was significantly lower than that of the HD-1 standard strain, with a toxicity that was three-fold greater than that of the standard. Additionally, the BR 83 isolate was approximately 22-fold more toxic than the S1269 isolate, which had the highest LC_{50}. Only the S1450 isolate presented a significant X^2; thus, the LC_{50} could not be estimated (Table 1).

Protein and molecular characterization of *B. thuringiensis* isolates pathogenic to *E. lignosellus*

The PCR technique using total DNA of isolates and specific primers for the detection of *cry1*, *cry2*, *cry3*, *cry4A*, *cry 4B*, *cry10*, and *cry11* resulted in the amplification of fragments of the expected sizes (Bravo et al., 1998; Céron et al., 1995; Ibarra et al., 2003; Vidal-Quist et al., 2009) and consequently, the determination of which *cry* genes were in the isolates of *B. thuringiensis*.

The amplicons produced with the greatest frequency corresponded to *cry1*, *cry2*, and *cry3* genes. The *cry1* gene was detected in all isolates, except BR52 and BR53, which also did not exhibit the *cry2* gene. The expected fragment for the *cry3* gene appeared in BR145, S1534, and S1302 isolates, whereas the fragment for the *cry11* gene occurred only in the BR53 isolate. Only the BR52 isolate did not show a PCR product consistent with the selected *primers*. The fragments of the expected size for the other *primers* used were not observed in all isolates, indicating the absence of *cry4A*, *cry4B*, and *cry10* genes in some isolates (Table 2). The protein

profile analysis of the spore-crystal mixtures of the isolates from the selective bioassays revealed bands of 60, 65, 70, 80, and 130 kDa. The isolate used as the standard, *B. thuringiensis* subsp. *kurstaki*, had a protein profile of 65 and 130 kDa (Höfte & Whiteley, 1989; Lereclus, Delécluse, & Lecadet, 1993). All data related to the protein and molecular characterization of *B. thuringiensis* isolates are presented in Table 2. An image of protein electrophoresis on 10% polyacrylamide gel (SDS-PAGE) illustrating the protein profile of some isolates is shown in Figure 1.

Table 2. Genetic protein profiles and morphological characterization of crystals of *Bacillus thuringiensis* isolates.

Isolate	Genetic profile	Protein profile (kDa)	Bipyramidal	Spherical	Cuboidal
BR83	*cry1, cry2*	130/70	+	+	+
BR145[a]	*cry1, cry2,cry3*	130/65-70	+	+	-
BR09[b]	*cry1, cry2*	130/65	+	+	+
BR78	*cry1, cry2*	65/80	+	+	+
S1534	*cry1, cry2, cry3*	130/70	+	+	-
S1302	*cry1, cry2, cry3*	70	+	+	-
HD1	*cry1, cry2*	65/130	+	-	+
BR52	-	70	+	-	-
BR38	*cry1, cry2*	130/70	+	+	-
BR53	*cry11*	70	+	-	-
S545	*cry1*	130	+	-	-
S1269	*cry1*	130	+	-	-
S1450	*cry1, cry2*	130/65	+	+	+

The "Morphological analysis" header spans the Bipyramidal, Spherical, and Cuboidal columns.

[a]Data from Ricieto et al. (2013); [b]Data from Santos et al. (2009).

Morphological characterization of *B. thuringiensis* isolates

Morphological analysis using optical and scanning electron microscopy showed protein inclusions of different forms. The BR83, BR09, BR78, and S1450 isolates showed three different crystalline protein inclusions: bipyramidal, spherical, and cuboidal. Isolates BR145, S1534, S1302, and BR38 exhibited bipyramidal and spherical crystals, whereas isolates S545, S1269, BR52, and BR53 contained only bipyramidal crystals (Table 2).

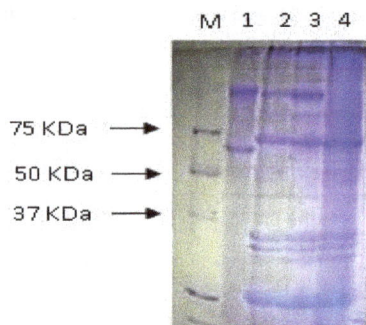

Figure 1. Protein profile produced by isolates toxic to *Elasmopalpus lignosellus* (Zeller, 1848) (Lepidoptera, Pyralidae). M, Rainbow molecular weight marker (GE); 1 - BR09; 2 - BR38; 3 - BR83; 4 - BR52.

Discussion

The selective and dose bioassays revealed that *B. thuringiensis* isolates were active against *E. lignosellus*. Of the isolates tested, 25.53% caused mortality above 85%; therefore, these isolates can be further tested as a possible alternative method for the management of populations of *E. lignosellus*. Based on dose bioassays, isolate BR83 had the lowest lethal concentration. Furthermore, BR83 had superior activity compared with that of the standard *B. thuringiensis* subsp. *kurstaki* HD-1 strain, although the activity was not significantly different from that of the BR145, BR09, BR78, and S1534 isolates. Those four isolates carried *cry1* and *cry2* genes, which are toxic to insects of the order Lepidoptera (Ricieto et al., 2013). In addition to these genes, BR145 and S1302 isolates also carried the *cry3* gene, with reported toxicity to Lepidoptera and Coleoptera (Brizzard & Whitley, 1988).

The *cry1* and *cry2* genes were found in almost all selected isolates, except for BR52, which did not yield an amplification product for any of the used primers, and BR53, which amplified only for the *cry11* gene. The genes in the Cry1 subfamily are the most abundant and occur in approximately half of the isolates identified to date. The *cry2* gene is also very common, particularly among isolates harbouring the *cry1* gene (Porcar & Juárez-Perez, 2003; Arrieta, Hernández, & Espinoza, 2004).

Isolates containing *cry1* and/or *cry2* genes were also the most abundant in the collection studied by Vidal-Quist et al. (2009), representing over 45% of the total. In studies with the BR37 isolate, Santos et al. (2009) identified eight genes of which seven were in the Cry1 group, showing the frequent occurrence of *cry1* genes in isolates of *B. thuringiensis* (Bravo et al., 1998).

Genes in the Cry1 and Cry2 protein families have been reported in the isolates used for the control of *E. lignosellus*, and the effects of Cry1 proteins in transgenic peanut and soybeans plants have been examined under field conditions. In peanuts, various levels of resistance to the lesser cornstalk borer were provided by the introduction of the *cry1Ac* gene, from complete larval mortality to 66% reduction in larval weight (Singsit et al., 1997). For soybean, plants expressing the *cry1Ac* gene had four-fold more resistance to *E. lignosellus* than that of the wild-type isolate (Walker, All, Mcpherson, Boerma, & Parrott, 2000).

Assessments of genetically modified corn crops revealed that hybrids containing *cry9C, cry1F,* and *cry1Ab* genes did not differ in resistance to *E. lignosellus*; however, transgenic plants were superior to non-transgenic hybrids (Vilella, Waquil, Vilela, Siegfried, & Foster, 2002). Additionally, in laboratory bioassays, isolates containing *cry2A* gene and the HD-1 standard isolate proved effective against *E. lignosellus* (Moar, Pusztai-Carey, & Mack, 1995).

These examples demonstrate the different methodologies used for the control of *E. lignosellus*, but the mortality rates are different between isolates harbouring many genes and those caused by genetically modified plants carrying only the primary gene responsible for the toxicity. However, to select new isolates that carry toxic genes to be tested individually against the pest, the bioassays used in this study must be conducted. Thereby, the selected gene, after many tests, can be inserted into plants for pest control and increase the possibilities for pest management.

The protein profiles of most of the tested isolates that amplified with the *cry1* and *cry2* genes (BR83, BR145, BR09, BR38, S1534, and S1450) showed bands of 130 and 65/70 kDa, which confirmed their specificity for Lepidoptera. Cry1 class proteins have a molecular weight of approximately 130-140 kDa (Höfte & Whiteley, 1989), whereas proteins of the Cry2 and Cry3 groups, which are active against Lepidoptera and Coleoptera, have values of 65-70 kDa (Bravo et al., 2004).

Although the BR09 isolate did not differ from the most toxic isolate (BR83) or the HD-1 standard, the LC_{50} of the BR09 isolate was approximately 1.8-fold lower than that of HD-1. Santos et al. (2009) obtained a similar value when testing the same isolate against *Spodoptera eridania*. Nevertheless, LC_{50} values may vary among species; for example, in studies against *Spodoptera cosmioides*, the LC_{50} of the HD-1 standard strain was two-fold lower than that of the BR09 isolate (Santos et al., 2009). Moreover, in a selection performed by Constanski et al., 2015, no differences were detected between the LC_{50} of the HD-1 standard strain and that of the other isolates tested against *S. eridania* and *S. cosmioides*.

The toxicological profile of the BR09 isolate that was effective against *E. lignosellus* can be explained by the expression of *cry1* and *cry2* genes, similar to the HD-1 strain, which has the *cry1Aa, cry1Ab, cry1Ac, cry2A,* and *cry2B* genes in its genome (Li et al., 2005). Furthermore, SDS-PAGE protein analysis of the spore and crystal mixture revealed two polypeptides with approximate molecular weights of 70-130 kDa. Santos et al. (2009) also tested the BR09 isolate and showed a profile with *cry1Aa, cry1Ab, cry1Ac,* and *cry2Aa* genes.

Although the S1450 isolate caused mortality of *E. lignosellus* larvae above 85%, in the selective bioassays, the X^2 value was significant, i.e., the data did not fit the Probit model. Thus, the LC_{50} could not be estimated for isolate S1450. This isolate is in the *kurstaki* serotype described in the literature as toxic to insects of the order Lepidoptera (Monnerat et al., 2007) and causing 100% mortality of *Agrotis ipsilon* (Lepidoptera: Noctuidae). Additionally, *cry1Aa, cry1Ab, cry1Ac, cry1,* and *cry2* genes were revealed by the molecular analysis, which confirmed the activity of the isolate against Lepidoptera (Menezes, Fiuza, Martins, Praça, & Monnerat, 2010).

The BR78 isolate, with an LC_{50} that did not differ from that of the most toxic isolates, also carried *cry1* and *cry2* genes. However, only two primary polypeptides of approximately 65 and 80 kDa were associated with this isolate, which are related only to the *cry2* gene. The result was similar for the S1302 isolate, which despite amplifying with *cry1, cry2,* and *cry3* genes, revealed only a band of 70 kDa. As a possible explanation, a poor performance or even the absence of the promoter led to low expression of the *cry1* gene, which prevented the display of a 130 kDa band.

According to Alper et al. (2014), some strains that harbour the same genes may not be as effective as other toxic strains that have those same genes, indicating that these genes could be poorly expressed because of a weak promoter in the strains. Thus, poor expression of the *cry* genes in the BR78 and S1302 isolates could explain the *cry* genes in the genetic profile but not in the protein profile.

Armengol, Escobar, Maldonado, and Orduz (2007), identified isolates toxic to *S. frugiperda* containing *cry1Aa, cry1Ab, cry1Ac, cry1B,* and *cry1D* genes, with protein profiles that revealed bands only at 60 kDa. According to these authors, the correlation between the identified protein profiles

and the *cry* genes cannot occur when the *cry* proteins are encoded by unknown genes or have not been amplified by the *primers* used. Additionally, the identified genes may encode proteins with low-level or inactive expression.

Although the LC_{50} did not differ from that of the HD-1 standard isolate, only the BR52 isolate failed to obtain amplification of the expected fragment sizes with the specific primers employed. Nevertheless, the optical and electron microscope observations revealed the production of protein inclusions, suggesting that the isolate might contain genes not covered by the set of primers used in the PCR analysis or *cry* genes not yet described. Additionally, the analysis of the protein profile showed a polypeptide of approximately 70 kDa that corresponded to the Cry2 class, possibly explaining the insecticidal activity of isolate BR52 against Lepidoptera.

Only isolate BR53 expressed the *cry11* gene. However, this gene was most likely not the cause of the toxicity of this isolate against *E. lignosellus*, because the gene is usually associated with activity against larvae of Diptera, such as *Simulium* spp., *Culex* spp. and *Aedes aegypti* (Vidal-Quist et al., 2009). Thus, isolate BR53 might contain some other gene toxic to *E. lignosellus* not disclosed by the set of primers used, requiring further investigation to search for new *cry* genes.

Based on the different shapes of protein crystals detected by optical and electron microscopy, the types of Cry proteins that composed the crystal could be inferred, which provided information on the insecticidal activity of the isolates (Lereclus, Delécluse, & Lecadet, 1993; Vilas-Boas et al., 2007).

The similarity between the morphological and molecular analyses was notable in this study. Isolates that expressed *cry1* and *cry2* genes contained primarily bipyramidal, cuboidal, and spherical shaped crystals. These three different shapes were observed in the BR09, BR78, and BR83 isolates, which contained the *cry1* and *cry2* genes, and in the S1450 isolate that harboured *cry1, cry2,* and *cry3* genes. The BR38 isolate with *cry1* and *cry2* genes and the S1302, S1534, and BR145 isolates with *cry1, cry2* and *cry3* genes all contained bipyramidal and spherical crystals. The crystal protein composition was studied in some of these isolates previously, and those studies confirm the shapes observed in this study (Praça et. al., 2004; Santos et al., 2009; Ricieto et al., 2013). Only bipyramidal crystals were produced by the isolates S545 and S1269, which contained only the *cry1* gene, and BR52 and BR53.

The selection of isolates containing Cry1 and Cry2 proteins and harbouring more than one Cry protein is essential in the prospecting for new virulent *cry* genes against lepidopterans such as *E. lignosellus*. Once identified, these are the isolates that can be used for pest management either through bioinsecticide formulations or as a source for transgenic plant manipulation.

Bacillus thuringiensis is the most successful pathogen agent used for insect control, representing almost 2% of the total insecticide market and up to 90% of bioinsecticide formulations. Currently, formulations based on *B. thuringiensis* are increasing, and this market is expected to continue to grow. Most of these products are based on spore-crystal preparations derived from a few strains such as *B. thuringiensis* var. *kurstaki* HD1. Thus, the detection of *B. thuringiensis* strains with new genes is extremely important to develop more efficient bioinsecticides (Bravo, Likitvivatanavong, Gill, & Soberón, 2011; Lemes et al., 2015).

For some products formulated with *B. thuringiensis*, their use in agriculture is limited, because Cry toxins are more specific for first larval instars, are sensitive to sun radiation and have limited activity against borer insects (Bravo et al., 2011). Therefore, the internal feeding behaviour of *E. lignosellus* inside stalks and stems makes management with pesticide and biologic product applications alone difficult. Thus, to more effectively *E. lignosellus* control, plants must be genetically modified with *B. thuringiensis* genes.

The discovery of these *cry* genes is of great interest for the production of new transgenic plants and also for genes tacking or pyramidalization. For example, insertion of different *cry* genes into genetically modified plants that use different receptors in the insect midgut membrane can help to extend the protection against more insect pests. Additionally, the beginning of resistance can be delayed or prevented, because more than one toxic protein would be acting against the same insect species (Sanahuja, Banakar, Twyman, Capell, & Christou, 2011; Hernández-Rodríguez, Hernández-Martínez, Van Rie, Escriche, & Ferré, 2013).

Researchers and commercial companies have recently used the technique of gene pyramidalization. As an example, the SmartStax® corn developed in cooperation between Monsanto and Dow Agro Sciences companies showed that the pyramidalization of *cry1A.105*, *cry2Ab*, *cry3Bb1*, *cry34Ab1*, *cry35Ab1*, and *cry1Fa2* genes was essential for managing pesticide-resistant pests, in addition to providing effective control against a long list of pests in Coleoptera and Lepidoptera, including *E. lignosellus* (Marra, Piggott, & Goodwin, 2010).

The isolates evaluated in this study have potential for biotechnological control of *E. lignosellus*. Additionally, the isolates can be a gene source for the production of new crops or the management of currently insect-resistant plant cultivars. The genome sequencing of the studied isolates ensured accurate quantification of *cry* genes for further selection and use against *E. lignosellus* through insertion into the genome of economically relevant crops attacked by this pest.

Conclusion

Among the 47 isolates studied, 12 caused mortality above 85%. Isolates BR145, BR09, BR78, S1534, and S1302 had the lowest LC_{50} values and did not differ from the standard HD-1. The protein profiles produced bands with molecular masses of 60-130 kDa. The molecular characterization showed the presence of *cry1*, *cry2*, *cry3*, and *cry11* genes.

The morphological analysis identified three different crystal inclusions: bipyramidal, spherical and cuboidal. As a result of these characterizations, these isolates have potential for biotechnological control of *E. lignosellus* and should be important candidates for more studies and the development of new biopesticides or genetically modified plants.

Acknowledgements

The authors wish to thank the National Council for Scientific and Technological Development (CNPq) for financial support.

References

Abbott, W. S. (1925). A method of computing the effectiveness of an insecticide. *Journal of Economic Entomology, 18*(2), 265-266.

Alper, M., Günes, H., Tatlipina, A., Çöl, B., Civelek, H. S., Özkan, C., & Poyraz, B. (2014). Distribution, occurrence of *cry* genes, and lepidopteran toxicity of native *Bacillus thuringiensis* isolated from fig tree environments in Aydın Province. *Turkish Journal of Agriculture and Forestry, 38*(6), 898-907.

Armengol, G., Escobar, M. C., Maldonado, M. E., & Orduz, S. (2007). Diversity of Colombian strains of *Bacillus thuringiensis* with insecticidal activity against dipteran and lepidopteran insects. *Journal of Applied Microbiology, 102*(1), 77-88.

Arrieta, G., Hernández, A., & Espinoza, A. M. (2004). Diversity of *Bacillus thuringiensis* strains isolated from coffee plantations infested with the coffee berry borer *Hypothenemus hampei* Ferrari. *Revista de Biología Tropical, 52*(3), 757-764.

Bertani, G. (1951). Studies on lysogenesis I. The mode of phage liberation by lysogenic *Escherichia coli*. *Journal of Bacteriology, 62*(3), 293-300.

Bravo, A., Gómez, I., Conde, J., Muñoz-Garay, C., Sánchez, J., Miranda, R., ... Soberón, M. (2004). Oligomerization triggers binding of a *Bacillus thuringiensis* Cry1Ab pore-forming toxin to aminopeptidase N receptor leading to insertion into membrane microdomains. *Biochimica et Biophysica Acta*, *1667*(3), 38-46.

Bravo, A., Likitvivatanavong, S., Gill, S. S., & Soberón, M. (2011). *Bacillus thuringiensis*: a story of a successful bioinsecticide. *Insect Biochemistry and Molecular Biology*, *41*(7), 423-431.

Bravo, A., Sarabia, S., Lopez, L., Ontiveros, H., Abarca, C, Ortiz, A., ... Quintero, R. (1998). Characterization of *cry* genes in a Mexican *Bacillus thuringiensis* strain collection. *Applied and Environmental Microbiology*, *64*(12), 4965-4972.

Brizzard, B. L., & Whiteley, H. R. (1988). Nucleotide sequence of an additional crystal protein gene cloned from *Bacillus thuringiensis* subsp. *thuringiensis*. *Nucleic Acids Research*, *16*(6), 2723-2724.

Céron, J., Ortíz, A., Quintero, R., Güereca, L., & Bravo, A. (1995). Specific PCR primers directed to identify *cry*1 and *cry*3 genes within a *Bacillus thuringiensis* strains collection. *Applied and Environmental Microbiology*, *61*(11), 3826-3831.

Constanski, K. C., Zorzetti, J., Vilas Bôas, G. T., Ricieto, A. P. S., Fazion, F. A. P., Vilas Boas, L. A., ... Neves, P. M. O, J. (2015). Seleção e caracterização molecular de isolados de *Bacillus thuringiensis* para o controle de *Spodoptera* spp. *Pesquisa Agropecuária Brasileira*, *50*(8), 730-733.

Downes, F. P., & Ito, K. (2001). *Compendium of methods for the microbiological examination of foods* (4th ed.). Washington, DC: American Public Health Association.

Finney, D. J. (1971). *Probit analysis* (3rd ed.). Cambridge, UK: Cambridge University Press.

Gallo, D., Nakano, O., Neto, S. S., Carvalho, R. P. L., Batista, G. C., Filho, E. B., ... Omoto, C. (2002). *Entomologia agrícola*. Piracicaba, SP: Fealq.

Gill, H. K., McSorley, R., Goyal, G., & Webb, S. E. (2010). Mulch as a potential management strategy for lesser cornstalk borer, *Elasmopalpus lignosellus* (Insecta: Lepidoptera: Pyralidae), in bush bean (*Phaseolus vulgaris*). *Florida Entomologist*, *93*(2), 183-190.

Greene, G. L., Leppla, N. C., & Dickerson, W. A. (1976). Velvetbean caterpillar: a rearing procedure and artificial medium. *Journal of Economic Entomology*, *69*(4), 487-488.

Hernández-Rodríguez, C. S., Hernández-Martínez, P., Van Rie, J., Escriche, B., & Ferré, J. (2013). Shared midgut binding sites for Cry1A, Cry1Aa, Cry1Ab, Cry1Ac and Cry1Fa proteins from *Bacillus thuringiensis* in two important corn pests, *Ostrinian ubilalis* and *Spodoptera frugiperda*. *PLoS ONE*, *8*(7), e68164.

Höfte, H., & Whiteley, H. R. (1989). Insecticidal crystal proteins of *Bacillus thuringiensis*. *Microbiology and Molecular Biology Reviews*, *53*(2), 242-255.

Ibarra, J. E., Rincón, M. C. D., Ordúz, S., Noriega, D., Benintende, G., Monnerat, R., ... Bravo, A. (2003). Diversity of *Bacillus thuringiensis* strains from Latin America with insecticidal activity against different mosquito species. *Applied and Environmental Microbiology*, *69*(9), 5269-5274.

Jham, G. N., Silva, A. A., Lima, E. R., & Viana, P. A. (2007). Identification of acetates in *Elasmopalpulus lignosellus* pheromone glands using a newly created mass spectral database and Kóvats retention indices. *Química Nova*, *30*(4), 916-919.

Lecadet, M. M., Chaufaux, J., Ribier, J., & Lereclus, D. (1992). Construction of novel *Bacillus thuringiensis* strain with different insecticidal activities by transduction and transformation. *Applied and Environmental Microbiology*, *58*(3), 840-849.

Lemes, A. R. N., Marucci, S. C., Costa, J. R. V., Alves, E. C. C., Fernandes, O. A., Lemos, M. V. F., & Desidério, J. A. (2015). Selection of strains from *B. thuringiensis* genes containing effective in the control of *Spodoptera frugiperda*. *Bt Research*, *6*(1), 1-8.

Lereclus, D., Delécluse, A., & Lecadet, M. M. (1993). Diversity of *Bacillus thuringiensis* toxins and genes. In: P. F. Enwistle, J. Cory, M. Bailey, S, Higgs (Eds.), *Bacillus thuringiensis, an environmental biopesticide: Theory and practice* (p. 37-69). Chichester, UK: John Wiley & Son Ltd.

Li, H., Oppert, B., Higgins, R. A., Huang, F., Buschman, L. L., & Zhu, K. Y. (2005). Susceptibility of Dipel-resistant and -susceptible *Ostrinia nubilalis* (Lepidoptera: Crambidae) to individual *Bacillus thuringiensis* protoxins. *Journal of Economic Entomology*, *98*(4), 1333-1340.

Ministério da Agricultura, Pecuária e Abastecimento [MAPA]. (1995). *Normas e exigências para execução de testes de produtos químicos para fins de registro no MAPA*. Brasília, DF: Ministério da Agricultura e Reforma Agrária.

Marra, M. C., Piggott, N. E., & Goodwin, B. K. (2010). The anticipated value of Smart Stax™ for US corn growers. *AgBio Forum*, *13*(1), 1-12.

Menezes, R. S., Fiuza, V. D., Martins, E. S.,Praça, L. B., & Monnerat, R. G. (2010). Seleção e caracterização de estirpes de *Bacillus thuringiensis* tóxicas a *Agrotisipsilon*. *Universitas Ciências da Saúde*, *8*(1), 1-13.

Moar, W. J., Pusztai-Carey, M., & Mack, T. P. (1995). Toxicity of purified proteins and the HD-1 strain from *Bacillus thuringiensis* against lesser cornstalk borer (Lepidoptera: Pyralidae). *Journal of Economic Entomology*, *88*(3), 606-609.

Monnerat, R. G., Batista, A. C., Medeiros, P. T., Martins, E., Melatti, V., Praça, L., Dumas, V., ... Berry C. (2007). Screening of Brazilian *Bacilus thuringiensis* isolates active against *Spodoptera frugiperda, Plutella xylostella* and *Anticarsia gemmatalis*. *Biological Control*, *41*(3), 291-295.

Pardo-López, L., Soberón, M., & Bravo, A. (2013). *Bacillus thuringiensis* insecticidal three-domain Cry toxins: mode of action, insect resistance and consequences for crop protection. *FEMS Microbiology Reviews*, *37*(1), 3-22.

Porcar, M., & Juárez-Pérez, V. (2003). PCR-based identification of *Bacillus thuringiensis* pesticidal crystal genes. *FEMS Microbiology Reviews*, *26*(5), 419-432.

Praça, L. B., Batista, A. C., Martins, E. S., Siqueira, C. B., Dias, D. G. S., Gomes, A. C. M. M., ... Monnerat, R. G. (2004). Estirpes de *Bacillus thuringiensis* efetivas contra insetos das ordens Lepidoptera, Coleoptera e Diptera. *Pesquisa Agropecuária Brasileira, 39*(1), 11-16.

Ricieto, A. P. S., Fazion, F. A. P., Carvalho Filho, C. D., Vilas-Boas, L. A., & Vilas-Bôas, G. T. (2013). Effect of vegetation on the presence and genetic diversity of *Bacillus thuringiensis* in soil. *Canadian Journal of Microbiology, 59*(1), 28-33.

Sanahuja, G., Banakar, R., Twyman, R. M., Capell, T., & Christou, P. (2011). *Bacillus thuringiensis*: a century of research, development and commercial applications. *Plant Biotechnology Journal, 9*(3), 283-300.

Santos, K., Neves, P. M. O. J., Meneguim, A. M., Santos, R. B., Santos, W. J., Vilas-Bôas, G. T., ... Monnerat, R. (2009). Selection and characterization of the *Bacillus thuringiensis* strains toxic to *Spodoptera eridania* (Cramer), *Spodoptera cosmioides* (Walker) and *Spodoptera frugiperda* (Smith) (Lepidoptera: Noctuidae). *Biological Control, 50*(2), 157-163.

Singsit, C., Adang, M. J., Lynch, R. E., Anderson, W. F., Wang, A., Cardineau, G., & Ozias-Akins, P. (1997). Expression of a *Bacillus thuringiensis* cryA(c) gene in transgenic peanut and its efficacy against lesser cornstalk borer. *Transgenic Research, 6*(2), 169-176.

Sun, Y., Fu, Z., Ding, X., & Xia, L. (2008). Evaluating the insecticidal genes and their expressed products in *Bacillus thuringiensis* strains by combining PCR with Mass Spectrometry. *Appliedand Environmental Microbiology, 74*(21), 6811-6813.

Viana, P. A. (2004). Lagarta-elasmo. In J. R. Salvadori, C. J. Ávila, M. T. B. Silva. (Ed.), *Pragas de solo no Brasil* (p. 379-408). Passo Fundo, RS: Embrapa Trigo; Dourados, MS: Embrapa Agropecuária Oeste; Cruz Alta, RS: Fundacep Fecotrigo.

Vidal-Quist, J. C., Castañera, P., & González-Cabrera, J. (2009). Diversity of *Bacillus thuringiensis* strains isolated from citrus orchards in Spain and evaluation of their insecticidal activity against *Ceratitis capitata. Journal of Microbiology and Biotechnology, 19*(8), 749-759.

Vilas-Bôas, G. T., Peruca, A. P. S., & Arantes, O. M. N. (2007). Biology and taxonomy of *Bacillus cereus, Bacillus anthracis* and *Bacillus thuringiensis. Canadian Journal of Microbiology, 53*(6), 673- 687.

Vilella, F. M. F., Waquil, J. M., Vilela, E. F., Siegfried, B. D., & Foster, J. E. (2002). Selection of the fall armyworm, *Spodoptera frugiperda* (J. E. Smith) (Lepidoptera: Noctuidae) for survival on Cry 1A(b)*Bt.* toxin. *Revista Brasileira de Milho e Sorgo, 1*(1), 12-17.

Walker, D. R., All, J. N., Mcpherson, R. M., Boerma, H. R., & Parrott, W. A. (2000). Field evaluation of soybean engineered with a synthetic *crylAc* transgene for resistance to corn earworm, soybean looper, velvetbean caterpillar (Lepidoptera: Noctuidae), and lesser cornstalk borer (Lepidoptera: Pyralidae). *Journal of Economic Entomology, 93*(3), 613-622.

23

Insecticide effects of *Ruta graveolens*, *Copaifera langsdorffii* and *Chenopodium ambrosioides* against pests and natural enemies in commercial tomato plantation

Flávia Silva Barbosa, Germano Leão Demolin Leite[*], Sérgio Monteze Alves, Aline Fonseca Nascimento, Vinícius de Abreu D'Ávila and Cândido Alves da Costa

*Instituto de Ciências Agrárias, Universidade Federal de Minas Gerais, Av. Universitária 1000, Cx. Postal 135, 39404-006, Montes Claros, Minas Gerais, Brazil. *Author for correspondence. E-mail: gldleite@ufmg.br*

ABSTRACT. The aim of this study was to evaluate the insecticide effect of watery leaf extracts of *Ruta graveolens* (Rutaceae), alcoholic leaf extracts of *Copaifera langsdorffii* (Caesalpinaceae) and *Chenopodium ambrosioides* (Chenopodiaceae) in the concentration of 5% under field conditions. The experiment design was randomized blocks with six replications. The parcels treated with plant extracts showed reduction in the population of pests when compared with the control parcels. The extract elaborated with *C. langsdorffii* presented greater insecticidal effect under *Bemisia tabaci* (Hemiptera: Aleyrodidae) and sum of pests. It was verified that after 24 hours of spraying, the parcels treated with the extract of *C. ambrosioides* presented minor numbers of adults of *Tuta absoluta* (Lepidoptera: Gelechiidae), followed by the parcels treated with extract of *R. graveolens*. There were smaller numbers of parasitoid eggs of lepidopterans *Trichogramma* sp. (Hymenoptera: Trichogrammatidae) and sum of natural enemies (predators + parasitoids) in the parcels that had received spraying with extracts from the plants of *C. langsdorffii* and *C. ambrosioides*, followed by *R. graveolens*, compared to the control. There were a smaller number of parasitoids from the family Eulophidae (Hymenoptera) attacking caterpillars of *T. absoluta* in plants treated with *R. graveolens*, followed by *C. langsdorffii* and *C. ambrosioides* than in the control.
Keywords: *Lycopersicon esculentum*, bioinsecticide, alternative control.

Introduction

Lycopersicon esculentum Mill. (Solanaceae) is one of the most important vegetables cultivated in the world, with regard to area as well as commercial value, being the second most produced vegetable in Brazil. The tomato is considered one of the few cultures for which pests and diseases are equally important, being a host plant to about 200 species of arthropods (CARVALHO et al., 2002).

Among the pests of tomato plants are the transmitters of viral diseases *Bemisia tabaci* Gennadius (Homoptera: Aleyrodidae), *Frankliniella schulzei* Trybom (Thysanoptera: Thripidae), *Myzus persicae* Sulzer and *Macrosiphum euphorbiae* Thomas (Hemiptera: Aphididae), leaf miners *Tuta absoluta* Meyrick (Lepidoptera: Gelechiidae) and *Liriomyza huidobrensis* Blanchard (Diptera: Agromyzidae), and fruit borers *T. absoluta*, *Neoleucinodes elegantalis*

Guenée (Lepidoptera: Crambidae) and *Helicoverpa zea* Bod. (Lepidoptera: Noctuidae) (GALLO et al., 2002). They can provoke high losses in tomato production, as verified by Picanço et al. (2007), in which some of these pests are capable of causing, directly or indirectly, a reduction of 58.39% in total income of culture.

For the control of these pests, successive applications of synthetic insecticides are used. This is undesirable due to elevated production costs, exposure of producers and consumers to harmful active ingredients, adverse effects on the environment, reduction of natural enemies, and selection of resistant individuals in the pest populations (THOMAZINI et al., 2000).

Therefore, there is a need to control pests without causing aggressions and unbalances in the environment through the search of alternative methods of pest control that favor the natural enemies that are indispensable to establish the biological balance and to reduce production costs, such as with the use of plant extracts (VIEGAS JÚNIOR, 2003). Thus, organic cultivation, since it uses environmentally safe technologies, constitutes a promising alternative (VENZON et al., 2006). Organic agricultural production seeks to maximize the social benefits, self-sustainment, the reduction or elimination of the dependence on inputs, non-renewable energy and preservation of the environment through optimization of the use of natural and socioeconomic resources available in a maintainable and rational way (HAMERSCHMIDT et al., 2000; PENTEADO, 2000).

Problems such as selectivity to natural enemies, persistence to the environment and toxicity to mammals can be minimized by the use of insecticide plants as an alternative method of suppression of the herbivore (VIEGAS JÚNIOR, 2003; TREVISAN et al., 2006). Barbosa et al. (2009) studied different extraction methods and concentrations of tree plants. This author observed satisfactory mortality of aqueous leaf extracts of *Ruta graveolens* L. (Rutaceae) and *Artemisia verlotorum* Lamotte (Asteraceae), alcoholic leaf extracts of *Petiveria alliacea* L. (Phytolaccaceae) against *Diabrotica speciosa* (Germar) (Coleoptera: Chrysomelidae), in laboratory conditions. However, it is necessary to evaluate the insecticide effect of those extracts on other pests found, the impact on natural enemies, and the persistence of those extracts after some time exposed to field conditions, as the natural enemies promote efficient biocontrol of pests and favor greater productivity of the culture by means of effect cascades (CARDINALE et al., 2003).

The objective of this work is to evaluate the insecticide effect of aqueous leaf extracts of *R. graveolens* and alcoholic leaf extracts of *C. Langsdorffii* and *C. ambrosioides* against pests and natural enemies in commercial plantations of *L. esculentum*.

Material and methods

The experiment was carried in the city of Montes Claros, state of Minas Gerais, Brazil, in April 2007. The analysis was accomplished in a commercial plantation of *L. esculentum* var. Santa Clara, cultivated in Red-Yellow Latosol, with four months of planting, and trellised obliquely. The last pulverization with organosynthetic pesticides had been done three weeks before.

The experimental block design was randomized, with six replications and four treatments. Each plot had five rows with 25 plants per rows. The 15 central plants of the medium row were used for data collection. The plants were spaced 0.50 m apart within rows and 1.0 m between rows. The treatments were: 1) control without pulverization, 2) alcoholic leaf extract of *C. langsdorffii*, 3) alcoholic leaf extract of *C. Ambrosioides*, and 4) aqueous leaf extract of *R. graveolens*. The extracts were used in a dilution of 5%. Due to this dosage, Barbosa et al. (2009) observed insecticide effect in *D. speciosa* in *R. graveolens*, and another two plants by preliminary testing.

The *R. graveolens* and *C. ambrosioides* plants were organically cultivated at the Horto Medicinal of ICA/UFMG, while *C. langsdorffii* adult trees, present at the campus, in dystrophic Red Latosol of average texture.

To obtain the aqueous leaf extract of *R. graveolens* with three months of age (reproductive phase), 200 g of fresh leaves plus 1 L of distilled water were used. It was triturated in an industrial blender until homogenization. Soon afterwards, the mixture was heated until the beginning of ebullition. The mixture was then left to cool down, filtered and stored in a tinted glass until use. To obtain the alcoholic extracts elaborated with leaves of *C. langsdorffii*, unkown age (reproductive phase), and *C. Ambrosioides*, with three months of age (reproductive phase), 250 g of fresh leaves of each plant plus 1 L of hydrated commercial ethyl alcohol were used. The leaves were cut in small pieces and put in tinted glass flasks. The alcohol was then added and the mixture was then briefly agitated twice a day for 15 days. After this period, the solution was filtered and stored in tinted glass flasks until application.

The insecticide effect of the plant extracts was verified 24 and 72 hours after application by evaluating the number of pests, natural enemies (predators and parasitoids) and ants in 15 plants per plot, using the beating tray method (MIRANDA et al., 1998).

This method consists of beating the first apical expanded leaf inside a 34 x 26 x 5 cm white tray and counting the insects inside. The number of mines of *T. absoluta* and *Liriomyza* sp. were evaluated by direct counting in the first expanded leaf of 15 plants per plot (PICANÇO et al., 1998).

To evaluate the possible effect of the extracts on the parasitoid of eggs *Trichogramma* sp. (Hymenoptera: Trichogrammatidae) by natural infestation in the field, two white cards (12.0 x 3.0 cm) with 3,500 eggs of *Ephestia kuehniella* Zell. (Lepidoptera: Pyralidae), not parasitized, were put in the apical part of the canopy height (1.6 m height) of tomato plants. In the plots that were sprayed with extracts, a pulverization jet was driven on the cards. After 72 hours, the cards were collected off the field, stored in transparent white plastic bags, sealed and transported to the laboratory. Rearing took place in an incubator at a constant temperature of 25°C. After 15 days, the cards were evaluated with a 40x magnifying lens, counting the number of adults of *Trichogramma* sp. emerged from the parasitized eggs, number of eggs predated, eggs parasitized by *Trichogramma* sp. not emerged, eggs not parasitized and total eggs in the cards.

Field and laboratory data were transformed by $\sqrt{x+0.5}$ and arcsin, respectively, and later submitted to analysis of variance and Tukey test ($p < 0.05$).

Results and discussion

The smallest number of *B. tabaci* adults, mines of *T. absoluta* and sum of pests (sum of total pests found) was observed in the treatments that had received the extracts of *C. langsdorffii*, *C. ambrosioides* and *R. graveolens* and *T. absoluta* adults in the plants sprayed with *C. ambrosioides* and *R. graveolens* extracts after 24 hours spraying (Table 1). There were no significant effects ($p > 0.05$) 24 hours after spraying in *F. schulzei* (Table 1). 72 hours after spraying, a smaller number of *B. tabaci* adults were observed on the tomato plants treated with *C. langsdorffii* and *C. ambrosioides* than on the plants treated with *R. graveolens* and in the control group (Table 1). A smaller number of *F. schulzei* was noted in Table 1.

The control group, followed by the plants that had received treatments with the extracts of *C. ambrosioides*, *C. langsdorffii* and *R. graveolens* (Table 1). A smaller number of the sum of pests was observed in the tomato plants sprayed with *C. langsdorffii*, followed by *C. ambrosioides*, *R. graveolens* and the control group (Table 1). There were no significant effects ($p > 0.05$) noted in the number of mines and adults of *T. absoluta* (Table 1).

Table 1. Effect of extracts of *Copaifera langsdorfii*, *Chenopodium ambrosioides* and *Ruta graveolens* in the number (±standard error) of adults of *Bemisia tabaci* and *Frankliniella schulzei* (adults + nymphs)/tray, of small, medium, big and total mines/leaf and adults of *Tuta absoluta*/tray and sum of pests/tray on the *Lycopersicon esculentum*. Montes Claros, Minas Gerais State, Brazil, 2007.

Evaluation (h)	Control	C. langsdorfii	C. ambrosioides	R. graveolens
	Bemisia tabaci			
24	18.25±1,72Aa	13.90±1.22Ba	13.18±0.93Ba	14.45±1.83Ba
72	11.15±1,09Ab	7.80±0.58Bb	8.51±0.61Bb	9.97±0.94ABb
	Frankliniella schulzei			
24	1.44±0.14Aa	1.39±0.16Aa	0.98±0.10Aa	1.36±0.18Aa
72	1.03±0.11Ba	1.75±0.21Aa	1.29±0.13ABa	1.75±0.24Aa
	Small mines of *Tuta absoluta*			
24	6.26±0.40Aa	4.62±0.35Ba	4.9±0.43Ba	4.84±0.33Ba
72	4.66±0.33Aa	4.58±0.45Aa	5.09±0.47Aa	5.42±0.48Aa
	Medium mines of *Tuta absoluta*			
24	3.79±0.29Aa	2.74±0.22Ba	2.91±0.28Ba	3.36±0.33ABa
72	3.45±0.41Aa	3.14±0.27Aa	3.41±0.36Aa	3.7±0.44Aa
	Big mines of *Tuta absoluta*			
24	5.66±0.53Aa	2.99±0.30Ba	4.01±0.44Ba	4.03±0.43Ba
72	4.33±0.50Aa	3.47±0.31Aa	3.89±0.38Aa	4.41±0.45Aa
	Total mines of *Tuta absoluta*			
24	15.71±0.91Aa	10.35±0.59Ba	11.83±0.92Ba	12.23±0.87Ba
72	12.44±0.96Aa	11.18±0.77Aa	12.39±0.93Aa	13.53±1.08Aa
	Adults of *Tuta absoluta*			
24	0.05±0.02Aa	0.01±0.01ABa	0.00±0.00Ba	0.00±0.00Ba
72	0.03±0.01Aa	0.01±0.01Aa	0.05±0.03Aa	0.03±0.01Aa
	Sum of pests			
24	21.17±1.77Aa	16.48±1.26Ba	15.6±0.98Ba	17.58±1.89Ba
72	14.08±1.19Ab	10.83±0.64Bb	11.32±0.64ABb	13.51±0.94Ab

The average followed by the same capital letter within a row or lowercase letter within a column did not differ in Tukey test ($p < 0.05$).

Other pests found were observed, however, apparently not affected by the pulverizations separately (not sum of pests), such as *Diabrotica speciosa* (0.35 ± 0.02/tray) (Germar) and *Cerotoma* sp. (0.0036 ± 0.0019/tray) (Coleoptera: Chrysomelidae); *M. persicae* and *M. euphorbiae* (0.02 ± 0.01/tray) (Aphididae), *Empoasca* sp. (0.03 ± 0.01/tray) (Cicadellidae), Pentatomidae (0.01 ± 0.01/tray); Miridae (0.001 ± 0.001/tray); caterpillars of Noctuidae (0.05 ± 0.04/tray) and Geometridae (0.0013 ± 0.0013/tray) and *Liriomyza* sp. adults (Diptera: Agromyzidae) (0.008 ± 0.003/tray).

The insecticide effect on *C. langsdorffii* extract in this work is probably due to the presence of coumarin in their constitution (VEIGA JUNIOR; PINTO, 2002), showing effect against *Aedes aegypti* L. larvae (Diptera: Culicidae) (CHAITHONG et al., 2006).

The insecticide action of the plant *C. ambrosioides* that occurred in this study is probably the result of the flavonoids and terpenoids present in its structure (CRUZ et al., 2007). Silva et al. (2005) verified that 1.0 and 2.0% (p/p) concentrations of the *C. ambrosioides* extracts showed satisfactory insecticide effect against *Sitophilus zeamais* Mots. (Coleoptera: Curculionidae). They achieved 90.3 and 90.1% mortality, respectively.

The *R. graveolens* plants have glycosides (rutin), aromatic lactones (coumarin, bergapten, xantotoxine, rutaretine and rutamarine), anthocyanins glycosides, alkaloids (rutamine, rutalidine, cocusaginine, esquiamianine and ribalinidine), methyl ketones (methylnonil ketone and methyheptil ketone), flavonoids (hesperidine), rutaline, rutacridone and terpenes (a-pinene, limonene, cineole) (MARTINS et al., 2005). These compounds are probably responsible for insecticide effects on *S. zeamais* (ALMEIDA et al., 1999), *Ctenocephalides canis* Curtis (Siphonaptera: Pulicidae) (LEITE et al., 2006) and *D. speciosa* (BARBOSA et al., 2009) as well as on several pests observed in the tomato plants in field conditions in this work.

Aqueous leaf-branch extracts of *Trichilia pallida* Swartz (Meliaceae) presented harmful effects on the development of *T. absoluta* (THOMAZINI et al., 2000) and aqueous extracts of *Prosopis juliflora* (S.w.) D.C. and *Leucaena leucocephala* Wit. (Leguminoseae) in variable concentrations between 3 and 10%, provoking mortality and alterations in the fertility parameters of *B. tabaci* (CAVALCANTE et al., 2006), therefore proving efficiency of vegetable extracts as easy elaboration in the suppression of important pests in tomato plantations.

After 24 hours a smaller number of adults of *Trichogramma* sp. and sum of natural enemies

(predators + parasitoids) per tray were observed on the tomato plants that had been sprayed with the extracts of *C. langsdorffii* and *C. ambrosioides* than on the plants that had been treated with *R. graveolens* and the control group (Table 2). A smaller number of Eulophidae (Hymenoptera, parasitoid of *T. absoluta* caterpillars) was noted on tomato plants sprayed with the extract of *R. graveolens* than on the ones that had been treated with *C. langsdorffii* and *C. ambrosioides* and the control (Table 2). A smaller number of sum of parasitoids than in the control was observed in the plots sprayed with the extracts studied (Table 2). On the other hand, a higher number of *Syrphus* sp. larvae (Diptera: Syrphidae, predators of aphids) than in the other treatments was noted on the tomato plants that had received the extracts of *R. graveolens* (Table 2). A higher number of sum of ants, basically *Crematogaster* sp. (Hymenoptera: Formicidae), than the other treatments was observed on the plants sprayed with extract of *C. langsdorffii*, followed by the control group (Table 2).

However, after 72 hours, significant effects were detected on *Chrysoperla* sp. larvae (Neuroptera: Chrysopidae, predator of aphids, nymphs of whitefly and eggs of lepidopterous), observing larger numbers on sprayed tomato plants with *C. langsdorffii* extracts (Table 2). No significant effects (p > 0.05) were observed in the sum of predators after 24 hours as well as after 72 hours like Carabidae (Coleoptera) (0.26 ± 0.02/tray), *Orius* sp. (Hemiptera: Anthocoridae) (0.05 ± 0.01/tray), Vespidae (Hymenoptera) (0.03 ± 0.01/tray) and spiders (0.28 ± 0.02/tray) as well as in parasitoids like *Encarsia* sp. (Aphelinidae) (0.01 ± 0.001/tray) and the parasitoid not identified (0.38 ± 0.03/tray) (Hymenoptera).

A higher number of eggs of *E. kuehniella*, parasitized and not emerged per total eggs, was observed in cards sprayed with *C. langsdorffii* extract, demonstrating a possible insecticide effect of this extract when compared with the extracts of *C. ambrosioides*, *R. graveolens* and control, as well as no preference of insect predators and parasitoids in the cards treated with *C. langsdorffii* (Table 3). It was verified that the cards with eggs of *E. kuehniella*, sprayed with the extract of the plant *R. graveolens*, presented a larger residual insecticide effect.

There was a larger number of parasitized eggs and without emergence of parasitized eggs, with approximately 75% of embryos dead, with a smaller number of *Trichogramma* sp. emerged compared with *C. langsdorffii*, *C. ambrosioides* and the control group (Table 3).

Table 2. Effect of extracts of *Copaifera langsdorfii*, *Chenopodium ambrosioides* and *Ruta graveolens* in the number (±standard error) of *Trichogramma* sp., Eulophidae, larvae of *Syrphus* sp., larvae of *Chrysoperla* sp., sum of ants, sum of parasitoids, sum of predators and sum of natural enemies/tray on the *Lycopersicon esculentum*. Montes Claros, Minas Gerais State, Brazil, 2007.

Evaluation (hours)	Control	C. langsdorfii	C. ambrosioides	R. graveolens
Trichogramma sp.				
24	0.37±0.06Aa	0.20±0.05Ba	0.16±0.03Ba	0.26±0.05ABa
72	0.14±0.03Aa	0.18±0.05Aa	0.10±0.03Aa	0.20±0.04Aa
Eulophidae				
24	0.52±0.09Aa	0.38±0.08ABa	0.34±0.07ABa	0.28±0.08Ba
72	0.72±0.06Aa	0.47±0.05Aa	0.48±0.06Aa	0.55±0.04Aa
Syrphus sp.				
24	0.02±0.01Ba	0.00±0.00Ba	0.02±0.01Ba	0.10±0.03Aa
72	0.02±0.01Aa	0.02±0.01Aa	0.01±0.01Aa	0.05±0.02Aa
Chrysoperla sp.				
24	0.04±0.02Aa	0.01±0.01Aa	0.00±0.00Aa	0.03±0.01Aa
72	0.00±0.00Ba	0.05±0.02Aa	0.01±0.01Ba	0.00±0.00Ba
Sum of ants				
24	0.01±0.01ABa	0.04±0.01Aa	0.00±0.00Ba	0.00±0.00Ba
72	0.00±0.00Aa	0.01±0.01Aa	0.00±0.00Aa	0.00±0.00Aa
Sum of parasitoids				
24	0.91±0.12Aa	0.58±0.10Ba	0.55±0.09Ba	0.54±0.10Ba
72	0.72±0.12Aa	0.47±0.07Aa	0.48±0.08Aa	0.55±0.11Aa
Sum of predadors				
24	0.63±0.07Aa	0.51±0.07Aa	0.58±0.09Aa	0.74±0.09Aa
72	0.67±0.09Aa	0.71±0.09Aa	0.59±0.08Aa	0.72±0.10Aa
Sum of natural enemies				
24	1.54±0.14Aa	1.08±0.13Ba	1.13±0.13Ba	1.27±0.13ABa
72	1.39±0.16Aa	1.17±0.11Aa	1.07±0.11Aa	1.28±0.16Aa

The average followed by the same capital letter within a row or lowercase letter within a column did not differ in Tukey test (p < 0.05).

Table 3. Effect of extracts of *Copaifera langsdorfii*, *Chenopodium ambrosioides* and *Ruta graveolens* on the percentages (±standard error) of eggs not parasitized, *Trichogramma* sp. emerged, eggs predated and eggs parasitized not emerged per total eggs and eggs parasitized not emerged per total eggs parasitized. Montes Claros, Minas Gerais State, Brazil, 2007.

| Characteristics | Treatments | | | |
	Control	C. langsdorfii	C. ambrosioides	R. graveolens
Eggs not parasitized (%)/total eggs	44.64±12.53A	67.34±9.36A	61.84±11.12A	43.15±11.03A
Trichogramma sp. emerged (%)/total eggs	0.55±0.37A	0.30±0.08A	0.27±0.13A	0.01±0.01A
Eggs predated (%)/total eggs	54.79±12.69A	32.29±9.42A	37.85±11.18A	56.81±11.04A
Eggs parasitized not emerged (%)/total eggs	0.02±0.00B	0.07±0.01A	0.04±0.01AB	0.03±0.01B
Eggs parasitized not emerged (%)/total eggs parasitized	17.59±11.13B	21.48±4.47B	15.49±4.92B	75.00±11.18A

The average followed by the same capital letter within a row did not differ in Tukey test (p < 0.05).

It is probable that the spots treated with *C. langsdorffii* extract presented selectivity to ants, and the spots treated with *R. graveolens* presented selectivity to Syrphidae in function of the dose and formulation of the extracts used, because the effect of the botanical insecticides on insects is variable (MOREIRA et al., 2006).

Works regarding alternative control of insects through vegetable extracts focus on their compatibility with other management tactics, mainly with biological control. However, the answers found in this experiment for low selectivity of the extract of *R. graveolens* and *C. langsdorffii* on insects of the genus *Trichogramma* were verified by Raguram and Singh (1999) and Reddy and Manjunatha (2000) by testing the insecticide effect of *Azadirachta indicates* A. Juss on natural enemies. Carvalho et al. (2001) studied the insecticide effect of commercial products used mainly in tomato plantations on *T. pretiosum*.

They verified, under laboratory conditions, that the chemicals clorfluazuron, teflubenzuron, triflumuron, cyromazine, benomyl, chlorothalonil, mancozeb, iprodione, dimetomorf, tebufenozide and pirimicarb were selective to the two lineages of *T. pretiosum*, as well as the biological commercial product *Bacillus thuringiensis*, reaching a mortality of up to 30%. However, it is known that those synthetic chemicals remain for a longer time in the environment than natural insecticides – a fact that should be considered at the moment of choosing the product to be applied, as well as the best application time.

Conclusion

We conclude that the extract from *C. langsdorffii* presents insecticide action after 24 hours and maintains a good residual effect until 72 hours after application. *C. langsdorffii* has a greater insecticide effect on natural enemies than *C. ambrosioides* and *R. graveolens*. The extract elaborated with *R. graveolens* shows to be more selective to natural enemies than *C. langsdorffii* and *C. ambrosioides*.

Acknowledgments

To Capes, CNPq and Fapemig for research support, and Maria de Fátima Prates for using her commercial tomato plantation.

References

ALMEIDA, F. A. C.; GOLDFARB, A. C.; GOUVEIA, J. P. G. Avaliação de extratos vegetais e métodos de aplicação no controle de *Sitophilus* spp. **Revista Brasileira de Produtos Agroindustriais**, v. 1, n. 1, p. 13-20, 1999.

BARBOSA, S. L.; LEITE, G. L. D.; MARTINS, E. R.; GUANABENS, R. E. M.; SILVA, F. W. S. Métodos de extração e concentrações no efeito inseticida de *Ruta graveolens* L., *Artemisia verlotorum* Lamotte e *Petiveria alliacea* L. a *Diabrotica speciosa* Germar. **Revista Brasileira de Plantas Medicinais**, v. 11, n. 3, p. 221-229, 2009.

CARDINALE, B. J.; HARVEY, C. T.; GROSS, K.; IVES, A. R. Biodiversity and biocontrol: emergent impacts of a multi-enemy assemblage on pest suppression and crop yield in an agroecosystem. **Ecology Letters**, v. 6, n. 9, p. 857-865, 2003.

CARVALHO, G. A.; PARRA, J. R. P.; BAPTISTA, G. C. Seletividade de alguns produtos fitossanitários a duas linhagens de *Trichogramma pretiosum* Riley, 1879 (Hymenoptera: Trichogrammatidae). **Ciência e Agrotecnologia**, v. 25, n. 3, p. 583-591, 2001.

CARVALHO, G. A.; REIS, P. R.; MORAES, J. C.; FUINI, L. C.; ROCHA, L. C. D.; GOUSSAIN, M. M. Efeitos de alguns inseticidas utilizados na cultura do tomateiro (*Lycopersicon esculentum* Mill.) a *Trichogramma pretiosum* Riley, 1879 (Hymenoptera: Trichogrammatidae). **Ciência e Agrotecnologia**, v. 26, n. 6, p. 1160-1166, 2002.

CAVALCANTE, G. M.; MOREIRA, A. F. C.; VASCONCELOS, S. D. Potencialidade inseticida de extratos aquosos de essências florestais sobre mosca-branca. **Pesquisa Agropecuária Brasileira**, v. 41, n. 1, p. 9-14, 2006.

CHAITHONG, U.; CHOOCHOTE, W.; KAMSUK, K.; JITPAKDI, A.; TIPPAWANGKOSOL, P.; CHAIYASIT, D.; CHAMPAKAEW, D.; TUETUN, B.; PITASAWAT, B. Larvicidal effect of pepper plants on *Aedes aegypti* (L.) (Diptera: Culicidae). **Journal of Vector Ecology**, v. 31, n. 1, p. 138-144, 2006.

CRUZ, G. V. B.; PEREIRA, P. V. S.; PATRÍCIO, F. J.; COSTA, G. C.; SOUSA, S. M.; FRAZÃO, J. B.; ARAGÃO-FILHO, W. C.; MACIEL, M. C. G.; SILVA, L. A.; AMARAL, F. M. M.; BARROQUEIRO, E. S. B.; GUERRA, R. N. M.; NASCIMENTO, F. R. F. Increase of cellular recruitment, phagocytosis ability and nitric oxide production induced by hydroalcoholic extract from *Chenopodium ambrosioides* leaves. **Journal of Ethnopharmacology**, v. 111, n. 1, p. 148-154, 2007.

GALLO, D.; NAKANO, O.; SILVEIRA NETO, S.; CARVALHO, R. P. L.; BAPTISTA, G. C.; BERTI FILHO, E.; PARRA, J. R. P.; ZUCCHI, R. A.; ALVES, S. B.; VENDRAMIM, J. D.; MARCHINI, L. C.; LOPES, J. R. S.; OMOTO, C. **Entomologia agrícola**. Piracicaba: Fealq, 2002.

HAMERSCHMIDT, I.; SILVA, J. C. B. V.; LIZARELLI, P. H. **Agricultura orgânica**. Curitiba: Emater, 2000.

LEITE, G. L. D.; SANTOS, M. M. O.; GUANABENS, R. E. M.; SILVA, F. W. S.; REDOAN, A. C. M. Efeito de boldo chinês, do sabão de côco e da cipermetrina na mortalidade de pulgas em cachorro doméstico. **Revista Brasileira de Plantas Medicinais**, v. 8, n. 3, p. 96-98, 2006.

MARTINS, A. G.; ROSÁRIO, D. L.; BARROS, M. N.; JARDIM, M. A.G. Levantamento etnobotânico de plantas medicinais, alimentares e tóxicas da Ilha do Combu, Município de Belém, Estado do Pará, Brasil. **Revista Brasileira de Farmácia**, v. 86, n. 1, p. 21-30, 2005.

MIRANDA, M. M. M.; PICANÇO, M.; LEITE, G. L. D.; ZANUNCIO, J. C.; CLERCQ, P. Sampling and non-action levels for predators and parasitoids of virus vectors and leaf miners of tomato plants in Brazil. **Mededelingen Faculteit Landbouwwetenschappe Universiteit Gent**, v. 63, n. 1, p. 519-523, 1998.

MOREIRA, M. D.; PICANÇO, M. C.; SILVA, M. E.; MORENO, S. C.; MARTINS, J. C. Uso de inseticidas botânicos no controle de pragas. In: VENZON, M.; PAULA JÚNIOR, T. J.; PALLINI, A. (Ed.). **Controle alternativo de pragas e doenças**. Viçosa: Epamig/CTZM, 2006. p. 89-120.

PENTEADO, S. R. **Introdução à agricultura orgânica**: normas e técnicas de cultivo. Campinas: Grafimagem, 2000.

PICANÇO, M. C.; BACCI, L.; CRESPO, A. L. B.; MIRANDA, M. M. M.; MARTINS, J. C. Effect of integrated pest management practices on tomato production and conservation of natural enemies. **Agricultural and Forest Entomology**, v. 9, n. 4, p. 327-335, 2007.

PICANÇO, M.; LEITE, G. L. D.; GUEDES, R. N. C.; SILVA, E. E. A. Yield loss in trellised tomato affected by insecticidal sprays and plant spacing. **Crop Protection**, v. 17, n. 5, p. 447-452, 1998.

RAGURAN, S.; SINGH. R. P. Biological effects of neem (*Azadirachta indica*) seed on an egg parasitoid, *Trichogramma chilonis*. **Journal of Economic Entomology**, v. 92, n. 6, p. 1274-1280, 1999.

REDDY, G. V. P.; MANJUNATHA, M. Laboratory and field studies on the integrated pest management of *Helicoverpa armigera* (Hübner) in cotton, based on pheromone trap catch threshold level. **Journal of Applied Entomology**, v. 124, n. 5, p. 213-221, 2000.

SILVA, G.; ORREGO, O.; HEPP, R.; TAPIA, M. Búsqueda de plantas con propiedades insecticidas para el controlde *Sitophilus zeamais* en maíz almacenado. **Pesquisa Agropecuária Brasileira**, v. 40, n. 1, p. 11-17, 2005.

THOMAZINI, A. P. B. W.; VENDRAMIM, J. D.; LOPES, M. T. R. Extratos aquosos de *Trichilia pallida* e a traça-do-tomateiro. **Scientia Agricola**, v. 57, n. 1, p. 13-17, 2000.

TREVISAN, M. T. S; BEZERRA, M. Z. B.; SANTIAGO, G. M. P.; FEITOSA, C. M. F. Atividades larvicida e anticolinesterásica de plantas do gênero *Kalanchoe*. **Química Nova**, v. 29, n. 3, p. 415-418, 2006.

VEIGA JÚNIOR, V. F.; PINTO, A. C. O gênero *Copaifera* L. **Química Nova**, v. 25, n. 2, p. 273-286, 2002.

VIEGAS JÚNIOR, C. Terpenos com atividade inseticida: uma alternativa para o controle químico de insetos. **Química Nova**, v. 26, n. 3, p. 390-400, 2003.

VENZON, M.; TUELHER, E. S.; BONOMO, I. S.; TINOCO, R. S.; FONSECA, M. C. M.; PALLINI, A. Potencial de defensivos alternativos para o controle de pragas do cafeeiro. In: VENZON, M.; PAULA JÚNIOR, T. J.; PALLINI, A. (Ed.). **Tecnologias alternativas para o controle de pragas e doenças**. Viçosa: Epamig/CTZM, 2006. p.117-136.

24

Mapping of the time available for application of pesticides in the state of Paraná, Brazil

Alessandra Fagioli da Silva[*], Rone Batista de Oliveira and Marco Antonio Gandolfo

*Setor de Engenharia e Desenvolvimento Agrário, Universidade Estadual do Norte do Paraná, Campus Luiz Meneghel, Rodovia BR-369, km 54, Vila Maria, Cx. Postal, 261, 86360-000, Bandeirantes, Paraná, Brazil, *Author for correspondence. E-mail: alefagioli@hotmail.com*

ABSTRACT. Spraying practices are highly sensitive to variations in climatic conditions, so it the spatial variability of the number of monthly hours available for the application of pesticides in the state of Paraná, Brazil, was determined and analyzed. An hourly time series of the climatic data obtained from 54 meteorological stations from the Paraná, for the years 2004-2014 was analyzed. To determine the number of monthly hours available to perform the applications, the following conditions were established: temperature < 30°C, relative air humidity > 55%, wind speed between 3 and 12 km h[-1], and rainfall less than 0.2 mm h[-1]. After being cross-referenced, these data were analyzed using descriptive statistics and geostatistics. A high variability of availability for the performance of the applications was found, with 46.59 hours in June and 285 hours in August. A spatial dependence existed in the time available to perform pesticide applications in the state of Paraná, with variation occurring as a function of the month of the year. The Paraná geographic regions of the Central East, Metropolitan Curitiba, and Southwest are less favorable to perform the application. The mapping allows the management of spraying practices.

Keywords: application technology; spraying; climatic conditions; geostatistics.

Introduction

The state of the atmosphere directly affects the evaporation, deposition of spray droplets and can increase the risk of losses through drift and run-off. Thus, the appropriate time for spraying can vary considerably with climatic variables from place to place and during the year in the same area of cultivation, constituting phenomena with spatio-temporal indexing.

Natural phenomena, such as climatic variables, often present with a certain structure in the variations between neighbors; therefore, the variations are not random, and thus, they have some degree of spatial and/or temporal dependence. Spatio-temporal variability can be studied using geostatistical tools, based on the theory of regionalized variables, where the values of a variable are related to its spatio-temporal arrangement.

To characterize the variability of climatic data, an analysis of their distribution is necessary. Spatial variability can be studied with geostatistical tools, based on the theory of regionalized variables, where the values of a variable are related to their spatial arrangement. The observations made over a short space resemble one another more than those made over larger spaces (Vieira, Nielsen, & Biggar, 1981).

For pesticide spraying to be successful, several factors related to the application technology must be considered, including climatic conditions (Cunha, Pereira, Barbosa, & Silva, 2016). Often, the active ingredient is lost due to environmental conditions and inappropriate application times. The results of 15 field experiments under varying climatic conditions, boom heights, and driving speeds indicated that for the most-evaluated spraying conditions, the most decisive factors influencing total spray drift were boom height and wind speed, followed by temperature, driving speed, and relative air humidity (Arvidsson, Bergström, & Kreuger, 2011).

The meteorological conditions considered favorable for spraying are widely reported in the literature and are characterized by temperatures between 15 and 30°C, relative air humidity greater than 55%, and wind speeds varying from 2 to 10 km h^{-1} (Ruedell, 2002; Minguela & Cunha, 2010; Raetano, 2011). However, in various situations, these requirements are not met due to the need to spray under unfavorable conditions. Previous knowledge about the spatial distribution of the monthly number of hours available to perform applications can be used in the decision-making process for the best time/condition for the application of pesticides.

The spatial analysis of the availability in hours to perform the application of pesticides under an ideal state of atmospheric conditions is fundamental for labor management, planning, and sizing of the sprayer fleet in agricultural properties. In addition, spatial analysis provides a more complex analysis to diagnose the regions that need more attention when choosing the application technology to reduce the risk of

environmental impact, increase efficiency, and achieve the expected benefits of spraying practices.

The objective of this study was to analyze the spatial variability of the time available under appropriate conditions to perform pesticide applications in the state of Paraná, Brazil.

Material and method

The main meteorological elements that affect the quality and loss of pesticides during spraying are rainfall, air temperature, relative air humidity, and wind speed. This work was conducted through the analysis of historical data schedules of the meteorological variables air temperature (°C), air humidity (%), rainfall (mm), and wind speed (km h^{-1}) collected by the weather stations of the State of Paraná in southern Brazil (Figure 1).

The study included the entire territory of Paraná State, located between 22°29′30″ and 26°43′00″ south latitude and 48°05′37″ and 54°37′08″ west longitude. According to the Köppen classification, the climate in the state includes the types Cfa (subtropical) in the north, west, southwest, and coastal regions and Cfb (temperate) predominantly in the southern and southeastern portions of the state (Caviglione et al., 2000).

A map of the state of Paraná with the spatial distribution of the meteorological stations that collected the data is presented in Figure 1. Data were obtained from 54 stations encompassing the Pioneer North (4), Central North (5), Central West (2), Central East (4), Central South (6), Northwest (7), West (11), Southeast (4), Southwest (5), and Metropolitan Curitiba (6) regions. The data cover the period from 2004 to 2014.

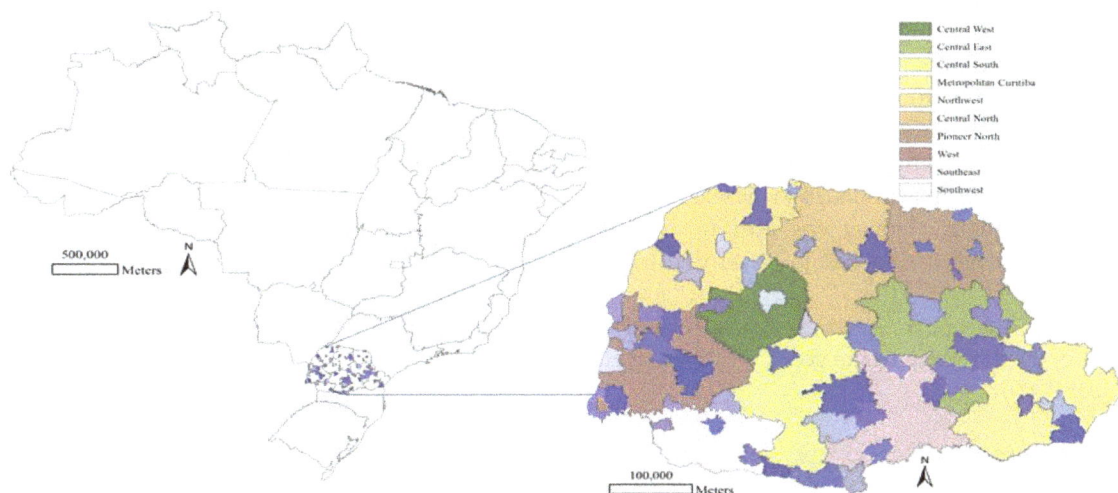

Figure 1. Spatial distribution of the agrometeorological data collection stations in the State of Paraná, Brazil.

Historical data on the studied variables were acquired from the National Institute of Meteorology (Instituto Nacional de Meteorologia – INMET) website (http://www.inmet.gov.br) and the Meteorological System of Paraná (Sistema Meteorológico do Paraná – SIMEPAR). The data were tabulated in an electronic spreadsheet (Excel®), followed by daily verification of the consistency of the available information. Data for those days whose information was inconsistent or had missing variables were eliminated.

With the aid of an electronic spreadsheet (Excel®) and using the ideal meteorological conditions for conducting pesticide applications as the parameter, a temperature below 30°C, a relative air humidity above 55%, wind speed between 3 and 12 km h^{-1} and rainfall less than 0.2 mm h^{-1} (ANDEF, 2004; Alvarenga et al., 2014), the data were cross-referenced to determine the times of day that all four weather conditions were within the intervals recommended in order to obtain better efficiency in the application. Next, the hours available for pesticide application were summed for each month, and the monthly mean for the 2004-2014 period was considered.

The data were analyzed using descriptive statistics of measures of central tendency (mean) and variability (standard deviation and coefficient of variation). Variographic analysis was applied to verify the existence and quantify the degree of spatial dependence by fitting the theoretical functions to the experimental variogram models, based on the assumption of stationarity of the intrinsic hypothesis, proposed by Vieira, Hatfield, Nielsen, and Biggar (1983), according to equation 1:

$$\gamma = \frac{1}{2N_h} \sum_{i=1}^{N_h} [Z(x_i) - Z(x_i + h)]^2 \qquad (1)$$

where: $\gamma(h)$ = experimental variogram or variogram of samples; $N(h)$ = number of experimental pairs of observations $Z(x_i)$, $Z(x_i + h)$, separated by a vector h; and $Z(x_i)$ and $Z(x_i + h)$ = pairs of data belonging to a distance class. The nugget effect (C_0), plateau ($C_0 + C$), and range (R_0) coefficients were determined from the fitting of the theoretical models to the experimental variograms. Ordinary kriging was used to estimate values at nonsampled sites and to create the isoline maps.

For the descriptive statistics and geostatistical analysis of the data, the program geoMS v. 1.0 was used (CMRP, 2000).

Result and discussion

Table 1 shows the results of the descriptive analysis of the number of hours available to carry out the

application of pesticides in each month for the period 2004-2014.

For these 54 stations, the central tendency represented by the mean indicates that the month of February had less time available for applications (113.32 hours) and that October had more time available (152.74 hours). High variability existed in the ability to carry out the applications, which were intrinsically influenced by the months and geographical location. The highest coefficient of variation (22.20%) was observed in the month of June and the lowest in the months of April and September (16.80%). The data presented normal distribution according to the Kolmogorov-Smirnov test (Table 1).

Table 1. Descriptive measures of hours available for pesticide application in each month in Paraná State, southern Brazil.

Variables	Mean	Standard deviation	Minimum	Maximum	CV (%)	K-S test Max D (p > 0.02)
January	142.74	31.46	56.48	264.18	22.00	0.08
February	113.32	20.79	53.74	197.80	18.30	0.09
March	129.84	28.28	46.65	253.22	21.70	0.08
April	120.25	20.18	54.19	206.07	16.80	0.10
May	125.47	23.10	55.11	205.82	18.40	0.10
June	118.85	26.38	46.59	259.34	22.20	0.09
July	132.03	24.76	57.49	267.64	18.75	0.09
August	131.22	24.18	54.57	285.00	18.43	0.08
September	140.42	23.62	65.58	282.12	16.80	0.07
October	152.74	25.93	77.69	242.31	16.98	0.09
November	152.11	31.55	60.61	272.14	20.74	0.07
December	144.04	29.13	64.53	248.75	20.23	0.09

The variographic analysis (Table 2) revealed that all months presented spatial dependence with the fitting of variograms to the spherical model, with a minimum range of 79,792.82 m (August) and maximum of 179,990.49 m (October). Several studies indicate this model is best adapted to describe the behavior of the variograms of soil attributes and of the different slope gradients and relief shapes (Trangmar, Yost, & Uehara, 1985; Sanchez, Marques Júnior, Pereira, & Souza, 2005; Lima, Oliveira, & Silva, 2012). The variogram shows the measure of the degree of spatial dependence among samples along a specific support (Landim, 2006), that is, the zone of influence or distance around a station to perform the mapping.

The month with the highest spatial dependence index (SDI) was August, with 65.96%. In practice, this indicates that the mean number of hours available for pesticide application is not random in the area, that the structure is spatially dependent, and that these variogram parameters allow the values in unmeasured locations to be estimated using ordinary kriging.

The availability to perform applications was not uniform over time or among stations and varied with geographic location, with the nearest stations presenting more similar results than the more distant stations (Figures 2 and 3). The same pattern was observed among the months, with the highest number of hours available for applying pesticides in all the months occurring in the Pioneer North, Central North, West, Central South, and Southeast geographical regions.

The greatest availability for pesticide application occurred in the months of March and April in the Pioneer North and Central North regions; in the months of May, June, July, and August in the West region; and in the months of September, November, December, and January in the Central South and Southeast regions of Paraná. Therefore, variation is necessary in the application technique and the number of machines to perform the sprays

at the appropriate time for control in the months with less availability and to meet the local demand of each region.

Table 2. Models and parameters of hours available for pesticide application in each month.

Variables	Model	C_0	$C+C_0$	Range (m)	SDI (%)
January	Spherical	3224.71	8957.61	172658.40	64.00
February	Spherical	2652.86	5478.53	159431.53	51.58
March	Spherical	3169.42	8319.76	122280.19	61.90
April	Spherical	5267.18	8778.64	125049.61	40.00
May	Spherical	3585.79	8366.61	87614.40	57.14
June	Spherical	3479.64	8654.36	102651.20	59.79
July	Spherical	3920.37	9025.73	91223.23	56.56
August	Spherical	2713.55	7970.89	79792.82	65.96
September	Spherical	2691.59	6930.75	92626.67	61.16
October	Spherical	4341.82	8857.11	179990.49	50.98
November	Spherical	3256.78	9118.59	143134.09	64.28
December	Spherical	4091.02	9304.64	169824.11	56.03

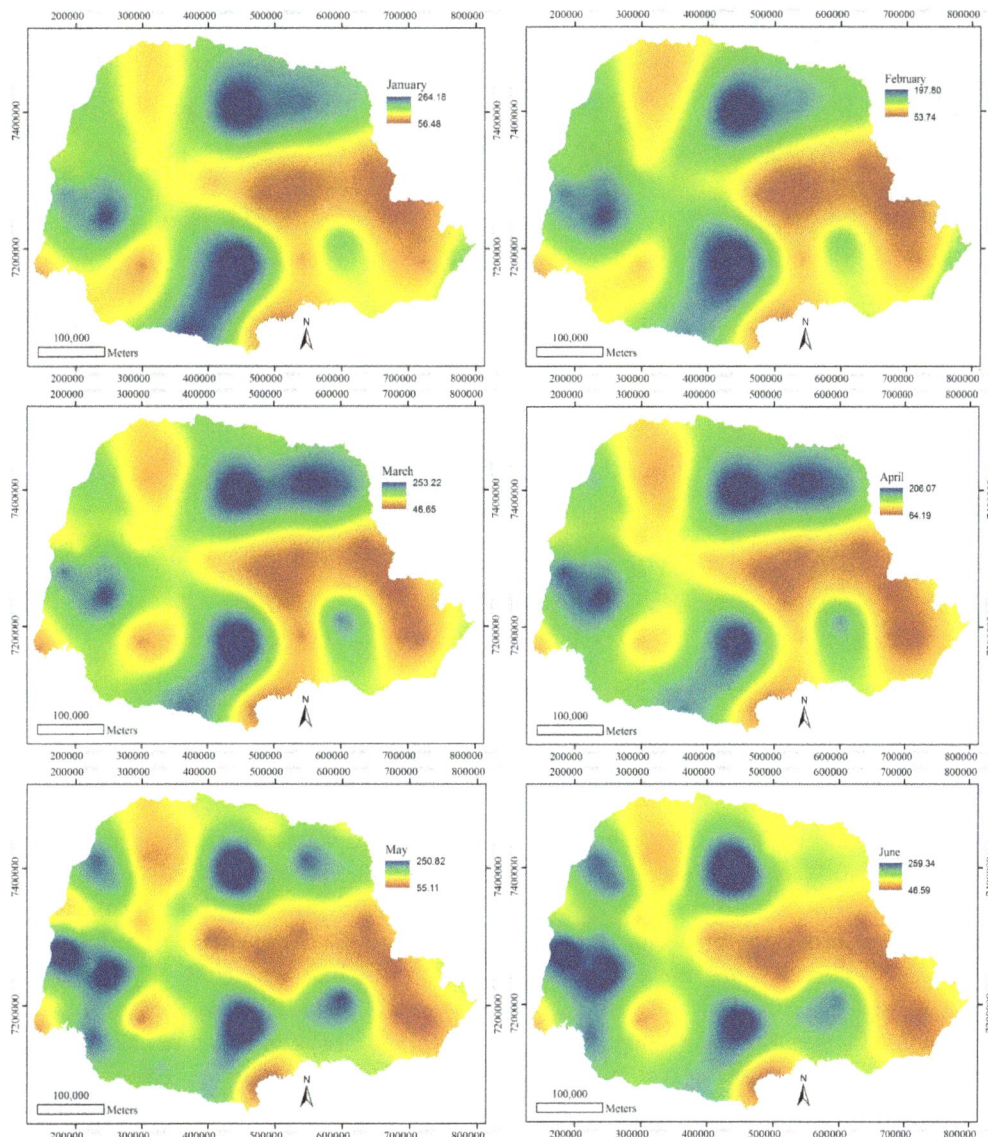

Figure 2. Spatial distribution of hourly availability for pesticide applications in the State of Paraná in the different months of the year.

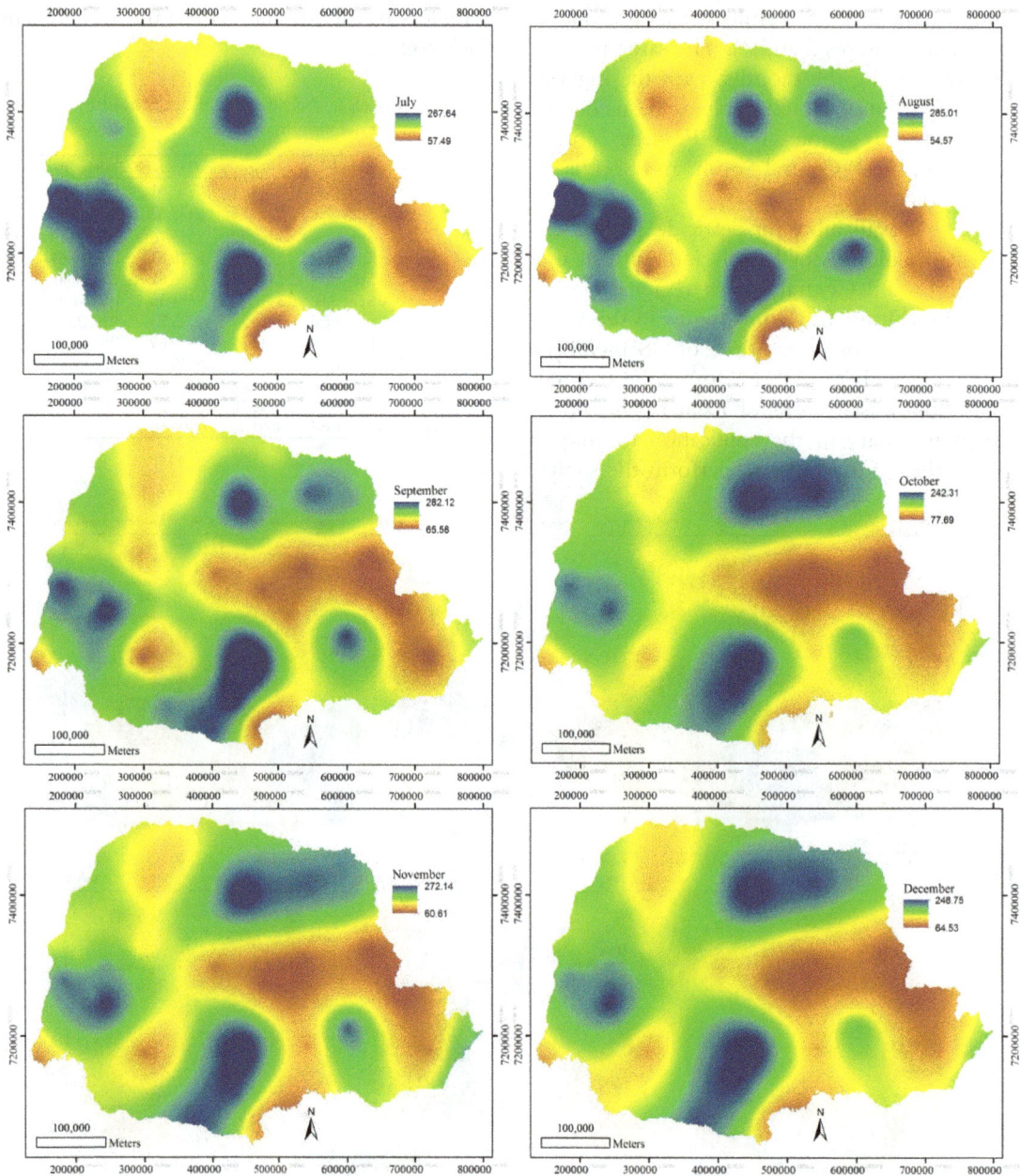

Figure 3. Spatial distribution of hourly availability for pesticide application in the State of Paraná in the different months of the year.

The availability in hours for application varies with the month, and the most appropriate months to apply pesticides are January, September, and November for the main summer crops, such as corn and soybean, and the least appropriate months are February and April, which is when the planting of the main winter crops begins (wheat and oats). The availability of hours for application also depends on the Brazilian region; in the southeast region, in the city of Uberlândia, state of Minas Gerais, a greater availability of times for pesticide application under the appropriate conditions were found for the

months of November, December, January, and February, and less availability of appropriate hours for pesticide application was found in the months of August and September (Cunha et al., 2016).

Due to the continental proportions of Brazil and the size of many agricultural properties in Brazil, however, the ideal time for pesticide spraying, considering the climatic conditions, is very variable in the different production regions (Alvarenga et al., 2014). In this case, the mapping and knowledge of the hourly availability to perform the applications under suitable climatic conditions allows the professionals involved with the application technology to plan ahead and size the fleet of machines and the most appropriate application techniques for each time of application in the month and region to achieve success and reduce the probability of failure when performed under unsuitable climatic conditions due to underestimating the size of the sprayer fleet needed.

The maps express regions with lower and greater restriction in the number of viable hours per month, indicating the need for variation in the pesticide application in the state of Paraná to meet the demand of each region and in the respective months in which spraying activities are conducted. Knowing the spatial variability allows analysis with more accuracy of the regional demands of the state and thus better definition of the type, timing, and movements of machines; an understanding of the technological levels; the selection of drift reduction and safety techniques for each region; and the adjustment of the droplet size and use of adjuvants to reduce the percentage of droplets smaller than 100 μm, which are the spray droplets that can cause environmental damage.

Wind speed showed a stronger correlation with the number of hours available to perform pesticide application; that is, wind speed is the climatic variable that most limited or affected the spatial distribution of availability for pesticide application in the state of Paraná in all months, with lower affects in the month of February (Table 3). Thus, wind speed serves as an alert, so that, in these regions and months, the application techniques and schedules can be adjusted to minimize the effects of very strong wind or absence of wind during spraying. In field experiments under variable climatic conditions, boom height and wind speed were the most decisive factors influencing total spray drift (Arvidsson et al., 2011).

Wind is very important, and pesticide application is not recommended in its absence and in situations of high speed (strong winds). In the absence of wind, very fine droplets remain suspended in the air; in very strong winds, the very fine droplets can be transported away from the target of action and dispersed with other pollutants up to hundreds of kilometers from the application site (Miller & Stoughton 2000). In addition to wind, other variables such as rainfall, temperature, and air humidity are also equally important for pesticide application.

Relative air humidity presented a weaker correlation with the number of hours available to perform the pesticide applications, thus indicating that in the state of Paraná, the relative air humidity does not limit pesticide application, which is more dependent on wind speed, temperature, and rainfall (Table 3).

Table 3. Pearson correlation between the number of hours available for pesticide application (NHAPA) and climatic variables.

NHAPA	Precipitation (mm)	Temperature (°C)	RH (%)	Wind (m s⁻¹)
January	0.33★	0.30★	0.23	0.99★
February	0.27★	0.27	0.12	0.31★
March	0.45★	0.36★	0.16	0.98★
April	0.50★	0.42★	0.24	0.99★
May	0.33★	0.38★	0.23	0.99★
June	0.31★	0.40★	0.19	0.99★
July	0.30★	0.38★	0.12	0.97★
August	0.25	0.27	-0.01	0.92★
September	0.19	0.15	-0.10	0.91★
October	0.24	0.19	-0.05	0.96★
November	0.15	0.16	-0.06	0.98★
December	0.21	0.26	0.14	0.99★

★ Significant correlation ($p < 0.05$).

Conclusion

Spatial dependence exists in the time available to perform pesticide applications in the state of Paraná, with variation occurring as a function of the month of the year.

The Pioneer North, Central North, West, Central South, and Southeast regions of Paraná are the most favorable to perform applications.

Wind speed is the climatic variable that most influences the mapping of the time available for the application of pesticides in the state of Paraná.

The mapping allows the management of spraying practices while taking into consideration the spatial variability of the availability in hours in each month and region of Paraná.

Acknowledgements

We acknowledge the Coordination for the Improvement of Higher Education (Coordenação de Aperfeiçoamento de Pessoal de Nível Superior – CAPES) for granting a scholarship to the first author. We also acknowledge INMET and SIMEPAR for providing the climatic data.

References

Alvarenga, C. B., Teixeira, M. M., Zolnier, S., Cecon, P. R., Siqueira, D. L., Rodriguês, D. E., ..., Rinaldi, P. C. N. (2014). Efeito do déficit de pressão de vapor d'água no ar na pulverização hidropneumática em alvos artificiais. *Bioscience Journal*, *30(1)*, 182-193.

Associação Nacional de Defesa Vegetal [ANDEF]. (2004). *Manual de tecnologia de aplicação*. São Paulo, SP: Linea Creativa.

Arvidsson, T., Bergström, L., & Kreuger, J. (2011). Spray drift as influenced by meteorological and technical factors. *Pesticide Management Science*, *67*(5), 586–598. doi: 10.1002/ps.2114

Caviglione, J. H, Kiihl, L. R. B, Caramori P. H., Oliveira, D., Galdino J., Borrozino, E., ..., Pugsley, L. (2000). *Cartas Climáticas do Estado do Paraná* [CD-ROM]. Londrina, PR: Instituto Agronômico do Paraná.

CMRP. (2000). *GeoMS - Geostatistical Modeling Software. v. 1.0.* Lisboa, PO: IST.

Cunha, J. P. A. R., Pereira, J. N. P., Barbosa, L. A., & Silva, C. R. (2016). Pesticide Application Windows In The Region Of Uberlândia-MG, Brazil. *Bioscience Journal*, *32*(2), 403-411. doi: 10.14393/BJ-v32n2a2016-31920

Landim, P. M. B. (2006). Sobre geoestatística e mapas. *Terra e Didática*, *2*(1), 19-33.

Lima, J. S. S., Oliveira, R. B., & Silva, S. A. (2012). Spatial variability of particle size fractions of an Oxisol cultivated with conilon coffee. *Revista Ceres*, *59*(6), 867-872. doi: 10.1590/S0034-737X2012000600018

Miller, P. C. H., & Stoughton, T. E. (2000). Response of spray drift from aerial application at a forest edge to atmospheric stability. *Agricultural and Forest Meteorology*, *100*(1), 49-58. doi:10.1016/S0168-1923(99)00084-2

Minguela, J. V., & Cunha, J. P. A. R. (2010). *Manual de aplicação de produtos fitossanitários*. Viçosa, MG: Aprenda Fácil.

Raetano, C. G. (2011). Introdução ao estudo da tecnologia de aplicação de produtos fitossanitários. In U. R. Antuniassi, & W. Boller (Ed.), *Tecnologia de aplicação para culturas anuais*. Passo Fundo, RS: Aldeia Norte, Botucatu, SP: FEPAF.

Ruedell, J. (2002). Tecnologia de aplicação de defensivos. *Plantio Direto*, *19*(6), 9-11.

Sanchez, R. B., Marques Júnior, J., Pereira, G. T., & Souza, Z. M. (2005). Variabilidade espacial de propriedades de Latossolo e da produção de café em diferentes superfícies geomórficas. *Revista Brasileira de Engenharia Agrícola e Ambiental*, *9*(4), 489-495. doi: 10.1590/S1415-43662005000400008

Trangmar B. B., Yost, R.S., & Uehara, G. (1985). Application of geostatistics to spatial estudies of soil properties. *Advances in Agronomy*, *38*(2), 45-93. doi: 10.1016/S0065-2113(08)60673-2

Vieira, S. R., Nielsen, D. R., & Biggar, J. W. (1981). Spatial variability of field-measured infiltration rate. *Soil Science Society of American Journal*, *45*(6), 1040-1048. doi: 10.2136/sssaj1981.03615995004500060007x

Vieira, S. R., Hatfield, J. L., Nielsen, D. R., & Biggar, J. W. (1983). Geostatistical theory and application to variability of some agronomical properties. *Hilgardia*, *51*(3), 1-75. doi: 10.3733/hilg.v51n03p075

25

Effect of bean genotypes, insecticides, and natural products on the control of *Bemisia tabaci* (Gennadius) biotype B (Hemiptera: Aleyrodidae) and *Caliothrips phaseoli* (Hood) (Thysanoptera: Thripidae)

Júlio Cesar Janini[1], Arlindo Leal Boiça Júnior[1*], Flávio Gonçalves Jesus[1], Anderson Gonçalves Silva[1], Sérgio Augusto Carbonell[2] and Alisson Fernando Chiorato[2]

[1]*Departamento de Fitossanidade, Faculdade de Ciências Agrárias e Veterinárias de Jaboticabal, Universidade Estadual Paulista "Júlio de Mesquita Filho", Via de Acesso Prof. Paulo Donato Castellane, s/n, 14884-900, Jaboticabal, São Paulo, Brazil.* [2]*Centro de Análise e Pesquisa Tecnológica do Agronegócio dos Grãos e Fibras, Instituto Agronômico de Campinas, Campinas, São Paulo, Brazil. *Author for correspondence. E-mail: aboicajr@fcav.unesp.br*

ABSTRACT. Effect of bean genotypes, insecticides, and natural products on the control of *Bemisia tabaci* (Gennadius) biotype B (Hemiptera: Aleyrodidae) and *Caliothrips phaseoli* (Hood) (Thysanoptera: Thripidae). The influence of bean genotypes associated with neem oil as insecticide was evaluated to control *B. tabaci* (Gennadius) biotype B and *C. phaseoli* (Hood) during the wet season sowing. The experimental design used was the randomized block arrangement in a 4x4x3 factorial scheme, represented by genotypes, neem oil and insecticides respectively, with three replications. The genotypes Carioca, IAC Harmonia, IAC Centauro and Pérola were used. The evaluations were done at 14 and 42 days after seedling emergence, by counting *B. tabaci* biotype B eggs and nymphs and *C. phaseoli* nymphs in the genotypes leaf. Conclusion: The *B. tabaci* biotype B eggs and nymphs number were smaller in IAC Centauro and higher in IAC Harmonia. The tested genotypes were similarly infested by *C. phaseoli*. IAC Centauro and IAC Harmonia genotypes associated with neem oil (highlighting the full dose – 1%) provided lower number of whitefly eggs and thrips nymphs. Neem oil at the full dose also reduced whitefly nymph number. In the tested genotypes the insecticide provided reduction in the number of whitefly eggs and nymphs as well in the thrips nymphs, with increase in the recommend dose.

Keywords: *Bemisia tabaci* biotype B, *Caliothrips phaseoli*, *Phaseoli vulgaris*, resistance of plants.

Introduction

Brazil has important position in the world bean's (*Phaseolus vulgaris* L.) production since it is considered the biggest consumer, being this plant the main vegetal-protein source. The bean is a traditional culture that is getting more and more space in the agribusiness.

It is consumed practically in all Brazilian States, cultivated along the year and its production comes almost 100% from the national lands. Bean can suffer insect attack and other pests that affect production before and post harvest. The estimative of yield losses by pests attack is from 33 to 86% (YOKOYAMA, 1998).

Among several factors that can cause low productivity of beans in Brazil, insects are harmful from seedling to post harvest, where the stocking can be damaged by them (MAGALHÃES; CARVALHO, 1998). There are several pests but whitefly – *Bemisia tabaci* (Gennadius) biotype B (Hemiptera: Aleyrodidae) and thrips – *Caliothrips phaseoli* (Hood) (Thysanoptera:Thripidae) are prominent because they attack the beans' leaves.

Amid the reasons for the high whitefly incidence, the soybean planting field expansion (its preferential host), the longer sowing time, and the successive and phased cultivations due the center pivot irrigation use (VIEIRA et al., 1998) can be cited. Large losses in vegetables, beans, soybeans, peanuts, cotton, and several ornamental plants (LOURENÇÃO; NAGAI, 1994; FRANÇA et al., 1996) are related to this pest attack spread all over the country.

According to Costa and Carvalho (1960) and Yokoyama (1998) the main damage caused by whitefly in the beans is the transmission of golden mosaic virus (GMV). It is a golden and shinning-type mosaic, turning the beans plant color in to an intense and generalized yellow. Causes economic losses that may vary from 30 to 100%, depending on the cultivar, stage of the plant, the vector population, presence of alternative hosts and environmental conditions (FARIA et al., 1996).

Thrips is a polyphagous species and the higher occurrence period is between November and April (GALLO et al., 2002). When the attack is intense, leaves became yellowish and fall, sometimes remain silver dots in leaves and pods (BOIÇA JÚNIOR et al., 2005).

Although Prabhaker et al. (1985) cited that the insect biological and behavioral characteristics as fast development, high fecundity, and big dispersion capacity are factors that increase the resistance to commercial insecticides from different chemical groups (DITTRICH et al., 1990). Torres et al. (2006) checked that among several aqueous extracts tested, the one based in neem oil affected the development of *Plutella xylostella* (L.) in almost all its cycle. Due to the problems caused by insecticides in the agroecosystem, alternative methods to control pest have been studied, as the use of resistant varieties to whitefly (BOIÇA JÚNIOR et al., 2000a).

This study examined the interaction between bean genotypes with plant-extract used as insecticides and chemical insecticides to control *B. tabaci* biotype B and *C. phaseoli* during the wet season.

Material and methods

The experiment was installed and carried out from October 2007 and January 2008 during the wet season in a clay dark-red Oxisoil at Phytosanity Department experimental field, Faculdade de Ciências Agrárias e Veterinárias da Universidade Estadual Paulista (Unesp)-Jaboticabal, São Paulo State (Phytossanity Department Unesp-Jaboticabal, São Paulo State, Brazil).

Plant spacing was 0.50 m between lines, 12 plants per linear meter density. Seedling fertilization was 430 kg ha^{-1}, formula 4-14-8 (NPK).

The statistic design used was the randomized blocks under a 4 x 3 x 3 factorial scheme (genotypes vs. insecticide plants vs. commercial insecticides). The experiment had three replications and thirty-six treatments as follows: Carioca, IAC Harmonia, IAC Centauro and Pérola genotypes, all of them with and without neem oil application (0.0, 0.5, and 1.0%) and thiametoxan 250 WG (0.0, 75, and 150 g ha^{-1}).

Each plot had four lines, five meters length each, summing up 10 m^2 total area and 5 m^2 useful area. Spacing adopted was 0.50 m between lines, sowing fifteen seeds per linear meter, with thinning ten days later, letting twelve plants per linear meter.

Weed control was done chemically pre-planting by incorporation trifluralin 450 EC under 3 L ha^{-1} dose and thirty days after plant emergence a hoed was done.

The treatment applications were done weekly, from seven to fifty-three days after emergence (DAE), using a 40 lb pol^{-2} pressure manual sprayer, discharge of 400 L ha^{-1}, trying to reach mainly under leafs.

Evaluation started at 14 DAE and they were done weekly, up to 42 DAE. In each evaluation ten leafs were collected by plot. Using a stereoscope *B. tabaci* biotype B eggs and nymphs and *C. phaseoli* nymphs were checked. The leaves collected were situated in the middle of the plant because according Rossetto et al. (1974), this is the favorite place where the insect oviposits.

Data related to the average number of *C. phaseoli* nymphs were transformed in $(x+0.5)1/2$. All data were submitted to analysis of variance by F test. If data were significant, averages were compared by Tukey test at 5% probability.

Results and discussion

Analyzing the average number of eggs from *B. tabaci* biotype B among evaluated treatments, we observed significant differences among them at 14 and 28 DAE, with both periods presenting IAC Centauro less infested than IAC Harmonia. Boiça Júnior and Vendramim (1986), working with different genotypes of beans observed that the cycle from egg to adult of *B. tabaci*, was 1.8 times lower in the period of "water" compared to "dry". Regarding the neem oil doses, the full dose (1.0%) reduced the number of eggs, differing from the treatment control. To the insecticides, differences were observed at 35 DAE, where the full dose (150 g ha⁻¹) reduced the whitefly oviposition, differing from the other treatments. It is possible to verify a significant interaction between genotypes vs. insecticide and between neem oil at 14 and 28 DAE.

Table 1. Average number of eggs of *B. tabaci* biotype B by ten leaves in bean genotypes associated or not with neem oil and insecticidal. Jaboticabal, São Paulo State, 2008.

Genotypes (G)	14 days	21 days	28 days	35 days	42 days
	Days after plant emergence¹				
IAC Harmonia	2.01 a	0.99 a	1.62 ab	2.17 a	1.76 a
IAC Centauro	1.22 bc	1.00 a	1.32 b	2.17 a	1.97 a
Pérola	0.97 c	0.87 a	1.61 ab	2.04 a	1.96 a
Carioca	1.43 b	1.03 a	1.96 a	2.34 a	2.09 a
F (G)	14.00**	0.88ᴺˢ	4.04 *	0.44ᴺˢ	0.65ᴺˢ
Neem (N)					
Control	1.61 a	1.08 a	1.86 a	2.51 a	2.30 a
Half Dose	1.51 a	0.88 a	1.60 ab	1.98 a	1.91 ab
Full Dose	1.10 b	0.96 a	1.43 b	2.02 a	1.62 b
F (N)	6.85**	2.55ᴺˢ	3.77 *	3.21ᴺˢ	5.62**
Insecticide (I)					
Control	1.57 a	1.08 a	1.69 a	2.26 ab	2.03 a
Half Dose	1.33 a	0.88 a	1.59 a	2.48 a	1.94 a
Full Dose	1.32 a	0.96 a	1.60 a	1.78 b	1.87 a
F (I)	1.97ᴺˢ	1.80ᴺˢ	0.22ᴺˢ	4.78*	0.31ᴺˢ
F (G x N)	6.72**	0.56ᴺˢ	3.48**	1.25ᴺˢ	1.58ᴺˢ
F (G x I)	3.50**	1.43ᴺˢ	2.37*	0.87ᴺˢ	1.31ᴺˢ
F (I x N)	0.97ᴺˢ	1.14ᴺˢ	0.87ᴺˢ	1.26ᴺˢ	0.60ᴺˢ
VC (%)	43.57	38.84	41.23	45.56	44.21

¹Averages followed by the same letters do not differ statistically by Tukey test at 5% probability. Data transformed in (x + 0.5)¹ᐟ².

Toscano et al. (2002) studying the oviposition of *B. tabaci* biotype B on tomato plants and Campos et al. (2005) in cotton plants observed that this pest prefers ovipositing in plants at younger stages. This is probably because the insect find chemical and morphological composition more favorable due to the plant age (WALKER; PERRING, 1994) and the stimuli involved between the insect and plant (LARA, 1991).

Through the unfolding of genotype vs. neem oil interaction in the evaluation of 14 and 28 DAE (Table 3) and considering the genotype effect into doses, we verified in general that IAC Centauro had less oviposition in the three doses tested related to other genotypes. To neem oil doses effect on the genotypes, the number of whitefly eggs observed was lower when neem oil dose was increased, mainly in the IAC Centauro and IAC Harmonia genotypes. Studying the *B. tabaci* biologic cycle in several bean genotypes Boiça Júnior and Vendramim (1986) concluded that Carioca cultivar was the one that grew under better conditions against *B. tabaci* and the higher incidence months were from November to March.

Table 2. Values of the unfolding analysis of the interaction between genotype versus neem oil and insecticidal versus genotypes, obtained from bean plants, at 14 and 28 days after plant emergence, for the average number of eggs of *B. tabaci* biotype B in ten leaflets. Jaboticabal, São Paulo State, 2008.

14 days after plant emergence (genotypes vs. neem oil)			
Genotypes	Control	Half dose (0.5%)	Full dose (1.0%)	F (G)
IAC Harmonia	2.66 aA	2.36 aA	1.02 bA	18.11★★
IAC Centauro	1.80 aB	1.17 abB	0.70 bB	7.22**
Pérola	0.80 aC	1.02 aB	1.08 aAB	0.51ᴺˢ
Carioca	1.18 aBC	1.50 aB	1.61 aB	1.17ᴺˢ
F (N)	15.59**	8.53**	3.40*	

14 days after plant emergence (genotypes vs. neem oil)			
Genotypes	Control	Half dose (0.5%)	Full dose (1.0%)	F (G)
IAC Harmonia	1.81 aA	1.81 aA	2.41 aA	2.88ᴺˢ
IAC Centauro	1.69 aA	1.02 abB	0.95 bB	3.95*
Pérola	0.88 aB	0.99 aB	1.02 aB	0.12ᴺˢ
Carioca	1.91 aA	1.44 abAB	0.95 bB	5.53**
F (I)	5.21**	3.56*	12.32**	

28 days after plant emergence (genotypes vs. neem oil)			
Genotypes	Control	Half dose (0.5%)	Full dose (1.0%)	F (G)
IAC Harmonia	2.28 aA	1.47 bA	1.12 bB	7.04**
IAC Centauro	1.03 aB	1.72 aA	1.20 aAB	2.59ᴺˢ
Pérola	2.16 aA	1.28 bA	1.39 bAB	4.57*
Carioca	1.97 aA	1.91 aA	1.99 aA	0.03 ᴺˢ
F (N)	6.41**	1.53ᴺˢ	3.08*	

28 days after plant emergence (genotypes vs. neem oil)			
Genotypes	Control	Half dose (0.5%)	Full dose (1.0%)	F (G)
IAC Harmonia	1.21 aB	1.94 aA	1.72 aAB	2.78ᴺˢ
IAC Centauro	1.51 aB	1.30 aA	1.14 aB	0.67ᴺˢ
Pérola	1.99 aB	1.49 aA	1.35 aB	2.28 ᴺˢ
Carioca	2.04 aA	1.64 aA	2.19 aA	1.61 ᴺˢ
F (I)	3.15*	1.44ᴺˢ	4.19**	

¹Averages followed by the same letters do not differ statistically by Tukey test at 5% probability. Data transformed in (x + 0.5)¹ᐟ².

Analyzing genotype vs. insecticides effect at 14 and 28 DAE, it is possible notice in general that IAC Centauro, Pérola and Carioca genotypes when under full and half doses there was a reducing tendency in the eggs number. To the doses into the genotypes effect noticed low influence in the oviposition, suggesting a lack of efficiency of the product against eggs.

Related to *B. tabaci* biotype B nymphs (Table 3), there are significant differences between genotypes at 28 and 35 DAE, when IAC Centauro presented a lower number of nymphs than IAC Harmonia and Carioca.

At 21 DAE checking the neem oil treatments under 0.5 and 1.0% concentration and at 35 DAE under 0.5% there were less infestations of *B. tabaci* biotype B where the infestation averages were 1.01 and 1.04 nymphs to the first evaluation and 1.87 to the second evaluation, being the higher ones the control treatment, 1.83 in the first evaluation and 2.49 in the second evaluation (Table 3).

When insecticide was sprayed (Table 3) was observed at 14 and 42 DAE differences between the evaluated doses, having less nymphs when half and full dose were used. These data evidence that the product used is efficient against nymphs, corroborating with Boiça Júnior et al. (2005) and Boiça Júnior et al. (2006) accounted that adding vegetal oil to insecticides there was good whitefly control in bean, as much in dry as in wet seasons.

By the interactions showed in Table 3 was observed a significant effect to genotypes vs. neem oil. Folding doses effect into genotypes (Table 4) in Control, Carioca was less infested than IAC Harmonia, while the other genotypes presented intermediate number of whitefly nymphs.

Table 3. Average number of nymph of *B. tabaci* biotype B by ten leaves in bean genotypes associated or not with neem oil and insecticidal. Jaboticabal, São Paulo State, 2008.

	Days after plant emergence				
Genotypes (G)	14 days	21 days	28 days	35 days	42 days
IAC Harmonia	1.27 a	1.35 a	1.40 ab	2.33 ab	1.19 a
IAC Centauro	1.42 a	1.32 a	1.31 b	1.77 b	1.59 a
Pérola	1.27 a	1.25 a	1.19 b	2.54 a	1.36 a
Carioca	1.32 a	1.39 a	1.94 a	2.16 ab	1.20 a
F (G)	0.37^{NS}	0.16^{NS}	4.51^{**}	4.43^{**}	2.75^{NS}
Neem (N)					
Control	1.38 a	1.83 a	1.60 a	2.49 a	1.33 a
Half Dose	1.27 a	1.01 b	1.26 a	2.25 ab	1.42 a
Full dose	1.31 a	1.14 b	1.52 a	1.87 b	1.26 a
F (N)	0.37^{NS}	11.56^{**}	1.75^{NS}	5.39^{**}	0.75^{NS}
Insecticide (I)					
Control	1.86 a	1.23 a	1.35 a	2.25 a	1.59 a
Half Dose	1.13 b	1.41 a	1.33 a	2.32 a	1.25 b
Full dose	0.98 b	1.34 a	1.71 a	2.03 a	1.17 b
F (I)	23.37^{**}	0.51^{NS}	2.51^{NS}	1.24^{NS}	5.26^{**}
F (G x N)	5.18^{**}	0.28^{NS}	2.13^{NS}	0.71^{NS}	0.88^{NS}
F (G x I)	0.74^{NS}	0.69^{NS}	1.52^{NS}	1.00^{NS}	1.56^{NS}
F (I x N)	2.35^{NS}	0.90^{NS}	2.48^{NS}	0.56^{NS}	0.12^{NS}
VC (%)	43.98	58.77	54.82	36.48	43.69

[1]Averages followed by the same low case letter in the line and capital letter in the column do not differ statistically by Tukey test at 5% probability. Data transformed in $(x + 0.5)^{1/2}$.

Table 4. Values of the unfolding analysis of the interaction between genotype versus neem oil, obtained from bean plants, at 35 days after plant emergence, for the average number of nymphs of *B. tabaci* in ten leaflets. Jaboticabal, São Paulo State, 2008.

Genotypes	Control	Half dose (0.5%)	Full dose (1.0%)	F (G)
IAC Harmonia	1.74 aA	0.93 bB	1.15 abA	4.63^{*}
IAC Centauro	1.54 aAB	1.21 aB	1.51 aA	0.89^{NS}
Pérola	1.28 aAB	0.97 aB	1.56 aA	2.32^{NS}
Carioca	0.98 bB	1.95 aA	1.02 bA	8.06^{**}
F (N)	2.85^{*}	5.97^{**}	1.91^{NS}	

[1]Averages followed by the same low case letter in the line and capital letter in the column do not differ statistically by Tukey test at 5% probability. Data transformed in $(x + 0.5)^{1/2}$.

Boiça Júnior et al. (2000a) evaluating the fertilization and insecticide effect in beans to control *B. tabaci* biotype B observed that IAPAR MD-806 and IAPAR MD-808 genotypes were less infected by whitefly nymphs from 21 to 35 DAE. Boiça Junior et al. (2005) evaluating vegetal oils associated to insecticides interaction observed that this combination was positive to control the pest and reduced the population from 7 to 42 DAE.

Lemos et al. (2003) evaluating the beans genotypes susceptibility to GMV verified an increase in the pest population in the genotypes without chemical treatment, and an increase of 87.5% when compared with the chemically treated genotypes. Boiça Júnior et al. (2000b) also obtained reduction in the pest population and the illness incidence using insecticide in the planting line.

The average number of *C. phaseoli* nymphs (Table 5) was similar in all evaluated genotypes in all treatments, suggesting all of them are susceptible to the insect. To nymphs average in the treatments with different neem oil concentrations, a significant difference was observed only at 35 DAE being the highest value observed in the control treatment, differing from the other treatments, evidencing the neem oil action against the insect.

The thiametoxan insecticide provided a reduction in the thrips nymphs number in all the evaluations (Table 5), increasing the efficiency in the full dose treatment (150 g ha⁻¹) characterizing a good control when related to the control treatment.

Among the interactions it is possible to register significance between genotypes vs. neem oil at 14 DAE and genotypes vs. insecticide at 14 and 42 DAE (Table 5). By the genotypes tested vs. neem oil interaction unfolding at 14 DAE (Table 5) we recorded the neem oil significant effect on the tested genotypes where the lower values of thrips nymphs number occurred at half and full doses to IAC Harmonia and IAC Centauro. To the neem oil effect on the genotypes of the control, we observed lower pest attack in the Carioca Genotype.

Table 5. Average number of nymphs of *C. phaseoli* by ten leaflets in genotypes of bean or not associated with oil and insecticide nin. Jaboticabal, São Paulo State, 2008.

	Days after plant emerging				
Genotypes (G)	14 days	21 days	28 days	35 days	42 days
IAC Harmonia	1.24 a	1.71 a	0.93 a	1.29 a	1.22 a
IAC Centauro	1.28 a	1.79 a	1.22 a	1.55 a	1.56 a
Pérola	1.34 a	1.93 a	1.05 a	1.71 a	1.27 a
Carioca	1.42 a	1.79 a	1.29 a	1.83 a	1.42 a
F (G)	0.83^{NS}	0.38^{NS}	1.61^{NS}	2.20^{NS}	2.40^{NS}
Neem (N)					
Control	1.83 a	1.80 a	1.15 a	1.89 a	1.44 a
Half Dose	1.40 a	1.84 a	1.21 a	1.43 b	1.36 a
Full dose	1.18 a	1.79 a	1.01 a	1.46 b	1.31 a
F (N)	2.47^{NS}	0.01^{NS}	0.88^{NS}	3.66^{*}	0.63^{NS}
Insecticide (I)					
Control	1.66 a	2.31 a	1.61 a	2.13 a	1.62 a
Half Dose	1.31 b	1.60 b	0.90 b	1.38 b	1.15 b
Full dose	1.00 c	1.50 b	0.85 b	1.27 b	1.33 ab
F (I)	18.51^{**}	11.66^{**}	14.96^{**}	11.77^{**}	7.69^{**}
F (G x N)	3.77^{**}	1.18^{NS}	0.58^{NS}	1.09^{NS}	1.18^{NS}
F (G x I)	5.54^{**}	0.33^{NS}	1.17^{NS}	0.72^{NS}	4.82^{**}
F (I x N)	0.26^{NS}	0.17^{NS}	1.99^{NS}	0.80^{NS}	1.60^{NS}
VC (%)	34.78	43.11	58.19	50.91	37.39

[1]Averages followed by the same low case letter in the line and capital letter in the column do not differ statistically by Tukey test at 5% probability. Data transformed in $(x + 0.5)^{1/2}$.

To the interactions genotypes vs. insecticides at 14 and 42 DAE (Table 6) we observed that in general the more the insecticide dose, the smaller the insect numbers in the tested genotypes. Between these and the control, again, the lower index was verified to genotype Carioca.

Table 6. Values of the unfolding analysis of the interaction between genotypes versus neem oil and insecticidal versus genotypes, obtained from bean plants, at 14 and 42 days after plant emergence, for the average number of nymphs of *C. phaseoli* in ten leaflets. Jaboticabal, São Paulo State, 2008.

14 days after plant emergence (genotypes vs. neem oil)				
Genotypes	Control	Half dose (0.5%)	Full dose (1.0%)	F (G)
IAC Harmonia	1.43 aAB	1.39 abA	0.90 bB	3.78*
IAC Centauro	1.72 aA	1.11 bA	0.99 bB	6.53**
Pérola	1.22 aAB	1.41 aA	1.40 aB	0.48^NS
Carioca	1.15 bB	1.68 aA	1.44 abA	3.00*
F (N)	2.82*	2.26^NS	3.30*	

14 days after plant emergence (genotypes vs. insecticides)				
Genotypes	Control	Half dose (75 g ha⁻¹)	Full dose (150 g ha⁻¹)	F (G)
IAC Harmonia	1.43 aA	1.28 aB	1.01 aA	1.92^NS
IAC Centauro	1.79 aA	0.87 bB	1.17 bA	9.16**
Pérola	1.45 aA	1.88 aA	0.70 bA	15.01**
Carioca	1.95 aA	1.22 bB	1.10 bA	9.04**
F (I)	2.77^NS	7.37**	1.78^NS	

42 days after plant emergence (genotypes vs. insecticides)				
Genotypes	Control	Half dose (75 g ha⁻¹)	Full dose (150 g ha⁻¹)	F (G)
IAC Harmonia	1.47 aB	1.14 aA	1.05 aB	1.70^NS
IAC Centauro	2.22 aA	1.38 bA	1.09 bB	11.69**
Pérola	1.47 aB	1.07 aA	1.27 aB	1.38^NS
Carioca	1.32 bB	1.01 bA	1.93 aA	7.38**
F (I)	5.51**	0.89^NS	5.64**	

¹Averages followed by the same low case letter in the line and capital letter in the column do not differ statistically by Tukey test, at 5% probability. Data transformed in (x + 0.5)^(1/2).

Boiça Júnior et al. (2008) studying the interaction of genotypes and insecticides to control *C. phaseoli* in the wet season verified satisfactory index in the pest control at 25, 32, 39 and 46 days after plant emergence in genotypes sprayed with insecticides, resulting in low average number of *C. phaseoli* nymphs.

Conclusion

The number of eggs and nymphs of *B. tabaci* biotype B was lower in IAC Centauro and higher in IAC Harmonia. The genotypes if are on the experiment were the some infestation for *C. phaseoli*. IAC Centauro and IAC Harmonia associated to neem oil, highlighting the ful dose (1.0%) provided lower whitefly and thrips eggs' number being that the last one acted reducing the whitefly nymphs' number. In the tested genotypes the insecticide provided reduction in the whitefly eggs and nymphs thrips nymphs' number, with increment in the recommended dose.

Acknowledgements

To the Conselho Nacional de Desenvolvimento Científico e Tecnológico–CNPq by the productivity and research scholarship to the second author.

References

BOIÇA JÚNIOR, A. L.; VENDRAMIM, J. D. Desenvolvimento de *Bemisia tabaci* em genótipos de feijão. **Anais da Sociedade Entomológica do Brasil**, v. 15, n. 2, p. 231-238, 1986.

BOIÇA JÚNIOR, A. L.; SANTOS, T. M.; MOÇOUÇAH, M. J. Adubação e inseticidas no controle de *Empoasca kraemeri* e *Bemisia tabaci*, em cultivares de feijoeiro semeados no inverno. **Scientia Agricola**, v. 57, n. 4, p. 635-641, 2000a.

BOIÇA JÚNIOR, A. L.; MUÇOUÇAH, M. J.; SANTOS, T. M.; BAUMGARTNER, J. G. Efeito de cultivares de fejoeiro, adubação e inseticidas sobre *Empoasca kraemeri* Ross & Moore, 1957 e *Bemisia tabaci* (Gennadius, 1889). **Acta Scientiarum. Agronomy**, v. 22, n. 4, p. 955-961, 2000b.

BOIÇA JÚNIOR, A. L.; ANGELINI A. L. M. R.; COSTA, G. M.; BARBOSA, J. C. Efeito do uso de óleos vegetais, associados ou não a inseticida, no controle de *Bemisia tabaci* (Genn.) e *Thrips tabaci* (Lind.), em feijoeiro, na época "das secas". **Boletin del Sanidad Vegetal Plagas**, v. 31, n. 32, p. 449-458, 2005.

BOIÇA JÚNIOR, A. L.; ANGELINI, M. R.; COSTA, G. M. Efeito do uso de óleos vegetais, associados ou não a inseticida, na eficácia de controle de *Bemisia tabaci* (Gennadius, 1889) e *Thrips tabaci* (Lind, 1888), em feijoeiro comum, na época "de inverno". **Bioscience Journal**, v. 22, n. 3, p. 23-31, 2006.

BOIÇA JÚNIOR, A. L.; JESUS, F. G.; CARBONELL, S. A. M.; PITTA, R. M.; CHIORATO, A. F. Efeito de genótipos de *Phaseolus vulgaris* associados ou não a inseticidas, no controle de *Bemisia tabaci* (Gennadius) biótipo B (Hemiptera: Aleyrodidae) e *Caliothrips phaseoli* (Hood) (Thysanoptera: Thripidae). **Boletín de Sanidad Vegetal Plagas**, v. 34, n. 1, p. 27-35, 2008.

CAMPOS, Z. R.; BOIÇA JÚNIOR, A. L.; LOURENÇÃO, A. L.; CAMPOS, A. R. Fatores que afetam a oviposition de *Bemisia tabaci* (Genn.) biótipo B (Hemiptera: Aleyrodidae) em algodoeiro. **Neotropical Entomology**, v.34, n. 5, p. 823-827, 2005.

COSTA, A. S.; CARVALHO, A. M. B. Comparative studies between abutilon and euphorbia mosaic viruses. **Phytopathology**, v. 38, n. 2, p. 129-152, 1960.

DITTRICH, V.; ERNST, G. H.; RUESCH, O.; UK, S. Resistance mechanisms in Guatemala, and Nicaragua. **Journal of Economic Entomology**, v. 83, n. 5, p. 1665-1670, 1990.

FARIA, J. C.; ANJOS, J. R. N.; COSTA, A. F.; SPERÂNCIO, C. A.; COSTA, C. L. Doenças causadas por vírus e seu controle. In: ARAUJO, R. S.; RAVA, C. A.; STONE, L. F.; ZIMMERMANN, M. J. O. (Coord.). **Cultura do feijoeiro comum no Brasil**. Piracicaba: Potafos, 1996. p. 731-760.

FRANÇA, F. H.; VILLAS BOAS, G. L.; CASTELO BRANCO, M. Ocorrência de *Bemisia argentifolii* Bellows & Perring (Homoptera, Aleyrodidae) no Distrito Federal. **Anais da Sociedade Entomológica do Brasil**, v. 25, n. 2, p. 369-372, 1996.

GALLO, D.; NAKANO, O.; SILVEIRA NETO, S.; BAPTISTA, G. C.; BERTI FILHO, E.; PARRA, J. R. P.; ZUCCHI, R. A.; ALVES, S. B.; VENDRAMIM, J. D.; MARCHINI, L. C.; LOPES, J. R. S.; OMOTO, S. **Entomologia agrícola**. Piracicaba: Fealq, 2002.

LARA, F. M. **Princípios da resistência de plantas a insetos**. São Paulo: Ícone, 1991.

LEMOS, L. B.; FORNASIERI FILHO, D.; SILVA, T. R. B.; SORATO, R. P. Suscetibilidade de genótipos de feijão ao vírus do mosaico dourado. **Pesquisa Agropecuária Brasileira**, v. 38, n. 5, p. 521-528, 2003.

LOURENÇÃO, A. L.; NAGAI, H. Surtos populacionais de *Bemisia tabaci* no estado de São Paulo. **Bragantia**, v. 53, n. 1, p. 53-59, 1994.

MAGALHÃES, B. P.; CARVALHO, S. M. Insetos associados à cultura. In: ZIMMERMAN, M. J. O.; ROCHA, M.; YAMADA, T. (Ed.). **Cultura do feijoeiro**: fatores que afetam a produtividade. Piracicaba: Potafós, 1998. p. 573-589.

PRABHAKER, N.; COUDRIET, D. L.; MEYERDIRK, D. E. Insecticide resistance in the sweetpotato whitefly, *Bemisia tabaci* (Homoptera: Aleyrodidae). **Journal of Economic Entomology**, v. 78, n. 4, p. 748-752, 1985.

ROSSETTO, C. J.; SANTIS, L.; PARADELA FILHO, O.

Z.; POMPEU, A. S. Espécies de tripes coletadas em cultura de feijoeiro. **Bragantia**, v. 33, n. 15, p. 9-14, 1974.

TORRES, A. L; BOIÇA JÚNIOR, A. L.; MEDEIROS, C. A. M.; BARROS, R. Efeito de extratos aquosos de *Azadirachta indica, melia azedarach* e *Aspidosperma pyrifolium* no desenvolvimento e oviposition de *Plutella xylostella*. **Bragantia**, v. 64, n. 3, p. 227-232, 2006.

TOSCANO, L. C.; BOIÇA JÚNIOR, A. L.; MARUYAMA, W. I. Fatores oviposition de *Bemisia tabaci* (Genn.) biótipo B (Hemiptera: Aleyrodidae) em tomateiro. **Neotropical Entomology**, v. 31, n. 6, p. 631-634, 2002.

VIEIRA, C.; PAULA JÚNIOR, T. J.; BORÉM, A. **Feijão**: aspectos gerais e cultura. Viçosa: UFV, 1998.

WALKER, G. P.; PERRING, T. M. Feeding and oviposition behavior of whiteflies (Homoptera: Aleyrodidae) interpreted from AC electronic feeding monitor waveforms. **Annals of the Entomological Society of American**, v. 18, n. 3, p. 363-374, 1994.

YOKOYAMA, M. Pragas. In: VIEIRA, C.; PAULA JÚNIOR, T. J.; BORÉM, A. (Ed.). **Feijão**: aspectos gerais e cultura no Estado de Minas. Viçosa: Universidade Federal de Viçosa, 1998. p. 357-374.

26

Carryover of tembotrione and atrazine affects yield and quality of potato tubers

Marcelo Rodrigues Reis[1], Leonardo Ângelo Aquino[1], Christiane Augusta Diniz Melo[1*], Daniel Valadão Silva[2] and Roque Carvalho Dias[3]

[1]Universidade Federal de Viçosa, Campus de Rio Paranaíba, Rodovia MG-230, Km 7, 38810-000, Rio Paranaíba, Minas Gerais, Brazil. [2]Departamento de Fitotecnia, Universidade Federal Rural do Semi-árido, Mossoró, Rio Grande do Norte, Brazil. [3]Universidade Estadual Paulista "Júlio de Mesquita Filho", Faculdade de Ciências Agronômicas, Botucatu, São Paulo, Brazil. *Author for correspondence. E-mail: chrisadinizmelo@yahoo.com.br

ABSTRACT. Crop rotation improves potato (*Solanum tuberosum* L.) crops; however, in some cases, it can have some negative effects due to the herbicides previously used in crops under rotation. These effects are the decline in the yield and the appearance of physiological disorders, including cracked tubers, which impair the quality and economic value of the potato. Two field experiments were performed with applications of tembotrione and atrazine, alone or in combination. In the Rio Paranaiba (Minas Gerais State) area, in clay soil, the herbicides were applied at post-emergence in corn (*Zea mays* L.). After corn harvest and soil preparation, the potatoes were planted. In the Serra do Salitre (Minas Gerais State) area, in the medium texture soil, the herbicides were applied and incorporated into the soil, and the potatoes were planted the day after. No injuries were found in the shoots. Decline in the potato yield was found only in the medium texture soil (Serra do Salitre). Atrazine did not affect the quality of potato tubers. In the two areas, tembotrione promoted cracks in "Atlantic" potato tubers.

Keywords: Atlantic; herbicide; physiological disorder; *Solanum tuberosum*.

Introduction

On the world stage, potato stands out among vegetables due to having the highest consumption and the largest quantity produced - 381 million metric tons in 2014 (FAO, 2017). Crop rotation is paramount for achieving the largest production and best quality of the potato (Campiglia, Paolini, Colla, & Mancinelli, 2009), mainly for interrupting life cycles of pests and diseases (Keiser Häberli & Stamp, 2012) and improving soil quality (Askari & Holden, 2015). However, the herbicides used in rotation crops such as corn and wheat can persist in the soil and impair growth and development of the potato plant.

Soils contaminated with herbicide residues reduce the yield of potato crops according to the cultivar used (Novo & Miranda Filho, 2006) due to the phenomenon known as herbicide carryover. Physiological disorders and the decline in the yield were observed due to sulfonylureas and imidazolinones (Eberlein & Guttieri, 1994; Eberlein, Westra, Harderlie, Withmore, & Guttieri, 1997; Novo & Miranda Filho, 2006), to auxin mimics

(Wall, 1994) and to glyphosate (Hutchinson, Felix, & Boydston, 2014), even at very low concentrations in the soil (for example, nicosulfuron 0.15 mg ha^{-1}) or due to treatment drift of areas adjacent to the potato crops.

Overall, the injuries caused by herbicide carryover are several patterns of chlorosis according to the herbicide, reductions in the growth and in the yield. In addition, these injuries can occur in tubers, causing physiological disorders such as multiple and deep cracks, tuber folding, tuber spiral, numerous side tubers connected to a single tuber or tuber chain (Eberlein et al., 1997; Thornton & Eberlein, 2001). The quality of tubers for fresh consumption or for industrial processing is reduced by the occurrence of any such physiological disorders.

Some Brazilian potato growers have reported that the tembotrione (2- {2-chloro-4-mesyl-3-[(2, 2, 2-trifluoroethoxy) methyl] benzoyl} cyclohexane-1, 3-dione) used in corn promotes physiological disorders in potatoes grown in succession, such as cracks in the tubers (personal communication). Those reports are substantiated by the 10-month restriction period information for the cultivation of potatoes after application of the product. Such information is included in the label for tembotrione (Laudis®), which is used in the United States of America (Bayer CropScience, 2015a). However, on the label of the tembotrione used in Brazil (Soberan®), Spain and Belgium (Laudis®), there is no information about a restriction period for the potato crop after using the product (Bayer CropScience, 2015 b; 2015c; 2015d).

Tembotrione is a 4-hydroxyphenylpyruvate dioxygenase (HPPD) - inhibitor herbicide that belongs to the class of triketones, first registered in 2007. This herbicide is highly mobile in the soil (PPDB, 2017a), pKa = 3.17, weak acid characteristic, $t_{1/2}$ soil = 10 -14 days under laboratory conditions and tends to remain more stable in soils with higher pH (Barchanska, Kluza, Krajczewska, & Maj, 2016). Little is known about the behaviour of tembotrione in soil under field conditions and its effects on succeeding crops.

Atrazine (6-chloro-N2-ethyl-N4-isopropyl-1, 3, 5-triazine-2, 4-diamine) is also a widely used herbicide in maize crop that has high mobility in soil, pka = 1.7, characterizing a very weak base and $t_{1/2}$ soil = 75 days under laboratory conditions (PPDB, 2017b). Atrazine carryover potential associated with other herbicides was investigated in cabbage and beet (Soltani, Sikkema, & Robinson, 2005), carrot (Bontempo et al., 2016), potato, broccoli, onion and cucumber (Robinson, 2008). According to Robinson (2008) the atrazine

accentuates the toxic effects caused by the mesotrione carryover in the soil on several vegetable crops. In a study completed in Brazilian soils, under tropical conditions, a reduction of the total yield of carrot was observed, grown in succession to corn, caused by the carryover of atrazine + tembotrione (Bontempo et al., 2016). This fact reinforces the need to also investigate the effects of atrazine associated with tembotrione in potato cultivation.

There may be considerable time from the start to the manifestation of physiological disorders, making it difficult to identify the causative factor and the time of initiation. Some disorders become an issue close to or during the harvest (Hiller, Koller, & Thorton, 1985). The identification of the causative factor becomes more complicated when the suspect factor may be present in the soil even before cultivation of potatoes, such as, for example, with herbicide carryover.

Thus, the objective of this study was to evaluate the carryover effects of tembotrione and atrazine on the quality and yield of potato tubers under field conditions.

Material and methods

Experimental details

Two experiments were evaluated, one in the summer of 2013/14 in Rio Paranaíba (19°12'29" S, 46°07'57" W and 1.136 m above sea level) and the other in the summer of 2014/15 in Serra do Salitre (19°6'41" S, 46°41'23" W and 1.203 m above sea level), both municipalities of Minas Gerais State, Brazil. In the area of Rio Parnaíba, the soil is classified as an Oxisol containing 42% clay, 51% silt and 7% sand, with a cation exchange capacity (CEC) of 5.79 cmol$_c$ dm^{-3}, 2.9% organic matter and pH of 5.5. In Serra do Salitre, the potatoes were grown in an Oxisol of medium texture with 19.4% clay, 2.7% silt and 77.9% sand, CEC of 5.79 cmol$_c$ dm^{-3}, 2.25% organic matter and pH of 6.2. The medium texture soil in Serra do Salitre was chosen because of its reduced sorption of herbicides that would enhance the phenomenon under study.

In the two experimental areas, herbicide treatments were applied in late spring, the season for corn planting in Brazil. The climatic data (rainfall, maximum and minimum temperatures) were collected daily in both areas (Figure 1A and B).

In both areas, the potato cultivar "Atlantic" was planted in a spacing of 0.87 m between rows and 0.22 m between plants. This cultivar was selected for being the most cultivated cultivar in the major potato-producing regions of Brazil. Cultural practices of fertilization, irrigation and pest management followed the standard management of potato-producing areas.

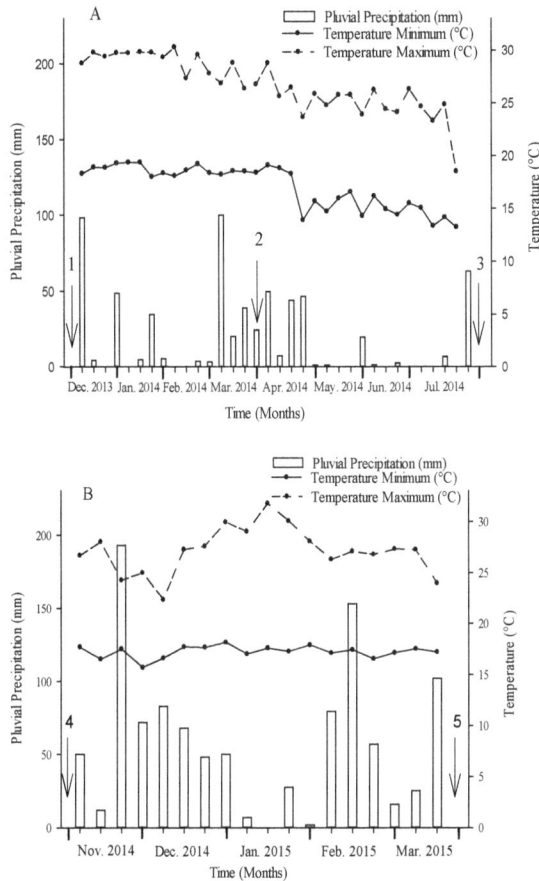

Figure 1. Pluvial precipitation (mm), maximal and minimal air temperature (°C) over experimental period in the experimental area of Rio Paranaíba (A) and Serra do Salitre (B). 1 - Herbicide application; 2 - potatoes planting; 3 - harvest; 4 - herbicide application and incorporation; and 5 - harvest.

Experimental design

In Rio Paranaíba, the treatments were tembotrione (50.4 and 100.8 g ha^{-1}), atrazine (2,000 g ha^{-1}), atrazine + tembotrione (2,000 + 100.8 g ha^{-1}) and a control with no application of herbicides that was manually weeded. The doses of herbicides are those recommended for corn on the labels of the products sold in Brazil. Treatments were applied at post-emergence in corn at the V3 (three fully expanded leaves) stage on December 12th, 2013. After corn harvest and soil preparation in the 0 - 40 cm layer, the potatoes were planted on March 28th, 2014.

In Serra do Salitre, the treatments were tembotrione (33.6 and 50.4 g ha^{-1}), atrazine (2,000 ha^{-1}), atrazine + tembotrione (2,000 + 50.4 g ha^{-1}) and a control with no application of herbicide that

was manually weeded. The herbicides were applied and incorporated into the soil layer of 0-40 cm on November 9th, 2014, and the potatoes were planted the day after to ensure greater herbicide-tuber contact. Half of the dose of tembotrione in the Rio Paranaíba area was used due to the higher bioavailability of the product in the medium texture soil, which could cause severe injury resulting in the death of plants. A sprayer pressurized with CO_2 operating at 200 kPa and spray volume of 200 L ha^{-1} were used in all treatments.

A randomized block experimental design with four replications was used in both areas. Each plot consisted of four potato rows with 8 m in length (27.8 m^2), and the useful area for evaluation consisted of two central rows of 6 m in length (10.5 m^2).

Evaluations

The injuries in the shoots of the potato plants were evaluated on days 15 and 30 after sprouting, when scores from 0 (no symptoms in the shoot) to 100% (total plant death) were assigned. The presence of cracks in tubers was evaluated in 20-day intervals on two potato plants by plot.

Harvest was carried out on July 18th, 2014 in Rio Paranaíba and on March 14th, 2015 in Serra do Salitre, fifteen days after the chemical desiccation of the shoots with paraquat (1,1-dimethyl-4,4'-bipyridinium dichloride) (400 g ha^{-1}). Tuber yield and percentage of tubers with cracks were evaluated.

Statistical analysis

Data were submitted to ANOVA ($p < 0.05$) using the SISVAR software for each experiment. The data were submitted to Tukey's test ($p < 0.05$).

Results

No symptoms caused by the herbicides could be seen on the shoots in the two experimental areas (data not shown).

In the Rio Paranaíba area, the potato tuber yield was not affected by tembotrione or by atrazine in the soil (Table 1).

Tembotrione promoted cracks in the tubers in both areas. The percentage of cracked tubers varied according to dose applied. Atrazine alone did not cause cracks in the tubers in the two areas (Table 1). In the area of Serra do Salitre, total tuber yield was reduced when tembotrione was applied in combination with atrazine or alone at both evaluated doses (Table 1).

Table 1. Total yield and yield of cracked potato tubers grown in Rio Paranaíba (Minas Gerais State) in clay soil and in the medium texture soil in Serra do Salitre (Minas Gerais State) with tembotrione and atrazine applications.

Treatments	Yield potato tubers (t ha⁻¹)		% C/T[1]
	Total	Cracked	
		Rio Paranaíba	
Control	35.4 a [2]	0 c	0.0
Atrazine (2,000 g ha⁻¹)	35.3 a	0 c	0.0
Tembotrione (50.4 g ha⁻¹)	34.9 a	2.0 b	5.7
Tembotrione (100.8 g ha⁻¹)	34.4 a	4.2 a	12.2
Atraz. + Temb. (2,000 + 100.8 g ha⁻¹)	36.0 a	2.4 ab	6.6
Variation Coefficient (%)	6.3	48.3	
		Serra do Salitre	
Control	33.8 a	0.0 c	0.0
Atrazine (2,000 g ha⁻¹)	34.0 a	0.0 c	0.0
Tembotrione (33.6 g ha⁻¹)	27.5 b	0.1 a	0.3
Tembotrione (50.4 g ha⁻¹)	26.4 bc	0.1 a	0.3
Atraz. + Temb. (2,000 + 50.4 g ha⁻¹)	22.8 c	0.03 b	0.1
Variation Coefficient (%)	5.4	22.1	

[1] % C/T: Mass percentage of cracked tubers in relation to the total tuber mass. [2] Means followed by the same letter do not differ by Tukey's test (p > 0.05).

The percentage of tubers with cracks was higher when tembotrione was applied in the medium texture soil (Serra do Salitre), and a lower percentage was found in the tembotrione and atrazine mixture (Table 1). In accordance with the results from Rio Paranaíba, atrazine did not cause cracks in the potatoes when applied alone.

Discussion

No symptoms were found in the shoots of potato plants planted in Rio Paranaíba and Serra do Salitre. Potato plants grown one year after application of atrazine (1,120 g ha⁻¹) and mesotrione (2-[4-(methylsulfonyl)-2-nitrobenzoyl]-1,3-cyclohexanedione) (280 g ha⁻¹) did not show injuries in the shoot (Robinson, 2008). Mesotrione is also an HPPD inhibitor and recommended for weed control in corn.

The reduction in potato yield in the presence of tembotrione only in the Serra do Salitre area can be attributed to the medium textured soil (19.4% clay), to the one-day interval from application to planting of potatoes, with a larger absorption of the herbicide by the plant and to the stress caused by the herbicide in the plant associated with higher maximum and minimum temperatures (Figure 1B) regarding the area of Rio Paranaíba (Figure 1A).

Atrazine alone did not reduce yield nor promote cracks in tubers in the two evaluated areas (Table 1). Robinson (2008) did not observe reduction in the yield or physiological disorders in potato tubers caused by the application of atrazine (1,120 g ha⁻¹). This result reinforces the observation that the reductions in the yields in Rio Paranaíba and Serra do Salitre and the increase in the percentage of cracked potatoes in Rio Paranaíba and Serra do

Salitre resulting from the application of the atrazine + tembotrione mixture had been caused by tembotrione.

Cracks in tubers of potatoes are associated with the presence of tembotrione in the soil, being more intensified in the area of Rio Paranaíba (Table 1), where tembotrione (100.8 g ha⁻¹) caused cracks in 11.4% of the total tubers. Cracked tubers are devalued by the industry due to low or no income from processing and by the consumer due to the poor appearance (Hiller et al., 1985).

The crack pattern observed was three branches from a single point regardless of the size of the tuber, with variable width, length and depth of cracks (Figure 2). This phenomenon has been associated with physiological disorders caused by herbicide drifting or in the soil (Eberlein et al., 1997), however, with no physiological explanations for the disorder. The identification of the causative factor of the physiological disorder may be induced or initiated at the beginning of the cycle and manifests only near or during harvest (Hiller et al., 1985). In experiments conducted in this work, no symptoms of crack initiation or induction in tubers of some plants monitored over the potato cycle were seen (data not shown).

There are reports associating the cracking of tubers to the presence of ALS-inhibiting herbicides (acetolactate synthase) such as nicosulfuron, metsulfuron and tribenuron, EPSPs-inhibiting herbicide (5-enolpyruvylshikimate-3-phosphate synthase) such as glyphosate and auxin mimics such as dicamba and clorpyralid (Eberlein et al., 1994; Wall, 1994; Novo & Miranda Filho, 2006; Hutchinson et al., 2014). No reports were found linking HPPD inhibitor herbicides to any physiological disorder in potato tubers.

The mass of cracked tubers accounted for 12.2% of the mass of total tubers with 100.8 g ha⁻¹ tembotrione in the area of Rio Paranaíba; however, this value did not exceed 0.3% in the area of Serra do Salitre (Table 1). The lower incidence of this injury in Serra do Salitre can be attributed to a greater leaching of tembotrione in the soil, explained by the weak acid characteristic and pKa of this herbicide, associated with rainfalls of 50 and 194 mm that occurred on the third day and on the third week after application of product, respectively (Figure 2). Weak acid characteristic herbicides in soils with pH (6.2) higher than the pKa (3.17) predominate in the ionic form, being less retained on soil colloids (Muller et al., 2014). Consequently, they remain in solution making them more prone to leaching in cases of high rainfall.

Figure 2. Cracks in potato tubers caused by tembotrione residues in the soil evidenced in Rio Paranaíba (Minas Gerais State).

In Serra do Salitre, it is likely that tembotrione was first absorbed by the plants, which promoted the stress and reduction in the yield. However, during the tuberization phase, the reduced amount of the product in the soil due to leaching was not sufficient to induce cracks in the tubers.

Conclusion

Tembotrione reduced the quality of "Atlantic" potato tubers since it promoted the occurrence of cracks. Atrazine did not promote cracks in tubers of "Atlantic" potatoes.

References

Askari, M. S., & Holden, N. M. (2015). Quantitative soil quality indexing of temperate arable management systems. *Soil & Tillage Research, 150*, 57-67. doi: 10.1016/j.still.2015.01.010

Barchanska, H., Kluza, A., Krajczewska, K. & Maj, J. (2016). Degradation study of mesotrione and other triketone herbicides on soils and sediments. *Journal of Soils and Sediments, 16*(1), 125-133. doi: 10.1007/s11368-015-1188-1

Bayer CropScience. (2015a). *Bayer CropScience United States: Laudis label*. Retrieved on Oct. 1, 2015 from https://www.bayercropscience.us/products/herbicides/laudis/label-msds.

Bayer CropScience. (2015b). *Bayer CropScience Brazil. Soberan®* [Herbicida]. Retrieved on Oct. 1, 2015 from http://www.bayercropscience.com.br/site/nossos produtos/protecaodecultivosebiotecnologia/DetalheDo Produto.fss?Produto=190.

Bayer CropScience. (2015c). *Bayer CropScience Spain. Laudis®*. Retrieved on Oct. 1, 2015 from http://www.cropscience.bayer.es/es-ES/Productos/Herbicidas/Laudis.aspx.

Bayer CropScience. (2015d). *Bayer CropScience Belgium. Laudis®*. Retrieved on Oct. 1, 2015 from http://www.cropscience.bayer.be/fr-FR/AllProducts/Herbicides/Laudis.aspx.

Bontempo, A. F., Carneiro, G. D., Guimarães, F. A., Reis, M. R., Silva, D. V., Rocha, B. H., ... Sediyama, T. (2016). Residual tembotrione and atrazine in carrot. *Journal Environmental Science Health - Part B, 51*(7), 465-468. doi: 10.1080/ 03601234.2016.1159458

Campiglia, E., Paolini, R., Colla, G., & Mancinelli, R. (2009). The effects of cover cropping on yield and weed control of potato in a transitional system. *Field Crops Research, 112*(1), 16-23. doi: 10.1016/j.fcr.2009.01.010

Eberlein, C. V., Westra, P., Harderlie, L. C., Withmore, J. C., & Guttieri, M. J. (1997). *Herbicide drift and carryover injury in potatoes*. Idaho, OR: Pacific Northwest Extension Publ. 498.

Eberlein, C.V., & Guttieri, M. J. (1994). Potato (*Solanum tuberosum*) response to simulated drift of imidazolinone herbicides. *Weed Science, 42*(1), 70-75.

Food and Agriculture Organization of the United Nations [FAO]. (2017). *FAOSTAT*: Crops. Retrieved on May 23, 2017 from http://faostat3.fao.org /browse/Q/QC/E.

Hiller, L. K., Koller, D. C., & Thorton, R. E. (1985). Physiological disorders of potato tubers. In P. H. Li (Ed.), *Potato Physiology* (p. 389-455). Orlando, FL: Academic Press Inc.

Hutchinson, P. J. S., Felix, J., & Boydston, R. (2014). Glyphosate carryover in seed potato: effects on mother crop and daughter tubers. *American Journal of Potato Research, 91*(4), 394-403. doi: 10.1007/s12230-013-9363-7

Keiser, A., Häberli, M., & Stamp, P. (2012). Quality deficiencies on potato (*Solanum tuberosum* L.) tubers caused by *Rhizoctonia solani*, wireworms (*Agriotes* ssp.) and slugs (*Deroceras reticulatum, Arion hortensis*) in different farming systems. *Field Crops Research, 128*, 147-155. doi: 10.1016/j.fcr.2012.01.004

Muller , K., Deurer, M., Kawamoto, K., Kuroda, T., Subedi, S., Hi, S., Komatsu, T., & Clothier, B. E. (2014). A new method to quantify how water repellency compromises soils' filtering function. *European Journal of Soil Science, 65*(3), 348-359. doi: 10.1111/ejss.12136

Novo, M. C. S. S., & Miranda Filho, H. S. (2006). Effect of sulfonylurea herbicides on tuberization of two potato cultivars. *Planta Daninha, 24*(1), 115-121. doi: 10.1590/S0100-83582006000100015

Pesticide Properties Database [PPDB]. (2017a). *Tembotrione*. Retrieved on May. 23, 2017 from http://sitem.herts.ac.uk/aeru/iupac/ Reports/ 1118.htm

Pesticide Properties Database [PPDB]. (2017b). *Atrazine*. Retrieved on May. 23, 2017 from http://sitem.herts.ac.uk/aeru/iupac/Reports/43.htm

Robinson, D. E. (2008). Atrazine accentuates carryover injury from mesotrione in vegetables crops. *Weed Technology, 22*(4), 641-645. doi: 10.1614/WT-08-055.1

Soltani, N., Sikkema, P. H., & Robinson, D. E. (2005) Effect of foramsulfuron and isoxaflutole residues on rotational vegetable crops. *HortScience, 40*(3), 620-622. doi: 10.7824/rbh.v15i1.434

Thornton, R. E., & Eberlein, C. V. (2001). Chemical injury. In W. R. Stevenson, Loria, R., Franc, G. D., & Weingartner, D. P. (Ed.), *Compendium of potato diseases*. (p. 92-94). Saint Paul, MI: American Phytopathological Society.

Wall, D. A. (1994). Potato (*Solanum tuberosum*) response to simulated drift of dicamba, clopyralid, and tribenuron. *Weed Science, 42*(1), 110-114.

27

The control and protection of cotton plants using natural insecticides against the colonization by *Aphis gossypii* Glover (Hemiptera: Aphididae)

Ezio dos Santos Pinto, Eduardo Moreira Barros, Jorge Braz Torres[*] and Robério Carlos dos Santos Neves

*Departamento de Agronomia-Entomologia, Universidade Federal Rural de Pernambuco, Rua Dom Manoel de Medeiros, s/n, 52171-900, Recife, Pernambuco, Brazil. *Author for correspondence. Email: jtorres@depa.ufrpe.br*

ABSTRACT. The cotton aphid, *Aphis gossypii* Glover (Hemiptera: Aphididae), is a key pest of cotton, irrespective of the use of conventional or organic management. In organic systems, however, the use of synthetic insecticides is not allowed, increasing the difficulty of controlling this pest. This work evaluated aphid control and the ability of products to prevent aphid infestation using natural insecticides compared to a standard synthetic insecticide. The control trial was conducted with four products [*Beauveria bassiana* (Boveril®), neem oil (Neemseto®), and cotton seed oil compared to thiamethoxam (Actara®)], and untreated plants served as the control group. The trial testing the efficacy of these products in preventing aphid infestation was conducted using the same products, excluding Boveril®. The evaluations were conducted 72 and 120h post-treatment for the efficacy and the protection against colonization trials, respectively. The aphid control by cotton seed oil, Neemseto®, and thiamethoxam was similar, with 100% control being achieved on the thiamethoxam-treated plants. Regarding the plant protection against aphid colonization, the insecticide thiamethoxam exhibited a better performance compared to the other tested products with steady results over the evaluation period. The natural products exhibited variable results with low protection against plant colonization throughout the evaluation period.

Keywords: Insecta, organic cotton, cotton aphid, alternative control, population growth rate.

Introduction

The cotton crop has made a considerable contribution to Brazilian agribusiness, with cotton production and the textile industry accounting for more than 16.4 million of direct and indirect employment in Brazil (VALDEZ, 2011). The cultivated area is expected to increase for the upcoming seasons, due to the worldwide demand for natural fibers (CONAB, 2011). In Brazil, cotton cultivation occurs primarily in the Cerrado and semiarid areas of the western and northeastern states. In the Cerrado, cotton is produced in large fields, adopting all available technologies, whereas it is produced by small growers in the semiarid areas (BARROS;

TORRES, 2010; FONTES et al., 2006). The cotton fields in the semiarid region consist of small areas, ranging from 0.3 to 8 hectares, using family labor and under a low input of technologies, such as chemical fertilization, mechanization, and pest control practices, and most of these fields can be classified as organic cotton because of the use of organic fertilizer and lack of pesticide utilization (BARROS; TORRES, 2010). Cotton plants, however, host a variety of herbivorous pests, and the lack of the proper adoption of pest control practices in organic cotton fields makes it difficult to produce a profitable crop.

Among the cotton pests, the cotton aphid *Aphis gossypii* Glover is an important pest that infests at the beginning of the crop season, delaying early plant development; the aphid infestation might extend through the development of the plants if control practices are not adopted. Plants infested with cotton aphids exhibit reduced development and curled leaves, especially the young leaves driving the growth of the main stem and the leaves of the reproductive branches (EBERT, 2008; LECLANT; DEGUINE, 1994). Beyond the damage caused directly to the plant due to the feeding behavior, aphids secrete honeydew on the leaves and the open lint, seriously endangering the cotton yield. In addition, the honeydew favors the development of black sooty mold fungus, which affects plant development and results in stick lint, causing problems during the spinning process at the textile mills (DEGUINE et al., 2000). Large colonies of cotton aphids are commonly produced due to the intrinsic biotic characteristics of the insects, such as rapid development and a parthenogenic mode of reproduction in the tropics, which are enhanced when coupled with high temperatures and plants under water stress (GODFREY et al., 2000; VAN EMDEN; HARRINGTON, 2007), both of which are common environmental conditions in semiarid regions.

The control of cotton aphids in cotton fields is primarily addressed with seed treatment or foliar spraying with systemic or contact broad-spectrum insecticides (ALMEIDA et al., 2008; TORRES; SILVA-TORRES, 2008). Nevertheless, the increased value of the fiber produced under organic systems has stimulated the small growers in the semiarid regions to adopt biorational pest control methods. The cotton cultivated under low environmental-impact production and family agriculture systems fetches a higher price, which compensates for the low yield commonly obtained. In addition, the recent cultivation of colored fiber

cottons (i.e., degrees of green and brown colors) under organic systems has a large opportunity to expand in the semiarid region, especially among the small growers. Thus, biorational methods of aphid control will be required where organic production systems are intended. Previous studies have shown that alternative products, such as entomopathogenic fungi (LOUREIRO; MOINO JÚNIOR, 2006; STEINKRAUS et al., 2002), natural oils, and plant product derivates (BAGAVAN et al., 2009; EL SHAFIE; BASEDOW, 2003; LIN et al., 2009; MAREGGIANI et al., 2008; SANTOS et al., 2004), have the potential to control cotton aphids. Based on the requirements for organic production, the utilization of natural insecticides is one way to control cotton aphid. Therefore, this study investigated the control of cotton aphids established on cotton plants and the ability of the treatment to prevent the colonization of treated plants. The tested products were the natural oil from seed cotton, a commercial formulation of neem oil, and a commercial formulation of *Beauveria bassiana* in comparison to thiamethoxam, a synthetic insecticide recommended to control cotton aphid in conventional cotton fields.

Material and methods

The experiments were conducted under greenhouse facilities and under controlled conditions in the Biological Control Laboratory of the Universidade Federal Rural de Pernambuco (UFRPE). Cotton plants of the variety BRS Verde were used for aphid-colony rearing and in the experiments. The plants were grown in plastic pots filled with 500 g of a mixture of soil and humus (3:2 by weight) and fertilized weekly with a 20 g L^{-1} of urea solution at rate of 20 mL per pot, beginning eight days after seedling emergence. Four seeds were planted per pot and were subsequently thinned to two plants per pot after five days from plant emergence. The aphid colony was maintained under greenhouse conditions with potted cotton plants kept inside cages of 1 x 1 x 0.8 m (WxLxH) that were covered with *voile* fabric to avoid other opportunist arthropods.

The study was conducted in the following two goals: (i) to evaluate the effect of the tested products in reducing the aphid population on infested plants and (ii) to verify the ability of the tested products in preventing the colonization of treated cotton plants exposed to aphid colonization.

Control of cotton aphid. To investigate the effect of the products in reducing aphid infestation, at 21 days after emergence, the cotton plants were subjected to aphid infestation by allowing them to contact other cotton plants inside the aphid-rearing cages for a period of 48h. After this colonization period, the plants were moved to the laboratory and randomly evaluated. The number of aphids was counted in the upper two fully expanded leaves of each plant using a bench 10x magnification lens to obtain the aphid infestation prior to the treatment. The petioles of these leaves were smoothly smeared with entomological glue (Cola Entomológica®, Biocontrole Ltda, São Paulo State, Brazil) to avoid aphid dispersion or hosting aphids dispersing from other parts of the plants. The experiment consisted of a completely randomized design with five treatments (four insecticides and control plants with no insecticide treatment) and five replications each. Each replication consisted of a pot with two plants. Thus, the replication mean was considered as the average of the aphids counted on four leaves (i.e., two leaves per plant x two plants per pot corresponding to 20 leaves counted from 10 plants per treatment).

The insecticides and concentrations tested were as follows: Boveril PM (*Beauveria bassiana* isolates ESALQ-PL63 and ESALQ-447 at 5×10^8 conidia g^{-1}) (Itaforte Bioprodutos, São Paulo State, Brazil); Neemseto® (azadirachtin) (Cruangi Neem do Brasil Ltda., Timbaúba, Pernambuco State, Brazil) at 1% concentration; cotton seed oil at 1%; and Actara 250 WG (thiamethoxam at 0.1 g of a.i. 200 mL^{-1}) (Syngenta S.A., São Paulo State, Brazil). Tween 20 at 0.02% was added to each dilution. For Boveril PM, the number of conidia counted in a 200 μL aliquot. Furthermore, the viability of the conidia in BDA and the pathongenicity against 3rd-instar larvae of *Diatraea saccharalis* (Fabr.) were tested. The results of these tests fit the standard values set by the manufacturer.

The plants were sprayed using an electric power Airbush set (Paasche Airsbush Co., Harwood Heights, IL, USA) under 15 lb pol^{-2} pressure. The volume of the insecticide dilution applied per plant was regulated to 1 mL to obtain a homogeneous plant covering according to a previous test. As aphids are also affected by water droplets, the infested plants comprising the control treatments were also sprayed with a dilution of water and Tween 20 at 0.02%. The treated plants were allowed to stand for 2h for the spray to dry the spray, and the plants were transferred to a climatic chamber at 27°C and a 12-h photophase. After 72h, the plants were evaluated by counting the number of live aphids per leaf.

The data of aphid per leaf pre- and 72h post-treatment were compared. In addition, the insectistatic effect of some of the tested products also relates to reproduction (DIMETRY; EL-HAWARY, 1995; ISMAN, 2006; NISBET, 1994); thus, the instantaneous population growth rate (r_i) was calculated based on Stark and Banks (2003) using the following formula: $r_i = \ln (N_f / N_0) / \Delta T$, where N_f indicates for the final number of individuals in the populations, N_0 indicates the initial number of individuals in the population, and Δt indicates the time (days) of observation post-treatment. In this case, Δt was three days. Positive values of r_i indicate a population increase during the observation period, and negative values of r_i indicate population decrease; values of $r_i = 0$ indicate no numerical change in the population.

The number of aphids per leaf and the values of r_i were subjected to normality and homogeneity tests with regard to the assumptions for the analysis of variance (SAS, 2001). The results were subsequently subjected to analysis of variance, and the means were compared using the Tukey HSD test at a 0.05 significance level (SAS, 2001).

Protection of cotton plants against aphid infestation. The products were also investigated with regard to their efficacy in protecting treated plants against aphid colonization; the product formulated with *B. bassiana* was excluded in this experiment, due to a low control efficacy. Cotton plants cultivated without aphid infestation and of the same age (~21 days after emergence) were sprayed using a procedure similar to that described above, and the treated and untreated (control treatment) plants were exposed to aphid colonization at two hours after spraying. To test the aphid colonization on the treated plants, the cotton plants were randomly placed between rows of potted cotton plants heavily infested with aphids, with the leaves touching to allow aphid colonization. A completely randomized design was established, with four treatments (three insecticides and control) and five replications each. Each replication mean was originated from two plants (two plants per pot). The evaluations consisted of whole-plant inspections for the presence of aphids at 24, 48, 72, and 120h after the exposure of the treated plants to the aphid-infested plants.

The average number of aphids per two plants was tested for normality and homogeneity of variance, and square root (x + 0.5) transformation was required to fit the assumptions for the analysis of variance. The data were then subjected to analysis of variance through a repeated measure procedure because the evaluations were conducted over time on the same plants using the SAS statistical package (SAS, 2001). To separate the means among the treatment, the Tukey HSD test was performed at 0.05 level of significance for each evaluation interval.

Results and discussion

Control of cotton aphid. The evaluation of aphid infestation on cotton plants prior to insecticide application resulted in a statistically similar average of aphids per plant ($p > 0.05$) (Figure 1), indicating that we could disregard the effect of the initial population on the final results across the treatments. Thus, we can conclude that the tested insecticides exhibited different control performances on cotton aphid ($F_{4, 20} = 26.89$, $p < 0.0001$). The aphid population on the untreated and plants treated with Boveril increased by 2.12- and 1.89-fold, respectively, during the 72h post-treatment period (Figure 1). The cotton plants treated with 1% cotton seed oil exhibited a slightly decrease of aphid population at 72h post-treatment that was statistically similar to the infestation prior to the treatment ($p > 0.05$). Among the natural products tested, the formulation of neem (Neemseto®) produced the greatest reduction in the aphid population with an average of 19.8 aphids per plant at 72h post-treatment, which was 12.9 times lower than the infestation prior to the treatment; the standard synthetic insecticide thiamethoxam exhibited 100% control of the aphids (Figure 1).

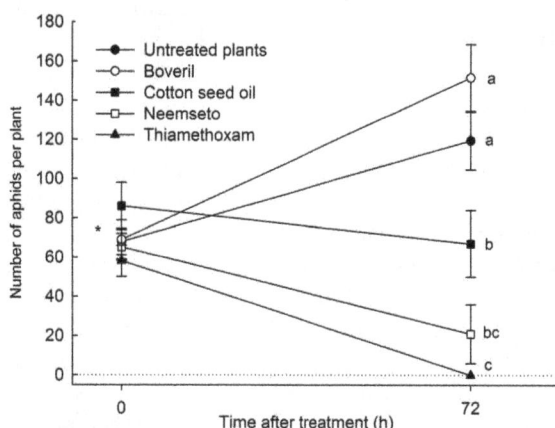

Figure 1. Control of cotton aphid *Aphis gossypii* with natural and synthetic insecticides. Note: *means (± SE) of initial infestation prior to insecticide application do not differ among treatments ($F_{4, 20} = 2.33$, $p = 0.0913$); while mean at 72h post insecticide application followed by different letter differ among treatments by Tukey HSD's test ($p < 0.05$).

The instantaneous population growth rate (r_i) calculated based on the final and initial number of aphids per cotton plant was not determined for the thiamethoxam treatment because of the 100% control at the final evaluation. However, based on r_i, the aphid population exhibited significant changes ($F_{3, 16} = 25.37$, $p < 0.0001$) with regard to untreated plants and those treated with Neemseto, cotton seed

oil, and Boveril (Figure 2). The untreated and Boveril-treated plants produced positive r_i values of 0.229 and 0.220, respectively, indicating similar population increases during the observation period. In contrast, the treatments with Neemseto and cotton seed oil resulted in significant reductions in the aphid population, with r_i values of -0.419 and -0.164, respectively. These results were similar between these two treatments, even though Neemseto resulted in 2.55-fold greater reduction (Figure 2).

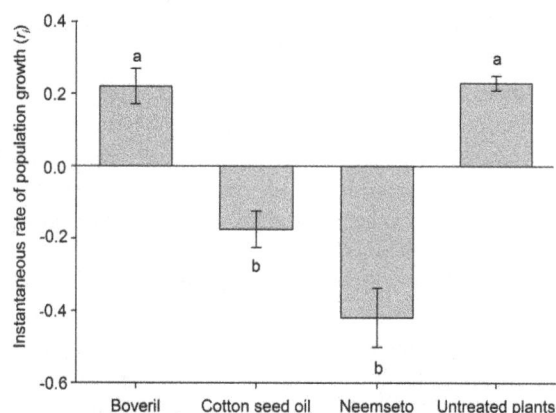

Figure 2. Cotton aphid population increase after application of natural insecticides on cotton plants. Bars holding similar letters do not differ by Tukey HSD's test ($p > 0.05$).

Protection of cotton plants against aphid infestation. The protection of the cotton plants against aphid colonization varied across the treatments and all of the evaluation intervals (Table 1). Among the treatments, the synthetic insecticide thiamethoxam exhibited the best performance in protecting the plants against aphid colonization, irrespective of the evaluation interval: the average population was 4.1 aphids per cotton plant at 120h post-treatment (Table 1). Among the natural products, there were no significant differences at the 120h-evaluation interval (Table 1). Between 24 and 120h for the untreated plants and after treatment with seed oil and Neemseto, the aphid population increased significantly at rates of 6.3-, 7.9-, and 10.9-fold, respectively (Table 1).

We found that the commercial formulation of *B. bassiana* performed differently from previous reports using different isolates of this fungus tested against cotton melon[-1] aphid. According to Loureiro and Moino Júnior (2006) and Araújo Júnior et al. (2009), *B. bassiana* was pathogenic to *A. gossypii*. In our study, the treatment of aphid-infested cotton plants with Boveril produced similar results as untreated plants: the observed aphid population growth was similar (Figures 1 and 2).

Table 1. Mean (± SE) number of cotton aphid per plant after application of natural and synthetic insecticides.

Treatments[1]	Time post-treatment (h)				Statistics
	24	48	72	120	
Untreated plants	33.8 ± 9.32 Ac	50.1 ± 5.04 Ac	112.2 ± 8.88 Ab	213.9 ± 19.59 Aa	$F_{3,12} = 48.33$, p < 0.0001
Cotton seed oil	36.8 ± 9.23 Ab	90.0 ± 28.87 Aab	199.9 ± 50.51 Aa	290.9 ± 88.75 Aa	$F_{3,12} = 10.44$, p = 0.0012
Neemseto®	21.1 ± 7.14 Ab	59.3 ± 10.41 Aab	142.7 ± 22.41 Aa	231.6 ± 42.07 Aa	$F_{3,12} = 15.31$, p = 0.0002
Thiamethoxam	1.4 ± 0.47 Ba	2.2 ± 1.37 Ba	2.2 ± 0.88 Ba	4.1 ± 1.92 Ba	$F_{3,12} = 0.86$, p = 0.4872
Statistics	$F_{3,12} = 19.36$, p < 0.0001	$F_{3,12} = 40.90$, p < 0.0001	$F_{3,12} = 89.24$, p < 0.0001	$F_{3,12} = 56.41$, p < 0.0001	

[1]Means followed by the same capital letters within column and small letters within rows do not differ by Tukey HSD's test (p > 0.05).

It is important to highlight that laboratory studies using isolates of *B. bassiana* are conducted in Petri dishes and under conditions that are usually favorable to the fungus, including high humidity. Confining aphids, phloem-sucking insects, on leaf discs can cause the aphids to move and/or stop feeding, resulting in stress to the insect, and the stress caused in the target pest is known to enhance the efficacy of the parasitism by *B. bassiana* (FURLONG; GRODEN, 2003; LORD, 2009). Another explanation that we can consider for the low efficacy of the tested formulation of *B. bassiana* can be the reduced post-treatment evaluation period. According to Tesfaye and Seyoum (2010), *B. bassiana* caused cumulative mortality of *A. gossypii* at 25°C from 73.3 to 93.3% but required ~5 days to produce 50% mortality. Furthermore, Vu et al. (2007) reported that among the fungi tested, *Lecanicillium lecanii, Paecilomyces farinosus, Beauveria bassiana, Metarhizium anisopliae, Cordyceps scarabaeicola,* and *Nomuraea rileyi, L. lecanii* performed as the best in controlling *A. gossypii*.

The aphid population on the plants treated with cotton seed oil demonstrated a slight decrease compared to the untreated plants, suggesting an effect on aphid reproduction. The negative outcome of the instantaneous rate of population growth supports this hypothesis of an effect on reproduction. Future studies using cotton seed oil should consider evaluation intervals that are sufficiently long to allow one aphid generation to ascertain the effect on reproduction.

The aphid control obtained with the commercial neem oil (Neemseto) was comparable to that obtained with the standard synthetic insecticide thiamethoxam (Figure 1). The topical effect of neem seems to predominate in this trial because the treatment consisted of spraying aphid-infested plants. Regarding the standard synthetic insecticide thiamethoxam, the results obtained fit those already reported with an excellent level of control and protection of cotton plants against colonization by aphids (TORRES; RUBERSON, 2004; TORRES; SILVA-TORRES, 2008). None of the natural products tested were able to protect the plants against aphid colonization; the final count varied, on average, from 231.6 to 290.9 aphids per plant at 120h post-treatment.

Although previous results suggest the systemic action of products with azadirachtin (SCHUMUTTERER, 1990; SOUZA; VENDRAMIM, 2005), only a delay in aphid infestation would be expected. Based on our results, although Neemseto showed efficacy in the control trial, the results were similar to the untreated plants when applied prior to infestation, evidencing no protection against aphid colonization (Table 1).

Conclusion

One application of Neemseto and Actara achieved similar control of cotton aphids at 72h after treatment followed by cotton seed oil, whereas the aphid populations on plants treated with Boveril and untreated plants were similar. Regarding protection against aphid infestation, only the insecticide Actara exhibited effectiveness in maintaining the aphid population near zero on cotton plants throughout the evaluation. However, further studies using more than one application and longer evaluation periods for cotton seed oil will be necessary to ascertain the efficacy of this natural product.

Acknowledgements

To the REDALGO project (Finep 0593/2007) and to the CNPq for grants.

References

ALMEIDA, R. P.; SILVA. C. D. A.; RAMALHO, F. S. Manejo integrado de pragas do algodoeiro no Brasil. In: BELTRÃO, N. E. M.; AZEVEDO, D. M. P. (Ed.). **O agronegócio do algodão no Brasil**. Brasília: Embrapa, 2008. p. 1035-1098.

ARAÚJO JÚNIOR, J. M.; MARQUES, E. J.; OLIVEIRA. J. V. Potencial de isolados de *Metarhizium anisopliae* e *Beauveria bassiana* e do óleo de nim no controle do pulgão *Lipaphis erysimi* (Kalt.) (Hemiptera: Aphididae). **Neotropical Entomology**, v. 38, n. 4, p. 520-525, 2009.

BAGAVAN, A.; KAMARAJ. C.; RAHUMAN. A. A.; ZAHIR. A. A.; PANDIVAN, G. Evaluation of larvicidal and nymphicidal potential of plant extracts against *Anopheles subpictus* Grassi. *Culex tritaeniorhynchus* Giles and *Aphis gossypii* Glover. **Parasitology Research**, v. 104, n. 5, p. 1109-1117, 2009.

BARROS, E. M.; TORRES, J. B. **Diagnóstico parcial da produção de algodão em Pernambuco**. Recife: UFRPE, 2010.

CONAB-Companhia Nacional de Abastecimento. **Acompanhamento de safra brasileira**: grãos. Safra 20102011. Quarto levantamento. Brasília: Conab, 2011.

DEGUINE, J. P.; GOZE, E.; LECLANT, F. The consequences of late outbreaks of the aphid *Aphis gossypii* in cotton growing in Central Africa: towards a possible method for the prevention of cotton stickiness. **International Journal of Pest Management**, v. 46, n. 1, p. 85-89, 2000.

DIMETRY, N. Z.; EL-HAWARY, F. M. A. Neem Azal-F as an inhibitor of growth and reproduction in the cowpea aphid *Aphis craccivora* Kock. **Journal of Applied Entomology**, v. 119, n. 1, p. 67-71, 1995.

EBERT, T. A. Melon aphid. *Aphis gossypii* (Hemiptera: Aphididae). In: CAPINERA, J. L. (Ed.). **Encyclopedia of entomology**. 2nd ed. Dordrecht: Springer, 2008. v. 1-4. p. 1374-1378.

EL SHAFIE, H. A. F.; BASEDOW, T. The efficacy of different neem preparations for the control of insects damaging potatoes and eggplants in the Sudan. **Crop Protection**, v. 22, n. 8, p. 1015-1021, 2003.

FONTES, E. M. G.; RAMALHO, F. S.; UNDERWOOD, E.; BARROSO, P. A. V.; SIMON, M. F.; SUJII, E. R.; PIRES, C. S. S.; BELTRÃO, N. E.; LUCENA, W. A.; FREIRE, E. C. The cotton agriculture context in Brazil. In: HILBECK, A.; ANDOW, D. A.; FONTES, E. M. G. (Ed.). **Environmental risk assessment of genetically modified organisms: methodologies for assessing Bt cotton in Brazil**. Wallingford: CABI Publishing, 2006. p. 21-66.

FURLONG, M. J.; GRODEN, E. Starvation induced stress and the susceptibility of the Colorado potato beetle, *Leptinotarsa decemlineata*, to infection by *Beauveria bassiana*. **Journal of Invertebrate Pathology**, v. 83, n. 2, p. 127-138, 2003.

GODFREY, L. D.; ROSENHEM, J. A.; GOODELL, P. B. Cotton aphid emerges as major pest in SJV cotton. **California Agriculture**, v. 54, n. 6, p. 26-29, 2000.

ISMAN, M. B. Botanical insecticides, deterrents, and repellents in modern agriculture and increasing regulated world. **Annual Review of Entomology**, v. 51, p. 45-66, 2006.

LECLANT, F.; DEGUINE, J. P. Aphids (Hemiptera: Aphididae). In: MATTEWS, G. A.; TUNSTALL, J. P. (Ed.). **Insect pests of cotton**. Wallingford: CAB International, 1994. p. 285-323.

LIN, C. Y.; WU, D. C.; YU, J. Z.; CHEN, B. H.; WANG, C. L. Control of silverleaf whitefly. cotton aphid and Kanzawa spider mite with oil and extracts from seeds of sugar apple. **Neotropical Entomology**, v. 38, n. 4, p. 531-536, 2009.

LORD, J. C. Efficacy of *Beauveria bassiana* for control of *Tribolium castaneum* with reduced oxygen and increased carbon dioxide. **Journal of Applied Entomology**, v. 133, n. 1, p. 101-107, 2009.

LOUREIRO, E. S.; MOINO JÚNIOR., A. Patogenicidade de fungos hifomicetos aos pulgões *Aphis gossypii* Glover e *Myzus persicae* Sulzer (Hemiptera: Aphididae). **Neotropical Entomology**, v. 35, n. 5, p. 660-665, 2006.

MAREGGIANI, G.; RUSSO. S.; ROCCA, M. *Eucalyptus globulus* (Mirtaceae) essential oil: efficacy against *Aphis gossypii* (Hemiptera: Aphididae). an agricultural pest. **Revista Latinoamericana de Química**, v. 36, n. 1, p. 16-21, 2008.

NISBET, A. J. The effects of azadirachtin-treated diets on the feeding behaviour and fecundity of the peach-potato aphid. *Myzus persicae*. **Entomologia Experimentalis et Applicata**, v. 71, p. 65-72, 1994.

SANTOS, T. M.; TORRES. A. L.; BOIÇA JUNIOR, A. L. Effect of neem extract on the cotton aphid. Pesquisa **Agropecuária Brasileira**, v. 39, n. 11, p. 1071-1076, 2004.

SAS-Statistical Analisys System. **SAS/STAT User's guide**. Version 8.02. TS level 2MO. Cary: SAS Institute Inc., 2001.

SCHUMUTTERER, H. Properties and potential of natural pesticides from the neem tree. *Azadirachta indica*. **Annual Review of Entomology**, v. 35, p. 271-297, 1990.

SOUZA, A. P.; VENDRAMIM, J. D. Efeito translaminar. sistêmico e de contato de extrato aquoso de sementes de nim Sobre *Bemisia tabaci* (Genn.) biótipo B em tomateiro. **Neotropical Entomology**, v. 34, n. 1, p. 83-87, 2005.

STARK, J. D.; BANKS, J. E. Population - level effects of pesticides and other toxicants on arthropods. **Annual Review of Entomology**, v. 48, p. 505-519, 2003.

STEINKRAUS, D. C.; BOYS, G. O.; ROSENHEIM, J. A. Classical biological control of *Aphis gossypii* (Homoptera: Aphididae) with *Neozygites fresenii* (Entomophthorales: Neozygitaceae) in California cotton. **Biological Control**, v. 25, n. 2, p. 297-304, 2002.

TESFAYE, D.; SEYOUM, E. Studies on the Pathogenicity of native entomopathogenic fungal isolates on the cotton/melon aphid, *Aphis gossypii* (Homoptera: Aphididae) Glover under different temperature regimes. **African Entomology**, v. 18, n. 2, p. 302-312, 2010.

TORRES, J. B.; RUBERSON, J. R. Toxicity of thiamethoxam and imidacloprid to *Podisus nigrispinus* (Dallas) (Heteroptera: Pentatomidae) nymphs associated to aphid and whitefly control in cotton. **Neotropical Entomology**, v. 33, n. 1, p. 99-106, 2004.

TORRES, J. B.; SILVA-TORRES, C. S. A. Interação entre inseticidas e umidade do solo no controle do pulgão e da mosca-branca em algodoeiro. **Pesquisa Agropecuária Brasileira**, v. 43, n. 8, p. 949-956, 2008.

VALDEZ, C. **Brazilian cotton - Briefing rooms**. Washington, D.C.: ERS-USDA, 2011.

VAN EMDEN, H. F.; HARRINGTON, R. **Aphids as crop pests**. Cambridge: CAB International, 2007.

VU, H.; HONG, S.; KIM, K. Selection of entomopathogenic fungi for aphid control. **Journal of Bioscience and Bioengineering**, v. 104, n. 6, p. 498-505, 2007.

28

Emergence of *Rottboellia exaltata* influenced by sowing depth, amount of sugarcane straw on the soil surface, and residual herbicide use

Núbia Maria Correia[*], Leonardo Petean Gomes and Fabio José Perussi

*Departamento de Fitossanidade, Universidade Estadual Paulista "Julio de Mesquita Filho", Via de Acesso Prof. Paulo Donato Castellane, s/n, 14884-900, Jaboticabal, São Paulo, Brazil. *Author for correspondence. E-mail: correianm@fcav.unesp.br*

ABSTRACT. Mechanical sugarcane harvest without burning and continuous straw on the soil surface may affect the *Rottboellia exaltata* infestation dynamics in sugarcane fields. Three greenhouse experiments were conducted with the aim of studying the effects of sowing depth (0, 2.5, 5, 7.5, and 10 cm), amount of sugarcane straw on the soil surface (0, 5, 10, and 15 ton ha^{-1}), and residual herbicide (clomazone, flumioxazin, imazapyr, isoxaflutole, and s-metolachlor) on the emergence of *Rottboellia exaltata*. For each experiment, a completely randomized design with four replicates was applied. The combination of mulch on soil surface (especially with larger amounts of straw) with deeper sowing depths provides less emergence and mass accumulation of *R. exaltata*. In bare soil, the sowing depth did not affect the weed dynamics. Clomazone and imazapyr were effective herbicides controlling *R. exaltata* regardless of the amount of straw on the soil surface. Flumioxazin was also effective in controlling *R. exaltata* but only under bare soil conditions. Even with 60 mm of accumulated rainfall over the 4 day period after application, the amount of flumioxazin leached to the soil was not enough to ensure the same control observed when applying the herbicide on bare soil.

Keywords: itchgrass, mulch, herbicide retention, weed dynamics.

Introduction

Rottboellia exaltata L.f. (synonym *R. cochinchinensis* (Lour.) Clayton) is an annual or perennial species depending on environmental conditions, and it is reproduced by seeds produced from stem fragments with budding nodes (KISSMANN, 1997). In addition to the damage caused by competition for water, light, nutrients, and space, decomposing plant residues from this species release totoxic compounds

in the soil that may inhibit germination and/or growth of adjacent species, including weeds and cultivated crops (KOBAYASHI et al., 2008; MEKSAWAT; PORNPROM, 2010). *R. exaltata* is the main weed for at least 18 crops in Africa, Central America, South America, Unites States, Australia, and Papua New Guinea (ANNING; GYAN-YEBOAH, 2007; HOLM et al., 1991; KISSMANN, 1997). This weed frequently occurs in Brazilian sugarcane crops in Rio de Janeiro (OLIVEIRA;

FREITAS, 2008) in focal infestations in São Paulo, Paraná e Mato Grosso do Sul. There are also reports of its occurrence in northern and central-western regions of Brazil (KISSMANN, 1997).

Mechanical sugarcane harvest without burning and continuous straw on the soil surface may affect the *R. exaltata* infestation dynamics in sugarcane fields. A reduced occurrence of weeds, mainly grasses has been observed in this harvest system. Correia and Durigan (2004) reported that sugarcane straw has an inhibitory effect on the emergence of *Brachiaria decumbens* and *Digitaria horizontalis* and reduced seed viability of such seeds since they do not germinate in the presence of straw or after its removal due to the physical, chemical, and/or biological effects of mulch. Thus, large amounts of sugarcane straw are effective in controlling *R. exaltata* (OLIVEIRA; FREITAS, 2009).

In contrast, straw can also compromise the ability of residual herbicide to reach the soil and, consequently, its ability to control weeds before their emergence. Depending on the physical and chemical attributes of herbicides, such as solubility, vapor pressure, and polarity, straw can have a differential influence on herbicide efficacy (RODRIGUES, 1993). After herbicide application, the amount and period when rainfall or irrigation occurs, as well as modifications occurring in the decomposing plant residues, are also important factors regarding the retention of the herbicide by the crop residue (CORREIA et al., 2007). If herbicides do not leach from the straw to the soil, they are exposed to losses by photodegradation, volatilization, and adsorption into plant residues. The degree of decomposition or aging of plant residues can also affect their capacity to absorb herbicides (MERSIE et al., 2006).

So far, information on the chemical control of *R. exaltata* in sugarcane crops is scarce, especially for herbicides applied pre-emergence. However, it is known that the chemical control of *R. exaltata* is costly because of the need to use up to six herbicide applications throughout the crop cycle (OLIVEIRA; FREITAS, 2009).

This study was conducted to test the hypothesis that (i) sowing depth does not affect the emergence of *R. exaltata*; (ii) large amounts of straw may inhibit the emergence of *R. exaltata* (iii) and impair herbicides applied pre-emergence, (iv) especially herbicides that are affected by the presence of sugarcane straw on the soil surface. The aim of this study was to study the effects of sowing depth, amount of sugarcane straw on the soil surface, and pre-emergence herbicide application on the emergence of *R. exaltata*.

Material and methods

Three experiments were conducted in pots inside a greenhouse between May 21, 2010 and May 7, 2011 at the Department of Phytosanitary Sciences at São Paulo State University (Universidade Estadual Paulista - UNESP), Jaboticabal Campus, São Paulo State, Brazil.

For the three experiments, a completely randomized design with four replicates was applied. In the first experiment (4 x 5 factorial design), four amounts of straw on the soil (0, 5, 10, and 15 ton ha^{-1}) and five sowing depths (0, 2.5, 5, 7.5, and 10 cm) of *R. exaltata* seeds were studied. In the second experiment (4 x 6 factorial design), four amounts of straw (0, 5, 10, and 15 ton ha^{-1}), five herbicides (clomazone at 1.20 kg ha^{-1}, flumioxazin at 0.25 kg ha^{-1}, imazapyr at 0.20 kg ha^{-1}, isoxaflutole at 0.225 kg ha^{-1}, and s-metolachlor at 2.88 kg ha^{-1}) applied pre-emergence, and an untreated control were studied. In the third experiment (4 x 3 + 4 factorial design), flumioxazin (250 g ha^{-1}) applied to four levels of soil cover by sugarcane straw (0, 5, 10, and 15 ton ha^{-1}) with simulated rain (20, 40, and 60 mm rainfall after application), and four untreated controls (0, 5, 10, or 15 ton ha^{-1} straw on the soil) were studied.

Each experimental unit consisted of 8 dm^{-3} plastic pots filled with substrate. The substrate was a mixture of soil, sand, and organic compost at a ratio of 3:1:1, respectively. The textural analysis of the substrate indicated 679, 261 and 61 g kg^{-1} of sand, silt and clay, respectively, as well as 28 g dm^{-3} of organic matter.

After mechanical harvest of sugarcane plants (var. RB 867515 third cut for the first experiment; var. RB 855453 sixth cut for the second experiment; and var. RB 835054 first cut for the third experiment), the straw remaining on the soil was collected and taken to the greenhouse, where it was dried completely.

In all experiments, 2.0 g of *R. exaltata* seeds was sown per pot. In the first experiment, a portion of the substrate in the pots was removed to plant the seeds at the depths indicated for each treatment (a ruler was used to measure the depths). In the second and third experiments, seeds were homogeneously distributed in each pot and sown at a 1 cm depth from the soil surface. For treatments with straw, a uniform layer of sugarcane straw in the respective quantity for each treatment was placed on soil surface of each pot after sowing. Straw was cut into smaller fragments less than or equal to the pot diameter.

The bottoms of the pots were sealed with a sheet of newspaper to prevent soil loss. Each pot was placed on a plastic container with a larger diameter and without holes to maintain the water regime. Soil moisture was monitored daily, and water in the containers was replaced as necessary.

In the second and third experiments, the herbicides were sprayed on pots prior to weed emergence. A backpack sprayer equipped with two flat-fan nozzles (XR 110015) spaced at 0.5 m and calibrated to deliver an equivalent of 200 L ha^{-1} at a constant pressure (maintained by CO_2) of 2.0 kgf cm^{-2} was used. The dates, times, and meteorological conditions for each application are shown in Table 1.

In the first and second experiments, rainfall depth equivalent to 25 mm was simulated after sowing or herbicide application. In the third experiment, rainfall equivalent to 20 mm depth was simulated in all treatments (after flumioxazin spraying), including untreated controls. For the 40 and 60 mm rainfall treatments, a new 20 mm simulation was conducted two days after the initial 20 mm rainfall, and the simulation was repeated once more two days after the 60 mm rainfall treatments. Thus, rainfall was cumulative for the 40 and 60 mm rainfall treatments (with additional 20 mm every two days).

The total number of emerged plants was counted at 14 and 35 days after herbicide application (DAA) for the second and third experiments or days after weed sowing (DAS) for the first experiment. At 35 DAA or DAS, the plants were harvested close to the soil, placed inside paper bags and dried in a convection oven at 50°C until a constant weight was achieved, and the shoot dry matter was then quantified.

The data were subjected to an F-test. In the first experiment, effects of straw amount and sowing depth or the interaction between them when significant were compared by polynomial fitting of the data. In the second experiment, significant effects of the herbicide treatments and mulch on the soil or the interaction between them were compared by Tukey's test at 5% probability (for herbicide treatments) or by polynomial fitting of the data (mulch on the soil). In the third experiment, significant effects of straw amount and rainfall or the interaction between them were compared by polynomial fitting of the data.

Results and discussion

Emergence of Rottboellia exaltata influenced by sowing depth and presence of sugarcane straw on the soil (first experiment)

Each factor (sowing depth and straw amount) and the interaction between them had a significant effect on R. exaltata weed density and dry matter.

When analyzing the depth x straw interaction, we observed that for all sowing depths studied, weed density linearly decreased with increasing amount of straw on the soil, except for superficial sowing at 14 DAA (Figure 1). Thus, emergence of R. exaltata was reduced with increasing amounts of sugarcane straw regardless of sowing depth. This effect was even more pronounced with increasing depths of seeds planted in the pots.

At depths of 2.5 and 10 cm, weed dry matter linearly decreased with increasing amounts of straw. The same trend occurred when the seeds were distributed on the soil surface (with polynomial fitting of the data). Thus, the least accumulated mass of R. exaltata weeds occurred with 15 ton ha^{-1} straw on the soil for the five depths studied. These data corroborated with report by Oliveira and Freitas (2009). Adding rice plant residues (4 and 6 ton ha^{-1}) to the soil surface also reduced the emergence of R. cochinchinensis (BOLFREY-ARKU et al., 2011).

Weed emergence was unaffected by sowing depth in bare soil (Figure 2). Moreover, weed density was unaffected by sowing depth in treatments with 5 ton ha^{-1} of straw. For dry matter, values decreased with increasing sowing depth. At 10 and 15 ton ha^{-1} straw, R. exaltata density and dry matter decreased linearly with increasing sowing depths.

A previous study has shown that the emergence of R. cochinchinensis is higher when the seeds are sown at the soil surface and that the emergence decreases with increasing sowing depths with no weeds emerging from seeds planted at a depth of 10 cm (BOLFREY-ARKU et al., 2011). The soil management conditions (conventional tillage, minimum tillage, and no tillage) for rice crops do not affect the emergence of this species (CHAUHAN; JOHNSON, 2009). Neither soil disturbances, which cause seed distribution throughout the soil profile, nor a complete absence of movement, which causes an accumulation of seeds at the soil surface, affect weed emergence dynamics or the level of infestation.

These results are at least partially explained by the ability of these seeds to germinate both in the light and in the dark (THOMAS; ALLISON, 1975). Although light is not a requirement for seed germination, a light/dark regime stimulates germination of R. cochinchinensis (BOLFREY-ARKU et al., 2011).

Therefore, there was less emergence and accumulation of dry mass with mulch on the soil, especially under higher amounts of straw, combined with deeper sowing depths. In bare soil, however, the sowing depth did not affect weed dynamics.

Table 1. Dates, times, and meteorological conditions for each herbicide application in the second and third experiments.

Experiments	Date	Time	Temperature (°C)		Air Relative humidity (%)	Wind speed (km h^{-1})	Cloudiness (%)
			Air	Soil			
Second	09/16/10	4:20 pm	35.7-34.5	35.7-40.4	22-24	zero	zero
Third	04/02/11	1:00 pm	29.4	29.1	66	3.5-4.0	90

Consequently, under bare conditions, there can be seedling emergence from seeds positioned up to a depth of 10 cm in the soil profile. This effect may interfere with the efficacy of herbicides applied during pre-emergence because herbicides should be distributed throughout the soil profile at depths up to 10 cm or be absorbed by seedling shoots. In addition, despite reduced seedling emergence observed with increasing straw amounts, this reduction was not enough to provide adequate control of *R. exaltata*, especially when the seeds were closer to the soil surface.

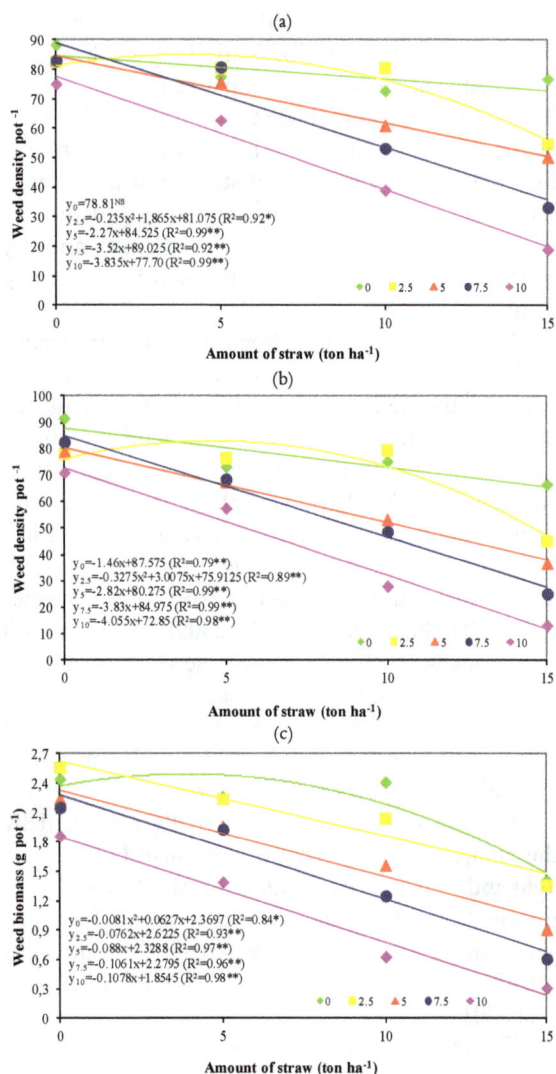

Figure 1. *Rottboellia exaltata* density at 14 (a) and 35 (b) days after sowing (DAS) and shoot dry matter at 35 DAS (c) as function of increasing amounts of sugarcane straw on the soil surface combined with increasing sowing depths.

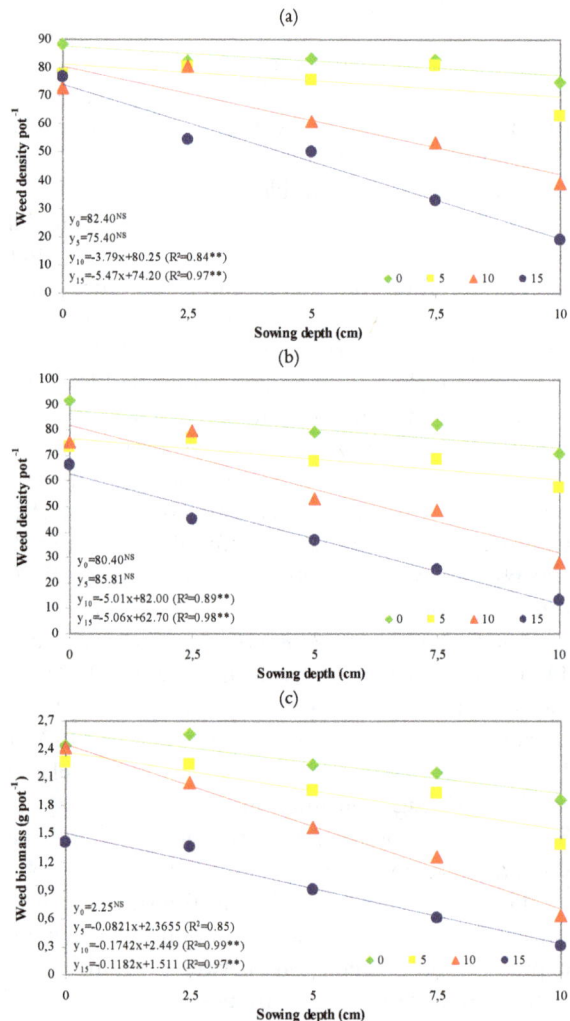

Figure 2. *Rottboellia exaltata* density at 14 (a) and 35 (b) days after sowing (DAS) shoot dry matter at 35 DAS (c) as function of increasing sowing depths combined with increasing amount of sugarcane straw on the soil surface.

Emergence of *Rottboellia exaltata* influenced by herbicide application during pre-emergence and the presence of sugarcane straw on the soil (second experiment)

The herbicide treatments and straw x herbicide interaction were significant for all variables evaluated. The amount of straw only significantly affected *R. exaltata* dry matter.

When analyzing the straw x herbicide interaction, we observed that in bare soil at 14 DAA, there was lower weed density when applying clomazone and flumioxazin compared to other herbicides (Table 2). At 5, 10, and 15 ton ha^{-1} of straw, clomazone was more effective than isoxaflutole combined with 10 ton ha⁻¹ of straw. At 35 DAA, clomazone provided decreases in weed density regardless of the presence of mulch on soil surface (Table 3). Combined with 10 ton ha⁻¹ of straw, flumioxazin, imazapyr, and isoxaflutole were also effective. Use of clomazone and imazapyr caused less shoot dry matter accumulation for all five straw amounts studied compared to other herbicides (Table 4). Flumioxazin and isoxaflutole were also effective in bare soil.

Although imazapyr did not reduce weed emergence, this herbicide inhibited weed growth and development as reflected in shoot dry matter (Table 4).

Table 2. *Rottboellia exaltata* density (plants pot⁻¹) at 14 days after herbicide application (DAA) pre-emergence with different amounts of sugarcane straw on the soil.

Herbicides/ Control	Doses (kg ha⁻¹)	Straw (ton ha⁻¹)			
		0	5	10	15
Clomazone	1.20	9.25 a¹	20.75 a	25.75 a	24.50 a
Flumioxazin	0.25	8.00 a	53.25 b	48.25 ab	57.50 bc
Imazapyr	0.20	69.25 bc	55.00 b	65.25 bc	80.25 c
Isoxaflutole	0.225	58.00 b	50.00 ab	31.25 a	44.25 ab
S-metolachlor	2.88	62.50 bc	71.25 b	84.25 c	53.25 abc
Control	-	90.00 c	79.00 b	85.00 c	77.00 c
LSD (in the column)		31.42			

¹Averages followed by lowercase letters in the columns indicate significant differences between the herbicide treatments within each amount of straw based on Tukey's test at 5% probability.

Table 3. *Rottboellia exaltata* density (plants pot⁻¹) at 35 days after herbicide application (DAA) pre-emergence with different amounts of sugarcane straw on the soil.

Herbicides/ Control	Doses (kg ha⁻¹)	Straw (ton ha⁻¹)			
		0	5	10	15
Clomazone	0.25	0.25 a¹	7.25 a	23.50 a	17.00 a
Flumioxazin	0.225	4.50 ab	61.25 bc	41.50 a	43.50 ab
Imazapyr	1.20	50.00 c	40.25 ab	38.75 a	47.00 ab
Isoxaflutole	0.20	38.00 bc	41.75 b	26.75 a	30.25 ab
S-metolachlor	2.88	58.75 cd	85.75 c	88.75 b	62.25 bc
Control	-	89.00 d	71.75 bc	87.25 b	84.00 c
LSD (in the column)		34.02			

¹Means followed by lowercase letters in the columns indicate significant differences between the herbicide treatments within each amount of straw based on Tukey's test at 5% probability.

Table 4. *Rottboellia exaltata* shoot dry matter (g pot⁻¹) at 35 days after herbicide application (DAA) pre-emergence with different amounts of sugarcane straw on the soil.

Herbicides/ Control	Doses (kg ha⁻¹)	Straw (ton ha⁻¹)			
		0	5	10	15
Clomazone	0.25	0.03 a¹	1.68 a	2.66 a	1.68 a
Flumioxazin	0.225	0.81 a	9.19 bc	7.96 bc	8.07 bc
Imazapyr	1.20	1.03 a	0.38 a	0.68 a	0.71 a
Isoxaflutole	0.20	4.71 a	5.51 ab	4.03 ab	3.48 ab
S-metolachlor	2.88	10.39 b	14.03 cd	13.16 cd	12.03 cd
Control	-	16.26 c	15.34 d	16.48 d	16.13 d
LSD (in the column)		5.20			

¹Averages followed by lowercase letters in the columns indicate significant differences between the herbicide treatments within each amount of straw based on Tukey's test at 5% probability.

For the control treatment and the clomazone, isoxaflutole and imazapyr herbicides, weed density and dry matter did not vary with increasing amounts

of straw on the soil surface (Figure 3). These results indicated that weeds did not germinate and/or emergence was inhibited by the presence of sugarcane straw and that herbicides were not affected by mulch on the soil. The 25 mm rainfall simulation after herbicide application was enough to leach herbicides from the straw to the soil, as the biological control of *R. exaltata* remained unaffected.

For flumioxazin and s-metolachlor, however, weed density varied with increasing amounts of straw (with polynomial fitting of the data), which resulted in less emergence in bare soil. The same behavior occurred for shoot dry matter when flumioxazin was applied. In contrast, there was no difference between the straw amounts when using s-metolachlor.

Among the factors that influence herbicide retention by the straw, the chemical and physical attributes of the molecules, especially their solubility and polarity, are essential for leaching herbicide to the soil via rainfall or irrigation. The solubility of the herbicides has been previously reported as follows: clomazone and imazapyr exhibit very high water solubility (1,100 and 11,272 mg L⁻¹, respectively); s-metolachlor and isoxaflutole (active metabolite DKN) exhibit high water solubility (480 and 326 mg L⁻¹, respectively); and flumioxazin exhibits low water solubility (1.79 mg L⁻¹) (RODRIGUES; ALMEIDA, 2011). The herbicide remaining on the straw (i.e., portion that was not leached to the soil) may be absorbed by surviving seedlings when they were growing up through the straw layer. However, this is possible only for herbicides that are absorbed by the mesocotyl and coleoptile of seedlings before they emerge above soil surface.

Although flumioxazin resulted in satisfactory control of *R. exaltata* under bare soil conditions, the level of weed control was compromised by mulch on the soil. The simulated 25 mm of rainfall after spraying was not enough to leach herbicide from the straw to the soil. This finding motivated us to conduct the third experiment to test the ability of flumioxazin to control *R. exaltata* with increasing rainfall intensities after application and with different straw amounts on the soil surface.

Control of *Rottboellia exaltata* by flumioxazin combined with rainfall intensity after application and presence of sugarcane straw on the soil (third experiment)

The amounts of straw and rainfall intensities significantly affected all of the variables evaluated. The straw x rainfall interaction was significant for weed density and shoot dry matter at 35 DAA.

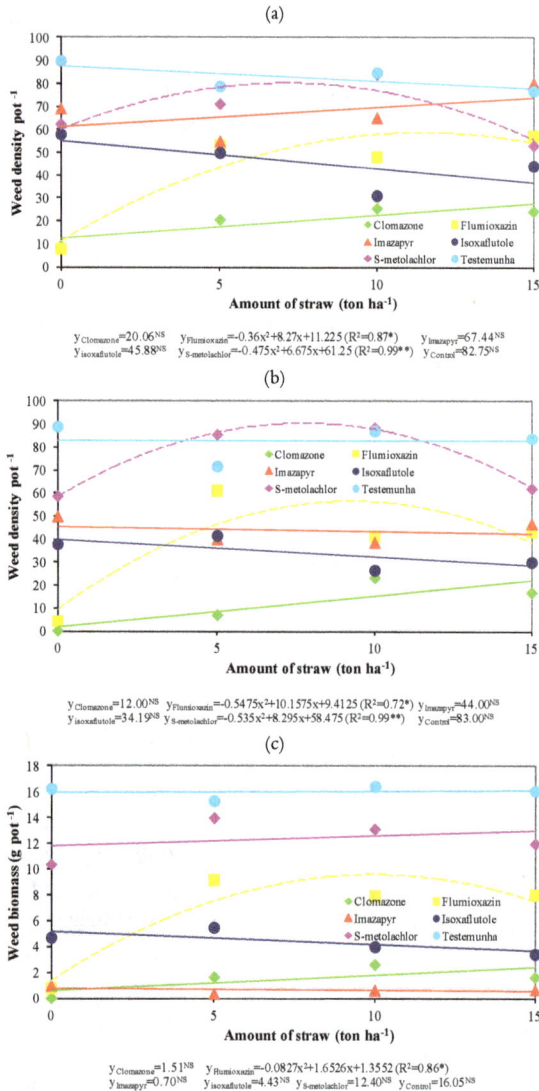

$y_{Clomazone}=20.06^{NS}$ $y_{Flumioxazin}=-0.36x^2+8.27x+11.225 (R^2=0.87*)$ $y_{Imazapyr}=67.44^{NS}$
$y_{isoxaflutole}=45.88^{NS}$ $y_{S-metolachlor}=-0.475x^2+6.675x+61.25 (R^2=0.99**)$ $y_{Control}=82.75^{NS}$

$y_{Clomazone}=12.00^{NS}$ $y_{Flumioxazin}=-0.5475x^2+10.1575x+9.4125 (R^2=0.72*)$ $y_{Imazapyr}=44.00^{NS}$
$y_{isoxaflutole}=34.19^{NS}$ $y_{S-metolachlor}=-0.535x^2+8.295x+58.475 (R^2=0.99**)$ $y_{Control}=83.00^{NS}$

$y_{Clomazone}=1.51^{NS}$ $y_{Flumioxazin}=-0.0827x^2+1.6526x+1.3552 (R^2=0.86*)$
$y_{Imazapyr}=0.70^{NS}$ $y_{isoxaflutole}=4.43^{NS}$ $y_{S-metolachlor}=12.40^{NS}$ $y_{Control}=16.05^{NS}$

Figure 3. *Rottboellia exaltata* density at 14 (a) and 35 (b) days after herbicide application (DAA) pre-emergence and shoot dry matter at 35 DAA (c) as function of increasing amounts of straw on the soil.

At 14 DAA, weed density increased with increasing amounts of straw on the soil surface with flumioxazin herbicide spraying (Figure 4). In contrast, weed emergence linearly decreased with increasing simulated rainfall intensities after application. This trend was observed for the five straw amounts studied where the straw x rainfall interaction was not significant.

At 35 DAA, weed density and dry matter varied with increasing amounts of straw for the three rainfall intensities (with polynomial fitting of the data) with less emergence and mass accumulation in bare soil (Figure 5).

In untreated controls with straw, higher weed density and amount of dry matter were obtained compared to the flumioxazin treatments regardless of the straw x rainfall combinations studied.

However, these data were not statistically analyzed and were only presented as graphs to visualize the potential weed infestation in that particular experimental period. The same set of treatments had already been studied in the first and second experiments but during different seasons and with straw from different origins. Thus, straw from different varieties may have influenced the weed dynamics by variations in allelopathic effects of the plant residues.

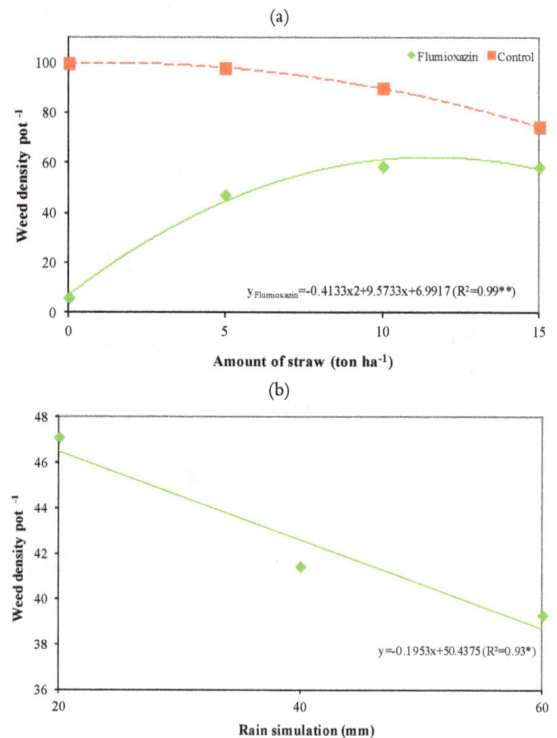

Figure 4. *Rottboellia exaltata* density at 14 days after flumioxazin herbicide application (DAA) with increasing amounts of sugarcane straw on the soil (a) and with increasing rainfall simulation intensities after application (b) in addition to untreated controls with four amounts of straw.

In bare soil, there was no difference between rainfall intensities for the variables evaluated (Figure 6). However, keeping 5, 10, and 15 ton ha^{-1} of straw on the soil linearly decreased weed dry matter with increasing rainfall simulation intensities after applying flumioxazin, thus resulting in less dry matter accumulation with 60 mm of water.

Increasing rainfall intensity caused increased leaching of herbicide retained in the straw into the soil, which favored weed control by flumioxazin under such conditions. Even with 60 mm of accumulated rainfall over the 4-days period after application, the amount of flumioxazin leached to the soil was not enough to ensure the same control observed in flumioxazin treatment on bare soil.

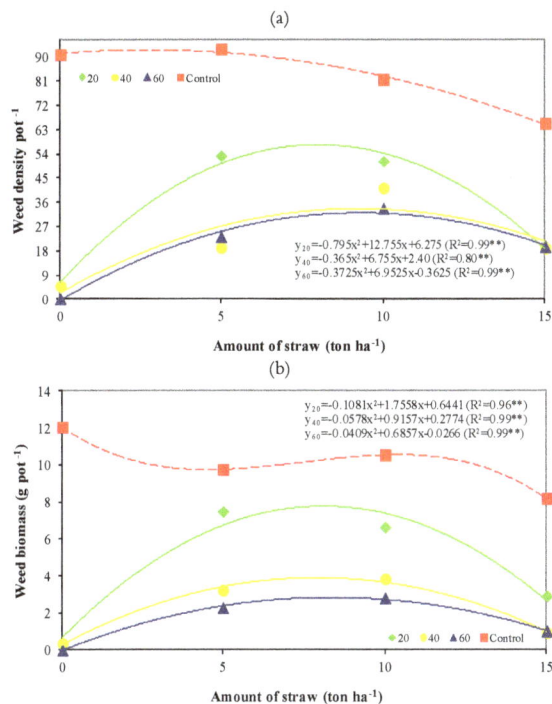

Figure 5. *Rottboellia exaltata* density (a) and shoot dry matter (b) at 35 days after flumioxazin herbicide application (DAA) as function of increasing amounts of sugarcane straw left on the soil combined with increasing rainfall simulation intensities after application in addition to untreated controls with four amounts of straw.

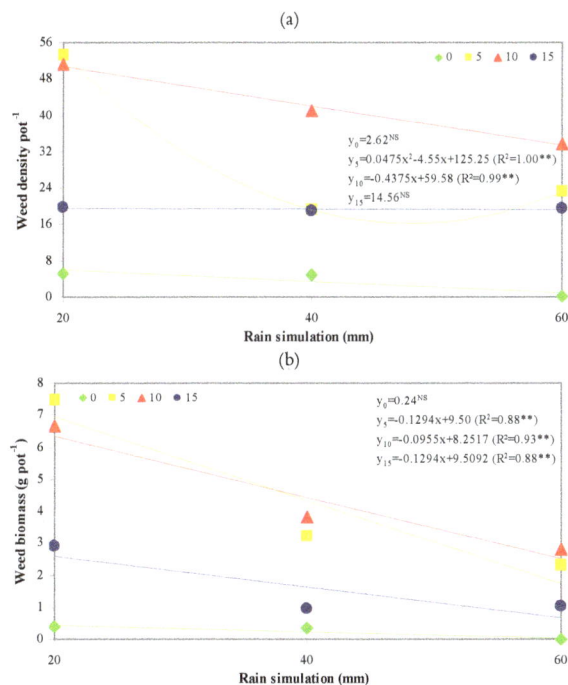

Figure 6. *Rottboellia exaltata* density (a) and shoot dry matter (b) at 35 days after flumioxazin herbicide application (DAA) as function of increasing rainfall simulation intensities combined with increasing amounts of straw left on the soil.

The chemical attributes of the flumioxazin (low solubility) would not contribute to its leaching to the soil in higher concentrations even when increasing the simulated rainfall intensity. However, no studies have been reported on the dynamics of flumioxazin in plant residues.

Conclusion

The combination of mulch on soil surface with deeper sowing depths provides less emergence and mass accumulation of *R. exaltata*. In bare soil, the sowing depth did not affect the weed dynamics.

Clomazone and imazapyr were effective herbicides controlling *R. exaltata* regardless of the amount of straw on the soil. Flumioxazin was also effective in controlling *R. exaltata* under bare soil conditions. Even with 60 mm of accumulated rainfall over the 4 day period after application, the amount of flumioxazin leached to the soil was not enough to ensure the same control observed when applying the herbicide on bare soil.

Acknowledgements

The authors would like to thank Bruno Daniel and Everton Henrique Camilo (Agronomy undergrad students) for their collaboration in some stages of this study.

References

ANNING, A. K.; GYAN-YEBOAH, K. Diversity and distribution of invasive weeds in Ashanti Region, Ghana. **African Journal of Ecology**, v. 45, n. 3, p. 355-360, 2007.

BOLFREY-ARKU, G. E.-K.; CHAUHAN, B. S.; JOHNSON, D. E. Seed germination ecology of itchgrass (*Rottboellia cochinchinensis*). **Weed Science**, v. 59, n. 2, p. 182-187, 2011.

CHAUHAN, B. S.; JOHNSON, D. E. Influence of tillage systems on weed seedling emergence pattern in rainfed rice. **Soil and Tillage Research**, v. 106, n. 1, p. 15-21, 2009.

CORREIA, N. M.; DURIGAN, J. C. Emergência de plantas daninhas em solo coberto com palha de cana-de-açúcar. **Planta Daninha**, v. 22, n. 1, p. 11-17, 2004.

CORREIA, N. M.; DURIGAN, J. C.; MELO, W. J. Envelhecimento de resíduos vegetais sobre o solo e os reflexos na eficácia de herbicidas aplicados em pré-emergência. **Bragantia**, v. 66, n. 1, p. 101-110, 2007.

HOLM, L. G.; PLUCKNETT, D. L.; PANCHO, J. V.; HERBERGER, J. P. **The world's worst weeds**: distribution and biology. Malabar: The University Press of Hawaii, 1991.

KISSMANN, K. G. **Plantas infestantes nocivas**. 2. ed. São Paulo: BASF, 1997.

KOBAYASHI, K.; ITAYA, D.; MAHATAMNUCHOKE, P.; PORNPROM, T. Allelopathic potential of itchgrass

(*Rottboellia exaltata* L.f.) powder incorporated into soil. **Weed Biology and Management**, v. 8, n. 14, p. 64-68, 2008.

MEKSAWAT, S.; PORNPROM, T. Allelopathic effect of itchgrass (*Rottboellia exaltata* L.f.) on seed germination and plant growth. **Weed Biology and Management**, v. 10, n. 14, p. 16-24, 2010.

MERSIE, W.; SEYBOLD, C. A.; WU, J.; McNAMEE, C. Atrazine and metolachlor sorption to switchgrass residues. **Communications in Soil Science and Plant Analysis**, v. 37, n. 3-4, p. 465-472, 2006.

OLIVEIRA, A. R.; FREITAS, S. P. Levantamento fitossociológico de plantas daninhas em áreas de produção de cana-de-açúcar. **Planta Daninha**, v. 26, n. 1, p. 33-46, 2008.

OLIVEIRA, A. R.; FREITAS, S. P. Palha de cana-de-açúcar associada ao herbicida trifloxysulfuron sodium + ametryn no controle de *Rottboellia exaltata*. **Bragantia**, v. 68, n. 1, p. 187-194, 2009.

RODRIGUES, B. N. Influência da cobertura morta no comportamento dos herbicidas imazaquin e clomazone. **Planta Daninha**, v. 11, n. 1-2, p. 21-28, 1993.

RODRIGUES, B. N.; ALMEIDA, F. L. S. **Guia de herbicidas**. 6. ed. Londrina: Edição dos autores, 2011.

THOMAS, P. E. L.; ALLISON, J. C. S. Seed dormancy and germination in *Rottboellia exaltata*. **Journal of Agricultural Science**, v. 85, n. 8, p. 129-134, 1975.

PERMISSIONS

All chapters in this book were first published in ASA, by Editora da Universidade Estadual de Maringá; hereby published with permission under the Creative Commons Attribution License or equivalent. Every chapter published in this book has been scrutinized by our experts. Their significance has been extensively debated. The topics covered herein carry significant findings which will fuel the growth of the discipline. They may even be implemented as practical applications or may be referred to as a beginning point for another development.

The contributors of this book come from diverse backgrounds, making this book a truly international effort. This book will bring forth new frontiers with its revolutionizing research information and detailed analysis of the nascent developments around the world.

We would like to thank all the contributing authors for lending their expertise to make the book truly unique. They have played a crucial role in the development of this book. Without their invaluable contributions this book wouldn't have been possible. They have made vital efforts to compile up to date information on the varied aspects of this subject to make this book a valuable addition to the collection of many professionals and students.

This book was conceptualized with the vision of imparting up-to-date information and advanced data in this field. To ensure the same, a matchless editorial board was set up. Every individual on the board went through rigorous rounds of assessment to prove their worth. After which they invested a large part of their time researching and compiling the most relevant data for our readers.

The editorial board has been involved in producing this book since its inception. They have spent rigorous hours researching and exploring the diverse topics which have resulted in the successful publishing of this book. They have passed on their knowledge of decades through this book. To expedite this challenging task, the publisher supported the team at every step. A small team of assistant editors was also appointed to further simplify the editing procedure and attain best results for the readers.

Apart from the editorial board, the designing team has also invested a significant amount of their time in understanding the subject and creating the most relevant covers. They scrutinized every image to scout for the most suitable representation of the subject and create an appropriate cover for the book.

The publishing team has been an ardent support to the editorial, designing and production team. Their endless efforts to recruit the best for this project, has resulted in the accomplishment of this book. They are a veteran in the field of academics and their pool of knowledge is as vast as their experience in printing. Their expertise and guidance has proved useful at every step. Their uncompromising quality standards have made this book an exceptional effort. Their encouragement from time to time has been an inspiration for everyone.

The publisher and the editorial board hope that this book will prove to be a valuable piece of knowledge for researchers, students, practitioners and scholars across the globe.

LIST OF CONTRIBUTORS

Lígia Helena de Andrade
Universidade Federal Rural de Pernambuco, Recife, Pernambuco, Brazil

José Vargas de Oliveira, Mariana Oliveira Breda and Edmilson Jacinto Marques
Área de Fitossanidade, Departamento de Agronomia, Universidade Federal Rural de Pernambuco, R. Dom Manoel de Medeiros, s/n, 52171-900, Recife, Pernambuco, Brazil

Iracilda Maria de Moura Lima
Laboratório de Entomologia, Departamento de Zoologia, Centro de Ciências Biológicas, Universidade Federal de Alagoas, Maceió, Alagoas, Brazil

Marcelo Gonçalves Balan and Arney Eduardo do Amaral Ecker
Departamento de Agronomia, Universidade Estadual de Maringá, Avenida Colombo, 5790, 87020-900, Maringá, Paraná, Brazil

Otavio Jorge Grigoli Abi Saab and Gustavo de Oliveira Migliorini
Departamento de Agronomia, Universidade Estadual de Londrina, Londrina, Paraná, Brazil

Flávia Garcia Florido and Patrícia Andrea Monquero
Centro de Ciências Agrárias, Universidade Federal de São Carlos, Rod. Anhanguera, km 174, Cx. Postal 153, Araras, São Paulo, Brazil

Ana Carolina Ribeiro Dias
Departamento de Produção Vegetal, Escola Superior de Agricultura "Luiz de Queiroz", Universidade de São Paulo, Piracicaba, São Paulo, Brazil

Valdemar Luiz Tornisielo
Laboratório de Ecotoxicologia, Centro de Energia Nuclear na Agricultura, Piracicaba, São Paulo, Brazil

Roni Paulo Fortunato, Paulo Eduardo Degrande and Paulo Rogério Beltramin da Fonseca
Instituto Federal do Mato Grosso do Sul, Campus Ponta Porã, Rua Intibiré Vieira, s/n, BR-463, km 4,5, 79900-972, Ponta Porã, Mato Grosso do Sul, Brazil

Flávio Lemes Fernandes, Maria Elisa de Sena Fernandes and Juno Ferreira da Silva Diniz
Instituto de Ciências Agrárias, Universidade Federal de Viçosa, Campus de Rio Paranaíba, M- 230, Km 7, s/n, 38810-000, Rio Paranaíba, Minas Gerais, Brazil

Elisângela Novais Lopes
Programa de Pós-graduação em Entomologia, Universidade Estadual Paulista, Jaboticabal, São Paulo, Brazil

Madelaine Venzon
Empresa de Pesquisa Agropecuária de Minas Gerais, Viçosa, Minas Gerais, Brazil

Luís Antônio dos Santos Dias
Departamento de Fitotecnia, Universidade Federal de Viçosa, Viçosa, Minas Gerais, Brazil

Samuel Julio Martins, Flávio Henrique Vasconcelos Medeiros and Ricardo Magela Souza
Departamento de Fitopatologia, Universidade Federal de Lavras, Campus Universitário, 3037, 37200-000, Lavras, Minas Gerais, Brazil

Laíze Aparecida Ferreira Vilela
Departamento de Ciências do Solo, Universidade Federal de Lavras, Lavras, Minas Gerais, Brazil

Talita Roberta Ferreira Borges Silva, André Cirilo de Sousa Almeida, Tony de Lima Moura, Anderson Rodrigo da Silva and Flávio Gonçalves Jesus
Instituto Federal Goiano, Campus Urutaí, Rodovia Professor Geraldo Silva Nascimento, km 2,5, 75790-000, Urutaí, Goiás, Brazil

Silvia de Sousa Freitas
Departamento de Química, Universidade Federal de Goiás, Campus de Catalão, Catalão, Goiás, Brazil

Flávia Silva Barbosa, Germano Leão Demolin Leite, Marney Aparecida de Oliveira Paulino, Denilson de Oliveira Guilherme, Janini Tatiane Lima Souza Maia and Rodrigo Carvalho Fernandes
Instituto de Ciências Agrárias, Universidade Federal de Minas Gerais, Av. Universitária, 1000, Cx Postal 135, 39404-006, Montes Claros, Minas Gerais, Brazil

Gustavo Mack Teló, Enio Marchesan and Maurício Limberger de Oliveira
Grupo de Pesquisa em Arroz Irrigado, Departamento de Fitotecnia, Universidade Federal de Santa Maria, 97105-900, Santa Maria, Rio Grande do Sul, Brazil

Renato Zanella, Sandra Cadore Peixoto and Osmar Damian Prestes
Laboratório de Análise de Resíduos de Pesticidas, Departamento de Química, Universidade Federal de Santa Maria, Santa Maria, Rio Grande do Sul, Brazil

José Romário de Carvalho, Dirceu Pratissoli, Leandro Pin Dalvi and Marcos Américo Silva
Centro de Ciências Agrárias, Departamento de Produção Vegetal, Núcleo de Desenvolvimento Científico e Tecnológico em Manejo Fitossanitário, Setor de Entomologia, Universidade Federal do Espírito Santo, Alegre, Espírito Santo, Brazil

Regiane Cristina Oliveira de Freitas Bueno
Departamento de Produção Vegetal Defesa Fitossanitária, Faculdade de Ciências Agrônomicas, Botucatu, São Paulo, Brazil

Adeney de Freitas Bueno
Embrapa Soja, Rodovia Carlos João Strass, 86001-970, Londrina, Paraná, Brazil

Marcus Alvarenga Soares
Programa de Pós-graduação em Produção Vegetal, Universidade Federal dos Vales do Jequitinhonha e Mucuri, Rodovia MGT-367, Km 583, 5000, 39100-000, Diamantina, Minas Gerais, Brazil

Germano Leão Demolin Leite and Cleidson Soares Ferreira
Insetário George Washington Gomez de Moraes, Instituto de Ciências Agrárias, Universidade Federal de Minas Gerais, Montes Claros, Minas Gerais, Brazil

José Cola Zanuncio and Silma Leite Rocha
Departamento de Biologia Animal, Universidade Federal de Viçosa, Viçosa, Minas Gerais, Brazil

Veríssimo Gibran Mendes de Sá
Faculdade de Engenharia, Universidade do Estado de Minas Gerais, João Monlevade, Minas Gerais, Brazil

Joanei Cechin and Dirceu Agostinetto
Faculdade de Agronomia Eliseu Maciel, Universidade Federal de Pelotas, Campus Universitário s/n., Cx. Postal 354, 96010-900, Pelotas, Rio Grande do Sul, Brazil

Leandro Vargas
Empresa Brasileira de Pesquisa Agropecuária, Passo Fundo, Rio Grande do Sul, Brazil

Fabiane Pinto Lamego
Empresa Brasileira de Pesquisa Agropecuária, Embrapa Pecuária Sul, Bagé, Rio Grande do Sul, Brazil

Franciele Mariani
Instituto Federal de Educação, Ciência e Tecnologia, Campus Sertão, Sertão, Rio Grande do Sul, Brazil

Taísa Dal Magro
Universidade de Caxias do Sul, Vacaria, Rio Grande do Sul, Brazil

Luciano Coutinho Silva and Renato Paiva
Departamento de Biologia, Setor de Fisiologia Vegetal, Universidade Federal de Lavras, Cx. Postal 3037, 372000-000, Lavras, Minas Gerais, Brazil

Rony Swennen
Division of Crop Biotechnics, Laboratory for Tropical Crop Improvement, Katholieke University Leuven, Leuven, Belgium
Bioversity International, Katholieke University Leuven, Leuven, Belgium

Edwige Andrè and Bart Panis
Bioversity International, Katholieke University Leuven, Leuven, Belgium

Antonia Railda Roel
Programa de Pós-graduação em Biotecnologia, Universidade Católica Dom Bosco, Av. Tamandaré, 6000, 78117 900, Campo Grande, Mato Grosso do Sul, Brazil

Melissa Gindri Bragato Pistori
Programa de Pós-graduação em Biotecnologia, Universidade Católica Dom Bosco, Av. Tamandaré, 6000, 78117 900, Campo Grande, Mato Grosso do Sul, Brazil
Laboratório de Entomologia de Plantas Forrageiras Tropicais, Empresa Brasileira de Pesquisa Agropecuária, Campo Grande, Mato Grosso do Sul, Brazil

José Raul Valério
Laboratório de Entomologia de Plantas Forrageiras Tropicais, Empresa Brasileira de Pesquisa Agropecuária, Campo Grande, Mato Grosso do Sul, Brazil

Marlene Conceição Monteiro Oliveira
Laboratório de Entomologia de Plantas Forrageiras Tropicais, Empresa Brasileira de Pesquisa Agropecuária, Campo Grande, Mato Grosso do Sul, Brazil
Agência de Desenvolvimento Agrário e Extensão Rural, Campo Grande, Mato Grosso do Sul, Brazil

Eliane Grisoto
Escola Superior de Agricultura "Luiz de Queiroz", Piracicaba, São Paulo, Brazil

Rosemary Matias
Programa de Pós-graduação em Meio Ambiente e Desenvolvimento Regional, Universidade Anhanguera, Campo Grande, Mato Grosso do Sul, Brazil

Flávio Neves Celestino
Instituto Federal de Educação, Ciência e Tecnologia do Espírito Santo, Rod. ES-130, km 01, 29980-000, Montanha, Espírito Santo, Brazil

Dirceu Pratissoli, Lorena Contarini Machado, Hugo José Gonçalves dos Santos Junior and Leonardo Mardgan
Departamento de Produção Vegetal, Universidade Federal do Espírito Santo, Vitória, Espírito Santo, Brazil

Vagner Tebaldi de Queiroz
Departamento de Química e Física, Universidade Federal do Espírito Santo, Vitória, Espírito Santo, Brazil

Germano Leão Demolin Leite, Manoel Ferreira Souza, Patrícia Nery Silva Souza and Márcia Michelle Fonseca
Insetário G.W.G. de Moraes, Instituto de Ciências Agrárias, Universidade Federal de Minas Gerais, Av. Universitária, 1000, Montes Claros, Minas Gerais, Brazil

José Cola Zanuncio
Departamento de Entomologia, Universidade Federal de Viçosa, Viçosa, Minas Gerais, Brazil

Elizandro Ricardo Kluge and Kathia Szeuczuk
Departamento de Agronomia, Setor de Ciências Agrárias e Ambientais, Universidade Estadual do Centro-Oeste, Rua Simeão Camargo Varela de Sá, Vila Carli, 03, Guarapuava, Paraná, Brazil

Marcelo Cruz Mendes
Departamento de Agronomia, Setor de Fitotecnia e Grandes Culturas, Universidade Estadual do Centro-Oeste, Guarapuava, Paraná, Brazil

Marcos Ventura Faria
Departamento de Agronomia, Setor de Genética e Melhoramento, Universidade Estadual do Centro-Oeste, Guarapuava, Paraná, Brazil

Leandro Alvarenga Santos
Departamento de Agronomia, Setor de Fitopatologia., Universidade Estadual do Centro-Oeste, Guarapuava, Paraná, Brazil

Heloisa Oliveira dos Santos
Departamento de Agricultura, Setor de Tecnologia de Sementes, Universidade Federal de Lavras, Lavras, Minas Gerais, Brazil

Núbia Maria Correia
Departamento de Fitossanidade, Universidade Estadual Paulista, "Julio de Mesquista Filho", Via de Acesso Prof. Paulo Donato Castellane, s/n., 14884-900, Jaboticabal, São Paulo, Brazil

Leonardo Petean Gomes and Fabio José Perussi
Curso de Graduação em Agronomia, Universidade Estadual Paulista, "Julio de Mesquista Filho", Jaboticabal, São Paulo, Brazil

Diones Krinski
Departamento de Ciências Biológicas, Faculdade de Ciências Agrárias, Biológicas, Engenharia e da Saúde, Universidade do Estado de Mato Grosso, Campus Universitário de Tangará da Serra, Rodovia MT 358, km 7, Jardim Aeroporto, Tangará da Serra, Mato Grosso, Brazil

Luís Amilton Foerster
Departamento de Zoologia, Setor de Ciências Biológicas, Universidade Federal do Paraná, Curitiba, Paraná, Brazil

Cicero Deschamps
Departamento de Agronomia, Setor de Ciências Agrárias, Universidade Federal do Paraná, Curitiba, Paraná, Brazil

Carita Liberato Amaral, Guilherme Bacarin Pavan, Fernanda Campos Mastrotti Pereira and Pedro Luis da Costa Aguiar Alves
Faculdade de Ciências Agrárias e Veterinárias, Universidade Estadual Paulista "Júlio de Mesquita Filho", Rod. de Acesso Prof. Paulo Donato Castellane, Km 05, Zona rural, 14884-900, Jaboticabal, São Paulo. Brazil

Germano Leão Demolin Leite, Farley William Souza Silva, Rafael Eugênio Maia Guanabens, Luiz Arnaldo Fernandes, Lourdes Silva Figueiredo and Leonardo Ferreira Silva
Laboratório de Entomologia Universitário, Instituto de Ciências Agrárias, Universidade Federal de Minas Gerais, Av. Universitária, 1000, Montes Claros, Minas Gerais, Brazil

Janaina Zorzetti, Ana Paula Scaramal Ricietto, Fernanda Aparecida Pires Fazion, Ana, Pedro Manuel Oliveira Janeiro Neves and Gislayne Trindade Vilas-Bôas
Universidade Estadual de Londrina, Rodovia Celso Garcia Cid, Km 445, 86047-902, Londrina, Paraná, Brazil

Maria Meneguim
Instituto Agronômico do Paraná, Londrina, Paraná, Brazil

Flávia Silva Barbosa, Germano Leão Demolin Leite, Sérgio Monteze Alves, Aline Fonseca Nascimento, Vinícius de Abreu D'Ávila and Cândido Alves da Costa
Instituto de Ciências Agrárias, Universidade Federal de Minas Gerais, Av. Universitária 1000, Montes Claros, Minas Gerais, Brazil

Alessandra Fagioli da Silva, Rone Batista dé Oliveira and Marco Antonio Gandolfo
Setor de Engenharia e Desenvolvimento Agrário, Universidade Estadual do Norte do Paraná, Campus Luiz Meneghel, Rodovia BR-369, km 54, Vila Maria Bandeirantes, Paraná, Brazil

Júlio Cesar Janini, Arlindo Leal Boiça Júnior, Flávio Gonçalves Jesus and Anderson Gonçalves Silva
Departamento de Fitossanidade, Faculdade de Ciências Agrárias e Veterinárias de Jaboticabal, Universidade Estadual Paulista "Júlio de Mesquita Filho", Via de Acesso Prof. Paulo Donato Castellane, s/n, 14884-900, Jaboticabal, São Paulo, Brazil

Sérgio Augusto Carbonell and Alisson Fernando Chiorato
Centro de Análise e Pesquisa Tecnológica do Agronegócio dos Grãos e Fibras, Instituto Agronômico de Campinas, Campinas, São Paulo, Brazil

Marcelo Rodrigues Reis, Leonardo Ângelo Aquino and Christiane Augusta Diniz Melo
Universidade Federal de Viçosa, Campus de Rio Paranaíba, Rodovia MG-230, Km 7, 38810-000, Rio Paranaíba, Minas Gerais, Brazil

Daniel Valadão Silva
Departamento de Fitotecnia, Universidade Federal Rural do Semi-árido, Mossoró, Rio Grande do Norte, Brazil

Roque Carvalho Dias
Universidade Estadual Paulista "Júlio de Mesquita Filho", Faculdade de Ciências Agronômicas, Botucatu, São Paulo, Brazil

Ezio dos Santos Pinto, Eduardo Moreira Barros, Jorge Braz Torres and Robério Carlos dos Santos Neves
Departamento de Agronomia-Entomologia, Universidade Federal Rural de Pernambuco, Rua Dom Manoel de Medeiros, s/n, 52171-900, Recife, Pernambuco, Brazil

Núbia Maria Correia, Leonardo Petean Gomes and Fabio José Perussi
Departamento de Fitossanidade, Universidade Estadual Paulista "Julio de Mesquita Filho", Via de Acesso Prof. Paulo Donato Castellane, s/n, 14884-900, Jaboticabal, São Paulo, Brazil

Index

* 9 7 8 1 6 4 7 4 0 0 6 1 3 *